ACTIVATED METALS
in
ORGANIC SYNTHESIS

NEW DIRECTIONS in ORGANIC and BIOLOGICAL CHEMISTRY

Series Editor
C.W. Rees, FRS
Imperial College of Science,
Technology and Medicine
London, UK

Activated Metals in Organic Synthesis
Pedro Cintas

Capillary Electrophoresis: Theory and Practice
Patrick Camilleri

Cyclization Reactions
C. Thebtaranonth and Y. Thebtaranonth

Mannich Bases: Chemistry and Uses
Maurilio Tramontini and Luigi Angiolini

Vicarious Nucleophilic Substitution and Related Processes in Organic Synthesis
Mieczyslaw Makosza

Radical Cations and Anions
M. Chanon, S. Fukuzumi, and F. Chanon

Chlorosulfonic Acid: A Versatile Reagent
R. J. Cremlyn and J. P. Bassin

Aromatic Fluorination
James H. Clark and Tony W. Bastock

Selectivity in Lewis Acid Promoted Reactions
M. Santelli and J.-M. Pons

Dianion Chemistry
Charles M. Thompson

Asymmetric Methodology in Organic Synthesis
David J. Ager and Michael B. East

Chemistry of Pyridoxal Dependent Enzymes
David Gani

The Anomeric Effect
Eusebio Juaristi

Chiral Sulfur Reagents
M. Mikołajczyk, J. Drabowicz, and P. Kiełbasiński

ACTIVATED METALS in ORGANIC SYNTHESIS

Pedro Cintas
Department of Organic Chemistry
University of Extremadura
Badajoz, Spain

CRC Press
Boca Raton Ann Arbor London Tokyo

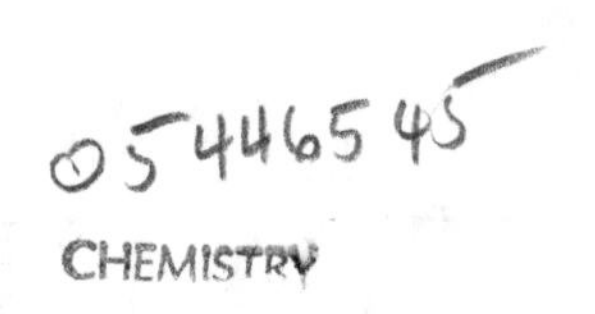

Library of Congress Cataloging-in-Publication Data

Cintas, Pedro
Activated metals in organic synthesis/Pedro Cintas.
p. cm.
Includes bibliographical references and index.
ISBN 0-8493-7863-X
1. Organic compounds—Synthesis. 2. Metals. I. Title.
QD262.C525 1993
547'.2—dc20 93-3570
CIP

This book represents information obtained from authentic and highly regarded sources. Reprinted material is quoted with permission, and sources are indicated. A wide variety of references are listed. Every reasonable effort has been made to give reliable data and information, but the author and the publisher cannot assume responsibility for the validity of all materials or for the consequences of their use.

Direct all inquiries to CRC Press, Inc., 2000 Corporate Blvd., N.W., Boca Raton, Florida 33431.

International Standard Book Number 0-8493-7863-X

Library of Congress Card Number 93-3570

Printed in the United States of America 1 2 3 4 5 6 7 8 9 0

Printed on acid-free paper

CONTENTS

PREFACE

Although the achievements of the past thirty years in pure organic chemistry have been quite remarkable, many chemists are convinced that only in the area of organometallic chemistry are there truly new reactions or novel processes waiting to be discovered. The importance of this discipline is clearly demonstrated by the increasing number of books, reviews, and articles dealing with specific reactions of particular metal derivatives, and their application in organic synthesis. The preparation of organometallic reagents represents therefore, an extremely important area of chemistry. Unfortunately, the high price, toxicity, and extreme sensitivity of many organometallic compounds constitute severe limitations and have added urgency to the search for other alternatives. An approach which has received considerable attention involves the use of activated metals. Thus, the reaction of a substrate at a metal surface, either in a catalytic fashion or in a stoichiometric reaction with consumption of the metal is a field that is receiving much current attention. Exceptional reactivity of the physically or chemically activated metal powders has been demonstrated by many reactions, including those of industrial importance. Several methods have been developed to increase the reactivity of the metal, in order to allow an easier access to some organometallic intermediates.

Most metals, with the exception of a few alkali and alkaline-earth elements, are relatively too inert to afford the desired organometallic species by direct metallation. Thus, reaction of an organic halide with a transition metal is known to be difficult. Transmetallation, which is a two-step process from organolithium or organomagnesium precursors and the corresponding metal halide is the next preference. Activation methods have made transmetallation possible in a single step. In some cases, this methodology proceeds with increased yields, under milder conditions, and in shorter times, and contrasts with classical transmetallation. Interestingly, some metal-mediated organic transformations are conducted in aqueous media and in the absence of an inert atmosphere.

Frequently a metal may not react, even if it is one of those known to be in principle highly reactive. The source of this absence of reactivity is often referred to as *passivation*, and the initial step in most organometallic preparations is metal activation or *depassivation*. The first part of this book provides a general account of the generation of activated metals by means of well-established procedures, and discusses how these highly reactive species can be used in the synthesis of some organometallic reagents. The concept of activation or *depassivation* may appear ambiguous and complex, and intricate phenomena of a physical or chemical nature may take place on the metal. In general, activation of the metal occurs by removing surface contaminants, usually a film of metal oxide, or more generally, increasing the reactive area. Importantly, this activation can be complemented in some instances by the transformation of the bulk solid to a dispersion, which greatly increases the surface area. For this reason, I have included a preliminary section on the basic forms of metals, from commercially available bulk solids usually referred to as massive metals, to those containing very small or fine particles and denoted as dispersed forms. This section is based mainly on reports by Klabunde, Bond, and others who have contributed significantly to this area. Emphasis has been placed on these dispersed metals, particularly on finely divided metal powders and metal slurries which are indeed organometallic species, since they are clusters of naked atoms stabilized by organic ligands.

The second chapter analyzes the principles and applications of a rather physical activation, the vaporization and subsequent co-condensation of metal atoms. Part one concludes with a description of the depassivating methods employed commonly in laboratory syntheses.

Part two describes some important metal-mediated organic reactions. These include representative reactions in which the activation of the metal plays a crucial role. In addition to metal-induced reductive methods and pinacolic, Reformatsky-, and Barbier-type reactions, I have added two relatively modern organic reactions: the McMurry ketone-olefin coupling and the Bernet-Vasella reaction. With a few exceptions, mechanisms are not generally discussed.

The reader should not expect full information on these topics because that is not the purpose of this book. My aim has been to bring out the synthetic uselfuness of this approach in some relevant organic transformations rather than to provide a comprehensive account; and it is obviously not possible to cover each topic in great depth. Nor would this be desirable, even if possible. Nevertheless, the sections will be a source of general information and will stimulate the further pursuit of individual topics.

In undertaking a task of this magnitude, particularly with regard to the large number of structural formulae, equations, and schemes, errors will be inevitable. I do hope, however, that serious as these may be, they will not detract from the essential interest of the book. Also, I apologize for any omissions or citations that have received less emphasis than deserved.

I am indebted to all my talented colleagues for their fruitful discussions and continuous help, and particularly to Professors Martin Avalos and Jose L. Jimenez for their advice and assistance. I am also grateful to my good friends Dr. Alois Fürstner (Graz, Austria) and Prof. Jean-Louis Luche (Grenoble) for their comments and valuable information on this subject. In addition, I would like to thank Professors Kenneth J. Klabunde (Manhattan, Kansas), Reuben D. Rieke (Lincoln, Nebraska), Andrea Vasella (Zürich), Helmut Bönnemann (Mülheim, Germany), and Dr. Bruno Bernet (Zürich) for their interest and gratifying encouragement in this project. Special thanks are due to Mr. Navin Sullivan (CRC Press Inc.) for his patience, advice, and assistance. Finally but not least, it is the sacrifice and understanding of my family that have made my writing possible.

It is hoped that the reader will find this contribution worth while and will not hesitate to suggest ways in which this material may be improved. Criticisms are always welcome.

Pedro Cintas
Badajoz, Spain

ABBREVIATIONS

Ac	Acetyl, acetate
AIBN	2,2'-azobisisobutyronitrile
Ar	Aryl
Bn	Benzyl
Boc	*tert*-Butyloxycarbonyl
Bz	Benzoyl
Bu	Butyl
Cat	Catalytic
Cbz	Carbobenzyloxy (Benzyloxycarbonyl)
COD	*cis,cis*-1,5-Cyclooctadiene
Cp	Cyclopentadienyl
DABCO	1,4-Diazabicyclo[2.2.2]octane
DBA	Dibenzylideneacetone
DBN	1,5-Diazabicyclo[4.3.0]non-5-ene
DBU	1,8-Diazabicyclo[5.4.0]undec-7-ene
DCC	*N,N'*-Dicyclohexylcarbodiimide
DDQ	2,3-Dichloro-5,6-dicyano-1,4-benzoquinone
DEAD	Diethylazodicarboxylate
de	Diastereomeric excess
DIBAH	Diisobutylaluminum hydride
Diglyme	Diethyleneglycol dimethyl ether
DMA	*N,N*-Dimethylacetamide
DMAP	4-Dimethylaminopyridine
DME	1,2-Dimethoxyethane
DMF	*N,N*-Dimethylformamide
DMSO	Dimethylsulfoxide
ee	Enantiomeric excess
EDA	Ethylendiamine (1,2-Diaminoethane)
Et	Ethyl
GIC	Graphite intercalation compound
Glyme	1,2-Dimethoxyethane
Gr	Graphite
HMPA	Hexamethylphosphoramide
LAH	Lithium aluminum hydride
LDA	Lithium diisopropylamide
M	Metal
M*	Activated metal

Abbreviations *(continued)*

MCPBA	*m*-Chloroperbenzoic acid
Me	Methyl
MEM	2-Methoxyethoxymethyl
MOM	Methoxymethyl
Ms	Methanesulfonyl
NBS	*N*-Bromosuccinimide
PCC	Pyridinium chlorochromate
PDC	Pyridinium dichromate
Ph	Phenyl
Pr	Propyl
Py	Pyridine
R_f	Perfluoroalkyl
SMAD	Solvated metal atom dispersion
TBDMS	*tert*-Butyldimethylsilyl
TBDPS	*tert*-Butyldiphenylsilyl
TFA	Trifluoroacetic acid
TfO	Triflate
THF	Tetrahydrofuran
THP	2-Tetrahydropyranyl
TMEDA	*N,N,N'N'*-Tetramethylethylenediamine
TMSCl	Trimethylsilyl chloride (Chlorotrimethylsilane)
Tr	Trityl (Triphenylmethyl)
Ts	*p*-Toluenesulfonyl
Δ	heat
))))	Ultrasound irradiation

PART ONE

METAL ACTIVATION

I. BASIC FORMS OF METALS

An in depth knowledge of the activation of metals requires a description of their structural forms. For the past fifty years, the properties and methods of preparation of highly dispersed transition metals have attracted considerable attention. Very small metallic particles exhibit special geometries as well as unusual magnetic, electrical, and chemical properties.[1-5] These features are particularly important in catalysis, metallurgy, metallography (the study of the structure and composition of metals and alloys), and in the preparation of new materials. Likewise, the study of formation, nucleation, and growth of small metal particles constitutes a current research activity of many theoretical and experimental scientists. By small particles are meant clusters of atoms or molecules of metals and alloys, ranging in size from <1 nm to almost 10 nm or having agglomeration numbers from <10 up to a few hundred, *i. e.*, species representing dimensions between single atoms or molecules and bulk materials.[1-4] Although these topics lie beyond the scope of this account, a brief note on the basic forms of metals will be first presented.

A. MASSIVE METALS

A general and structural division of metals into massive and dispersed forms is frequently presented for comparative purposes.

Massive or bulk metals are characterized by having a well-ordered infinite lattice of metal atoms. Several forms of bulk metals such as dusts, foils, pellets, or wires among others are commonly available, and have been used as catalysts. In general, a preliminary study of massive metals in catalysis research is performed in order to determine the role of the metal type and the effect of its surface. This can be cleaned by several mechanical or chemical procedures, which cause relevant changes in the properties of the metal, and therefore in catalytic activity. However, the most important form is the single crystal, which can be crystallographically characterized. In this case, several parameters can be studied as well as the effect of metal atoms in different positions of a metallic lattice. Moreover, single crystals are easily altered by cutting which opens up further catalytic possibilities.

Metal surfaces examined with the unaided eye or a simple binocular microscope at magnifications less than 10 diameters can reveal valuable information about crystalline, chemical, and mechanical heterogeneity. Crystalline heterogeneity is known as grain. Chemical heterogeneity arises from impurities, chemical segregation, and nonmetallic inclusions. Mechanical heterogeneity involves a local deformation of structure, distortions, and regions of chemical segregation, resulting from cold processes.

A further microscopic examination of polished or etched surfaces at magnifications from about 100 to 1500 diameters can yield additional information about the size and shape of grains, structural phases, nonmetallic inclusions, etc. Metallographic etching, that is, the action of a corrosive reagent on the polished surface can show a successive destruction.[6] This occurs because of the different rates of dissolution of the structural components under the attack of the etching agent. Several techniques are actually employed to reveal grain and other structural forms, examine oxide surfaces, and identify metallic phases. Particularly, the use of polarized light, scanning electron microscope photographs, and X-ray diffraction are valuable tools for this purpose.

Another interesting form of massive metal is that of thick films, which are also employed in catalysis. These films can be deposited on a wide variety of materials and substrates (glasses, alkali halides, etc.). A massive film must reach a thickness of about 50 nm,[7] since thinner films show different properties and should be considered as dispersed forms. The most important property of thick films is their porosity, which is demonstrated by the ratio of the surface area to the geometric surface area. Single metal crystals and massive metals have ratios ~1, while in metal films ratios are greater than unity. Metal films can be prepared by three general methods,[5,7] a) thermal evaporation, b) sputtering, and c) chemical deposition.

Powders with very large particles (>1 μm) are generally considered as massive metals. These powders can be activated mechanically or chemically. In mechanical cold working the metal is hardened and undergoes structural rather than chemical change. In hot working, the metal is softened by heat. This process changes the crystallinity of the metal. Moreover, mechanical activation can be achieved by pressing, rolling, filing and similar processes, although these methods have little effect on catalytic activity.

A remarkable effect on activation is exercized by sintering, which has considerable importance in dispersed forms of metals.[5] Sintering or diffusion bonding by heating is the increase in particle size above the equilibrium size obtained during preparation. Although sintering may have a thermal, chemical, or mechanical origin, only thermal sintering is particularly important, since the increase of metal particles can be expected to occur at higher temperatures than the preparation temperature.[7] Powder metallurgy uses sintering, since under this thermal effect the compressed shape of powders becomes consolidated and strengthened. The result is improved mechanical and physical properties of the material. In general, powders are heated (sintered) in inert atmospheres or in a vacuum at a temperature below the melting point of the metal or, in alloys, of the metal with the highest melting point. As the sintering proceeds, the total surface area decreases, which may alter the catalytic specific activity. Sintering can be prevented or inhibited by several methods,[5] in order to maintain the catalytic effect of powders.

B. DISPERSED METALS

Metals with very small metallic particles are usually described as dispersed metals. The unusual properties of dispersed forms have already been mentioned, and therefore the preparation of highly dispersed metals is stressed in order to ensure the best chemical and catalytic results. A concise classification of dispersed metals is relatively complicated. Davis and Klabunde[5] have established four general categories: a) thin metal films, b) metal powders, blacks, and colloids, c) porous and skeletal metals, and d) the so-called pseudoorganometallic powders.

1. Metal Films

Dispersed metal films or ultrathin metal films are composed of very small isolated metal crystals, which are often defined as crystallites. The discontinuity is, therefore, the main characteristic of these films in contrast to the continuous thicker films. Likewise, ultrathin films present small crystallite sizes (<0.5 nm) and high surface areas (from about 150 cm^2 to several m^2).[7,8] In some cases, the film can be as thin as a monolayer of atoms, which constitutes an ideal model of a highly dispersed metal catalyst.

Ultrathin films can be prepared mainly[9] by a) thermal evaporation, b) cathodic sputtering, and c) chemical technology. The latter includes electrodeposition and chemical vapor deposition methods. Despite their considerable importance in catalysis, these films have severe limitations. Thus, sintering is an unavoidable process when they are used in catalytic reactions at higher temperatures than at which they were deposited. The results are an increase in crystallite sizes and other structural changes. Furthermore, due to their inherent physical nature, they are easily contaminated.

2. Metal Powders

Metal powders represent the most important dispersed form under a chemical perspective, and are frequently employed in catalysis research. The main chemical methods for the preparation of finely divided metal powders will be treated in a further section. It should be pointed out that the concept of fine metal powder does not imply the formation of a dispersed metal in all cases. Some activated metal powders with high reactivity fall into the bulk metal category, which has been already described. In any event, an increase of reactivity can be expected to occur in metal powders with very small particles.

In contrast with metal films, metal powders are not easily poisoned initially. Several factors[5] have been suggested, although the greater porosity of metal powders must cause a retarded diffusion of contaminant molecules. However, once contamination occurs it is often irreversible. Furthermore, attempts to remove the contaminants may lead to sintering. For this reason, many metal powders are prepared *in situ* and used immediately. The main advantage of metal powders over single crystals and films is their much larger surface areas (0.1-20 m^2/g). This enables extreme experimental conditions, such as larger amounts of reactants or higher pressures. Moreover, large amounts of reaction products can be formed, since adsorption phenomena on other reaction surfaces (vessel, electrodes, etc.) are reduced.

With metal powders, sintering as well as nucleation and growth of metal particles seem to be very sensitive to the preparative conditions, and to the nature of the starting salt to produce the metal powder by thermal decomposition.[5] Metal powders are produced by either chemical or mechanical means. In chemical powdering, either a compound of the metal is reduced by a chemical agent or a liquid solution containing the metal is electrolyzed. Metal powders used in catalysis are generally prepared by hydrogen reduction of metal oxide powder.[7]

Table 1.1

Preparation of Metal Powders by Reductive Precipitation and Hydrogenation

Metal	Salt	Reducing Agent	H_2, T (°C)	Surface Area (m^2/g)
Pd	$(NH_3)_2PdCl_4$	$NaBH_4$	300	12.8
Pt	$H_2PtCl_6.6H_2O$	$NaBH_4$	300	3.7
Rh	$RhCl_3$	$NaBH_4$	300	30.5
Ru	$RuCl_3.3H_2O$	$NaBH_4$	300	26.4
Ru	$RuCl_3.3H_2O$	HCHO/KOH	200	4.0
Ru	$RuCl_3.3H_2O$	HCHO/KOH	200	33.0
Ir	$IrCl_4$	HCHO/KOH	200	11.0
Rh	$RhCl_3$	HCHO/KOH	200	7.0

(Reprinted with Permission from Davis, S. C.; Klabunde, K. J. *Chem. Rev.* **1982**, *82*, 153. Copyright 1982 American Chemical Society)

The oxide powder is usually formed by precipitation of the metal as hydroxide, carbonate or basic carbonate, or by thermal decomposition of the appropriate salt. Cold processes give a better metal dispersion than higher temperature treatments. Table 1.1 summarizes the effects of some preparative methods on dispersions of metal powders.[10-12]

These include reductive precipitation of metal salts in aqueous media followed by hydrogen reduction of the precipitate. The data of this table reveal important differences in surface areas depending on preparative procedures. Thus, Ru particles prepared by reduction of ruthenium trichloride in aqueous sodium borohydride gave a powder with a surface area of 26.4 m^2/g, but curiously, reduction of the same salt in aqueous formaldehyde afforded powder with surface areas of 4 and 33 m^2/g by two different groups.[11,12] It is very important, therefore, to examine a great number of experimental parameters that can affect the nature of the metal powder.

3. Organometallic Powders

Organometallic powders constitute an interesting dispersed form of metals, which have been prepared some years ago.[13-15] Under a more precise terminology, they must defined as pseudoorganometallic powders,[5] consisting of metal clusters/particles in organic media (substrates or solvents).

In principle, it is quite difficult to ascertain the exact nature of these powders. Actually, metal clusters are divided into the so-called naked and ligand-stabilized metal clusters. A naked cluster is essentially an aggregate of metal atoms without any protecting ligand shell. They are usually produced by laser evaporation from massive metals. Since the outer atoms of naked clusters are electronically unsaturated, they are extremely reactive compounds. However, these clusters can be stabilized on supports and then can serve as efficient and valuable catalysts.

In contrast, ligand-stabilized metal clusters can be formed by means of several, well-established chemical procedures. The properties of these molecular clusters are similar to those of transition metal complexes, and therefore determined by the nature and properties of ligands.

Under these general considerations, pseudoorganometallic powders can be visualized as naked clusters or particle aggregates, which are stabilized by another molecular fragment at low temperatures. In fact, hydrocarbon fragments support the metal particles and numerous transition metals, including nickel, cobalt, and iron can be stabilized.

The cycle of vaporization and further condensation of metal atoms (see section II) offers an attractive route for the preparation of highly active metal slurries. When nickel atoms (vapor) were codeposited with pentane vapor at 77 K (-196 °C), a black metal-alkane matrix resulted. Upon warming to pentane melting point (-130 °C) a slurry of very small nickel particles (or clusters) in pentane was formed. After 2 h at

-130 °C, metallic particles reacted extensively with pentane, which was cleaved mainly into C_1 species. The corresponding Ni-alkane slurry was evacuated by vacuum pumpoff to give the Ni-alkane pseudoorganometallic powder. This was quite thermally stable (>200 °C), but released methane and other hydrocarbon fragments when treated with water or hydrogen.[13-15] The C_1 and other alkane species remained bound to the nickel particle surfaces, although the nature of these interactions and the mechanism of formation of metal clusters will not be discussed here.[5,16,17] Scheme 1.1 describes the formation of these powders by using this procedure.

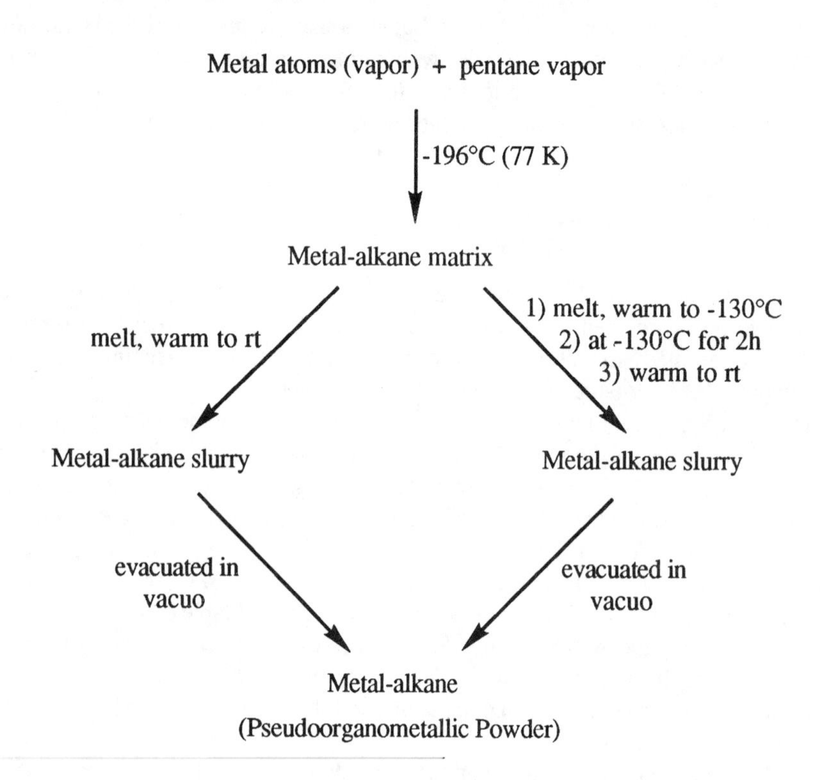

Scheme 1.1

(Reprinted with permission from Davis, S. C.; Klabunde, K. J. *Chem. Rev.* **1982**, *82* , 153. Copyright 1982 American Chemical Society.)

4. Other Dispersed Forms

Several concepts have been also proposed to denote other dispersed forms of metals. Metallic sponges are commonly referred to as blacks, albeit this denomination is not completely unambiguous. The term sponge denotes a porous aspect of particle aggregates as seen by an electron microscope. By using this technique, some metal powders are also defined as spongelike materials. Blacks or metallic sponges are easily prepared by the Willstätter method,[8] that is, reduction of metal salt by formaldehyde in alkaline solution. The term metal glass is often used to denote an amorphous metal powder.

Colloidal metals are fine dispersions of metal particles with diameters from about 10^3 to 10^4 Å, although colloids with smaller sizes (<100 Å) can be also obtained.[4] These dispersed forms, which are known as protecting colloids, can be held in a colloidal state by addition of a stabilizer[7]. They include substances such as gelatine, peptides or proteines, and macromolecules like poly(vinyl alcohol). Colloids are formed in aqueous media and a stabilizer is added so that hydrophilic colloidal suspensions are obtained. In the absence of a stabilizer, even with very small particles, colloids are easily coagulated and precipitate out of solution. Depending on the experimental conditions, metal colloids of different particle size are formed. The general methods for the preparation of colloidal metals have been reviewed.[4,5,7]

Porous metals can be stabilized with a minor structural component which prevents agglomeration of the metal particles.[5] In fact many common transition metals, such as iron, nickel or cobalt, are used in this form. Coprecipitation is the usual method for the preparation of stabilized porous metals. In this technique, metal hydroxides and the stabilizer (usually another metal, *e.g.*, Mg) are simultaneously precipitated from an aqueous solution. The dry precipitate is further reduced by hydrogen. The precipitates can be impregnated with structural promotors prior to hydrogen reduction. This addition enables the formation of catalysts with high surface areas and more stable toward sintering. Usually, the structural promotor can be a support such as silica, kieselguhr or molecular sieves. Thus, stabilized porous metals should be considered as supported catalysts in many cases.

Skeletal metals are commonly prepared by leaching a non-transition metal from an alloy or precipitate. Two salient examples of this dispersed form are Raney metals (Raney nickel and related catalysts), and the so-called Urushibara catalysts.[5] The former can be prepared by leaching aluminum from Ni-Al, Co-Al, or Fe-Al alloys with an aqueous base. Urushibara catalysts are generated from metal halides by precipitation with zinc or aluminum, followed by further digestion of the zinc or aluminum with an acid or alkali.

These intermetallic systems are important and useful reducing agents. One aspect worth mentioning is the fact that alloys of aluminum with nickel and other transition metals (from which Raney metal catalysts are prepared), are also reducing agents. Thus, the applications of nickel-aluminum alloy as a reductant in organic chemistry have been recently surveyed.[18] This alloy is commercially available as a gray powder consisting of almost equal proportion of aluminum and nickel, with traces of iron. The particle size is of ~65 μm, and therefore this form must be considered as a massive metal. The alloy is usually used in hydroxylic media such as an aqueous solution of NaOH, KOH, acid solutions (water with a weak organic acid), or mixtures of aqueous alkali with organic solvents. Although it is insoluble in water, reductions can be performed in this pure solvent.[18]

A Raney nickel alloy is generated from a Ni-Al alloy in sodium hydroxide aqueous solution in order to digest (leaching) the aluminum. The resulting mixture is decanted, washed with water under a hydrogen atmosphere, washed with aqueous ethanol, and stored in absolute ethanol.[5] The hydrogen atmosphere ensures a reduced state, which is essential for the activity of the catalyst. The main impurity in Raney nickel catalyst is due to the presence of aluminum oxide. In general, this oxide does not affect the catalytic activity and makes the catalyst resistant to sintering, probably due to a passivating effect. Furthermore, the crystallites of aluminum oxide on nickel surfaces increase the thermal stability of the catalyst, which undergoes a minor reduction in surface area.

Another interesting form of skeletal metal is the Urushibara catalyst,[19,20] introduced by this author in 1951. These catalysts are highly active reducing agents and often competitive with Raney nickel catalysts. They are prepared by precipitation of the catalyst metal (nickel, cobalt, or iron) from an aqueous solution of its salt (usually the chloride) by zinc or aluminum. The precipitated metal is then digested with an acid or an alkali, generally acetic acid or sodium hydroxide. Initially, Urushibara[19] tried to prepare a milder catalyst than Raney nickel, to reduce estrone into 17β-estradiol. Thus, zinc dust was added to a solution of nickel chloride which precipitated nickel metal. This was then added to an alkaline solution of estrone. However, this process can be carried out under two different conditions. Aluminum powder is added to reduce the estrone with the nascent hydrogen generated (eq. 1-1) and the nickel as a catalyst. Likewise, reduction is also successful when hydrogen was bubbled through the solution, but without adding aluminum.

$$2\,Al + 2\,OH^- + 6\,H_2O \longrightarrow 2\,[Al(OH)_4]^- + 3\,H_2 \qquad (1.1)$$

In the absence of alkalis the reduction failed. This fact can be attributed to the formation of zinc hydroxide chloride during the precipitation. This hydroxide is insoluble in water but soluble in alkali and acid, which remove Zn(OH)Cl and generate a very active catalyst. The Urushibara catalysts (U-catalysts) are comparable to Raney catalysts, and can be used for hydrogenation of alkynes and alkenes to alkanes, of carbonyl compounds to alcohols, and of aromatic nitro compounds to amines. Also, they effect reductive desulfurization, and can be used as dehydrogenation catalysts.[19,20]

Nickel-based Urushibara catalysts are denoted as U-Ni-A, with a crystallite size from 37 to 41 Å, and U-Ni-B (66-88 Å).[19] These and other acronyms are used to designate the method of preparation. U-Ni-A catalyst implies zinc as a reducing agent and further acid digestion with acetic acid. U-Ni-B involves digestion with base. U-

Ni-CA and U-Ni-CB catalysts are formed by precipitation at low temperatures followed by acid and base digestion, respectively. U-Ni-AA and U-Ni-BA are generated as U-Ni-A and U-Ni-B, but aluminum is now utilized as the reducing agent.[5,19]

References

1. Klabunde, K. J. *Chemistry of Free Atoms and Particles*; Academic Press: New York, 1980.
2. Benneman, K. H.; Koutecky, J. *Small Particles and Inorganic Clusters*; North-Holland: Amsterdam, 1985.
3. Alonso, J. A.; March, N. H. *Electrons in Metals and Alloys*; Academic Press: New York, 1989.
4. Henglein, A. *Chem. Rev.* **1989**, *89*, 1861.
5. For an excellent review on small metal particles: Davis, S. C.; Klabunde, K. J. *Chem. Rev.* **1982**, *82*, 153.
6. For a representative study on the corrosion of metallic magnesium by alkyl halides: Hill, C. L.; Sande, J. B. V.; Whitesides, G. M. *J. Org. Chem.* **1980**, *45*, 1020 and references cited therein.
7. Anderson, J. R. *Structure of Metallic Catalysts*; Academic Press: New York, 1975.
8. Bond, G. C. *Chem. Soc. Rev.* **1991**, *20*, 441 and references cited therein.
9. Chopra, K. L. *Thin Film Phenomena*; McGraw-Hill: New York, 1969.
10. Carter, J. L.; Cusumano, J. A.; Sinfelt, J. H. *J. Catal.* **1971**, *20*, 223.
11. Kobayashi, M.; Shirasaki, T. *J. Catal.* **1973**, *28*, 289.
12. Sárkány, A.; Matusek, K.; Tétényi, P. *J. Chem. Soc., Faraday Trans. 1* **1977**, *73*, 1699.
13. Davis, S. C.; Klabunde, K. J. *J. Am. Chem. Soc.* **1978**, *100*, 5973.
14. Davis, S. C.; Severson, S.; Klabunde, K. J. *J. Am. Chem. Soc.* **1981**, *103*, 3024.
15. Klabunde, K. J.; Efner, H. F.; Murdock, T. O.; Ropple, R. *J. Am. Chem. Soc.* **1976**, *98*, 1021.
16. Phillips, J. C. *Chem. Rev.* **1986**, *86*, 619.
17. Morse, M. D. *Chem. Rev.* **1986**, *86*, 1049.
18. Keefer, L. K.; Lunn, G. *Chem. Rev.* **1989**, *89*, 459.
19. Hata, K. *New Hydrogenating Catalysts: Urushibara Catalysts*; Wiley: New York, 1971; p 88.
20. Hata, K.; Motoyama, I.; Sakai, K. *Org. Prep. Proced. Int.* **1972**, *4*, 180.

II. METAL VAPOR CHEMISTRY

A. GENERAL CONSIDERATIONS

The outstanding activation method mentioned in Chapter I for the preparation of pseudoorganometallic powders is known as the metal vapor technique. This approach utilizes the high potential of metal atoms in vapor phase, which are co-condensed with an inorganic or organic substrate at low temperature.

Low temperature condensation of high temperature species is an old topic and has been extensively utilized.[1] Rice and Freamo[2] studied the thermal decomposition of hydrazoic acid. The gaseous imine radical formed condensed on very cold surfaces (at -196 °C) to a bright blue solid, which upon warming to room temperature gave ammonium azide. This preliminary and stimulating result marked the beginning of low temperature condensation reactions. A special feature of this vapor chemistry is that high temperature unstable species can be conveniently trapped under these cryogenic conditions.[3] However, a spectacular development of this technique was due to Skell and Westcott,[4] who vaporized carbon from a carbon arc under high vacuum. The carbon vapor produced was co-condensed with several organic compounds at 77 K (-196 °C) to provide new carbonaceous species, that is, compounds incorporating C_1, C_2, C_3, and other carbon vapor species.

It is noteworthy that high vacuum conditions must be maintained in order to prevent intermolecular collisions, which cause carbon atom aggregations. A vacuum apparatus was designed by the authors, and the surrounding walls were cooled with liquid nitrogen. Under this external refrigeration, the vapor of another organic compound was passed into the vacuum chamber and condensed with the carbon vapor on the walls. This work represents the practical beginning of this technique. Since then, atomic vapor chemistry has experienced a considerable advance and has been mainly developed by Skell, Timms, Klabunde, and Ozin among others, to name a few conspicuous researchers in this field.

Some general features of species and reaction conditions are discussed first. Species are generally vaporized at high temperatures (above 1000 °C) at low pressures (below 1 torr). It should be emphasized that many atomic and molecular species are thermodynamically stable only at high temperatures. From a more practical viewpoint, two requirements are necessary . Thus, high temperature species should be also reactive at low temperatures, and the process must occur at appreciable reaction rates. The latter is often a serious limitation in many organic reactions.

A great variety of species can be vaporized and then co-condensed with another compound. The simplest class of species as raw material for this purpose are the elements. Furthermore, gaseous atoms of elements (except the noble gases) fulfill the two basic criteria mentioned above. Atoms react faster than their corresponding elements due to a minor steric hindrance and the major availability of electrons in atomic orbitals. Likewise, vaporized atoms are species of higher energy than the elements in their ground state. Unfortunately, atoms of some elements vaporize at low temperatures in molecular forms, even if they are relatively stable in the gas phase with long lifetimes. For these atoms the conventional methods of vaporization, such as thermal dissociation and particularly electrical discharges are not recommended, since the discharge causes further polymerization or decomposition. Hydrogen and nitrogen atoms can be generated as long-lived atomic species by discharges remote from the reaction site.[5,6] In contrast, metal atoms show short lifetimes in the gas phase. However, recombination of metal atoms is also possible at low temperatures, and will be discussed later. It is obvious, therefore, that a pressure of a few torr is an essential requirement of this technique in order to minimize atom polymerization. Moreover, a cryogenic temperature ensures a good vacuum during the process. Experimentally, this must be a temperature where the vapor pressure is $<10^{-5}$ torr.

Since vaporization produces excited states of atoms, a convenient population of these species is required to allow a practical experiment. This should be performed on a scale which enables the isolation and further characterization of compounds produced. By thermal methods, even at 3000 °C, the population of excited states for some atoms can be relatively low. Nevertheless, low pressure electrical discharges and photochemical methods (photolysis) can generate considerable quantities of excited atoms.[1] In contrast, the rate of formation of species is usually lower than in thermal techniques.[1,7]

The additional energy of the gaseous atoms compared with their normal states[8] is clearly attributed to the inherent activation of the elements at high temperatures. At cryogenic temperatures (usually -196 °C), the enthalpies of formation will not be greatly affected by co-condensation reactions.[7] Then, chemical behavior of gaseous atoms runs parallel to the chemistry expected for the elements in their normal states. This is true for all the atoms having enthalpies below 40-50 kcal/mole. Larger values favor endothermic processes, which are not likely from the normal states of the elements. With the exception of halogen atoms and alkali and alkaline-earth metal atoms, which have quite low enthalpies of formation, the rest of the elements and particularly transition metal atoms possess a considerable intrinsic energy.

This feature can be further illustrated by considering the vaporization temperature of metals.[1] Thus, vaporization of the first-row elements takes place at lower temperatures than for the second- and third-row elements. This is particularly important for some heavy metals such as Nb, Mo, Ru, Os, or Ir, which require temperatures above 2500 °C to be vaporized. Therefore, the practical use of these atomic vapors is scarce and large-scale syntheses are not easily conducted. In contrast, other important elements, Cr, Mn, Fe, Co, Ni, Pd, Cu, Ag, and Au are vaporized at temperatures from 1100 to about 1600 °C. Likewise, the rate of formation for some gaseous atoms (Cu, Ag, Mn, Co, or Ni) can be as high as 1 mole/h. The group I and II metals (with the exception of beryllium which has an enthalpy of formation of 78 kcal/mole) can be easily vaporized at temperatures below 1000 °C under vacuum.

In any event, the formation of high temperature species is highly dependent on the vaporization method. Timms[1] has conspicuously reviewed the scope and limitations of different techniques for this purpose, including nonconventional methods like plasmas or photolysis. The simplest apparatus consists of a crucible made from a metal wire, which is heated by an electric current. The so-called resistive heating is frequently employed for the vaporization of metals. Solid or molten metal is heated until its vapor pressure becomes ~10^{-2} torr, so that a rapid evaporation takes place. The crucible can be mounted inside a large vacuum chamber and the walls are cooled at low temperature (liquid nitrogen). Then, metal vapors and the vapor of substrate are codeposited on the inside walls of the reactor.

By using crucibles, a wide variety of metals are successfully vaporized. These include Cr, Mn, Fe, Co, Ni, Cu, Ag, Au and Sn among others. Unfortunately, the use of crucibles is limited up to a temperature of about 1800 °C. From a practical viewpoint, higher temperatures constitute a serious limitation for many laboratory experiments.

Also, electrical discharges have been employed for the formation of high temperature species.[5] However, this method is not usual in synthesis, since discharges favor repolymerization of gaseous atoms. Other disadvantages are poor yields of products and unpredictable reactions.

Besides enthalpies of formation and vaporization temperatures, other factors must be considered in low temperature condensations. Thus, the lifetimes of the electronically excited states of atoms should be long enough to ensure their survival between formation and reaction with another substrate on a cold surface. Some excited states of transition metal atoms have very short lifetimes (a few nanoseconds). An additional complication of this matter arises when different electronic states of vapor species exhibit different reactivities, even at -196 °C. This aspect has been studied by Skell and co-workers[9-11] with carbon vapor species generated by using an arc.

Other cryogenic liquids can be used for the external refrigeration, such as liquid hydrogen (20 K) or even liquid helium (4 K). However at these temperatures, the activation enthalpy (or entropy) of such processes is dramatically affected. Much lower reaction rates constitute a severe handicap under these conditions. As Timms has remarked,[1] an increase of the temperature of the cold surface minimizes the differences in reaction rates due to small differences in activation enthalpies. Accordingly, the process of condensation should be conducted at higher temperatures (*e.g.*, at -78 °C). But, other limitations should be kept in mind, such as increases of vapor pressures, repolymerization of atoms, and inherent instability of products. Therefore, a plethora of thermodynamic and kinetic factors should be examined when a low temperature condensation experiment is envisaged.

Until now, we have considered the formation of atomic species at high temperatures. However, a great variety of molecular species can be easily generated in vaporization processes.

Transition metals vaporize as mainly monoatomic species.[12] Although some metal dimers have been also detected,[13] these species constitute a very small fraction (traces) under an efficient vaporization. The formation of molecular species occurs by the presence of a great number of gas molecules, which is strongly favored at high temperatures and low pressures.

Molecular species derived from metals can be useful reagents in low temperature condensations. Several chemical processes allow molecular species to be formed in the vaporization.[1] In general, vaporization of condensed phases generates species which will condense again on a cold surface. Condensed phases can vaporize both as monoatomic and molecular species. The latter are favored when the heat of polymerization is less than the corresponding heat of vaporization of the gaseous atom. Obviously, polymers possessing strong bonds will be stable in the vapor phase at high temperatures. In fact, halides, oxides, sulfides, and hydrides among others can be formed at high temperatures and then may react with inorganic and organic compounds at low temperatures. A special peculiarity of some of these species is that they contain elements in unstable or anomalous oxidation states, such as BF, CF_2, $SiCl_2$, AlF, ScF, LaCl, or polycarbonaceous species (C_2, C_3, or C_4). The main advantage of such compounds is derived from their inherent tendency to the normal oxidation state by reaction with other substrates. In these conditions, the combination with other molecules would compete with polymerization or disproportionation processes. Unfortunately, this desirable behavior occurs only with a few reactive species, like boron monofluoride.[14] Similarly, high temperature molecular species having elements in their normal oxidation states can be very unstable compared with these molecules at ordinary or lower temperatures, so that chemical reactions of those species are of possible synthetic value.

It has been already mentioned that metals vaporize as monomers, and because of metal-metal bonds have energies relatively low compared with the enthalpies of vaporization of atoms.[15,16] Bond dissociation energies are particularly low for alkali and alkaline-earth metals. These values rule out the practical existence of these diatomic species, which are exclusively detected by mass spectrometry and other spectroscopic techniques.

In general, the products of low temperature condensations are obtained in poor or moderate yields. Nevertheless, the yields of these reactions can be improved by using a large excess of the substrate, since this minimizes the possibility of collisions between atoms or molecules of the high temperature species.

Metal particles can be regarded as naked clusters, so that the highly extreme reactivity of these species constitutes an unavoidable problem. In order to prevent polymerization or disproportionation, high temperature species can be relatively stabilized as pseudoorganometallic powders (Chapter I). In a series of excellent works, Skell and his co-workers[9-11,17] isolated small quantities of carbon vapor species in neopentane matrices at 77 K (-196 °C), which further reacted with different substrates. Later, Brewer[18] was unsuccessful in isolating metal atoms in a sulfur hexafluoride matrix at -196 °C. At much lower cryogenic temperatures (-250 °C), and using a very large excess of sulfur hexafluoride, naked atoms were then obtained. They remained as isolated particles on warming to -196 °C.

Klabunde and his associates[19-22] have made valuable contributions to the preparation of pseudoorganometallic powders. Metal atoms were vaporized and then codeposited at -196 °C with another inert substrate (usually a hydrocarbon) to form a rigid matrix. Metal-alkane matrices were obtained, which can react with other compounds. However, pseudoorganometallic powders are not indefinitely stable at low temperatures. Thus, at higher temperatures (*e.g.*, the hydrocarbon melting point), metallic particles react with hydrocarbon molecules to produce several carbonaceous species.

Matrix isolation experiments are particularly important in co-condensation reactions. High temperature species are isolated in noble gas matrices at cryogenic temperatures (as low as 20 or 4 K). The technique is frequently employed in spectroscopy as a useful method.[23] Species isolated in an inert matrix can be surrounded by another compound. Condensation is expected to occur by diffusion at higher temperatures.[1,24] Reactions of metal and nonmetal atoms or clusters, with some inorganic and organic substrates at low temperatures under matrix isolation conditions are currently utilized in this research area.[25-29]

Metal vapor chemistry is an important tool for the preparation of many organometallic compounds. In fact, for a number of fundamental organometallics, co-condensation processes represent the exclusive way of access. Excellent surveys dealing with this topic have appeared in the literature,[1,30-44] and they constitute a detailed and valuable body of information on this technique.

Although metal vaporization should be considered as a physical rather than chemical activation, the scope of this strategy will be discussed in relation to the formation of active pseudoorganometallic powders. Likewise, these powders have proved to be highly active catalysts in several processes such as hydrogenation, hydrogenolysis, or dehydrogenation,[45-47] which have potential industrial importance.

B. ALKALI AND ALKALINE-EARTH METAL VAPORS

Despite their comparatively low enthalpies of vaporization, the use of alkali and alkaline-earth metals in co-condensation processes at low temperatures is rare compared to transition metals. This is due, in part, to the minor tendency of group I and II metals to form complexes with relevant ligands, such as olefins or aromatic rings.

The reaction of alkali metal vapors with organic halogen compounds generates mono- and diradical species. Thus, reaction between methyl bromide and sodium vapor into a vessel containing iodine vapor produces methyl iodide.[48] The process has been studied extensively from a kinetic viewpoint.[49] Product analysis suggests the intermediacy of alkyl, haloalkyl, vinyl, or dihaloalkyl radicals.[50-54] In a detailed study, Skell and Doerr prepared the 1,3-diradical trimethylenemethane by the reaction of potassium vapor with 2-iodomethyl-3-iodopropene.[51,54] By using high temperatures (from about 228 to 267 °C), some Wurtz-type syntheses were successfully carried out. The products consisted of 2-methyl propene, methylenecyclopropane, 1,4-dimethylenecyclohexane, *p*-xylene, and a mixture of butanes. When the reaction was performed with 1,3-diiodo-2-methylpropane, the major product was methylcyclopropane and compounds due to dimerization were not detected. This scarce tendency to undergo dimerization processes was also confirmed with other saturated diradicals formed from dihalides.

The process was conducted on a sodium-potassium alloy spray apparatus,[53,54] wherein a 200-mm helium pressure was maintained. The diiodide and a fine spray of Na-K (containing 88% K) were combined for 18 min with the aid of a helium stream. The temperature of the reaction zone varied between 232 and 236 °C. The products were passed successively through two traps at -196 °C and analyzed.

Mile[55] reviewed the formation of free radicals at low temperatures. In these studies, the reaction between a halohydrocarbon and alkali atoms (usually sodium atoms) in a rotating cryostat, was utilized to generate hydrocarbon radicals.

A rotating cryostat is simply a stainless steel drum, which contains liquid nitrogen and is rotating (~2400 rpm) under high vacuum (less than 10^{-5} torr). Vapors of both sodium metal and halohydrocarbon are then directed onto the outer surface of the rotating drum from the opposite sides. In this way, each reactant is deposited on a freshly formed layer of the other. Hydrocarbon radicals are rapidly generated and trapped by the next layer of halohydrocarbon molecules. The final result is the formation of a deposit of free radicals and halohydrocarbon molecules. The radicals produced are immediately transferred, at -196 °C under vacuum, into a tube for ESR experiments.

In addition, an excellent survey on the reaction of alkali metal vapors with halides of hydrocarbons and silahydrocarbons has appeared recently in the literature.[56]

Reactions between alkali metals and some ligands afford really charge-transfer complexes rather than clusters or organometallic complexes. Thus, lithium atoms co-condense with acetylene vapor to form complexes of the type Li(HC≡CH) and Li_2(HC≡CH). Additionally, other alkali metal atoms such as Na, K, or Cs are co-condensed with acetylene vapor to give similar complexes.[43] These can be represented satisfactorily as $[M]^+[HC{\equiv}CH]^-$. Species were isolated in inert matrices at low temperatures. A detailed study on the reaction of sodium atoms with acetylene was reported by Kasai.[57] In his experiments, variable quantities of both reactants were codeposited in argon matrices at 4 K. With a ratio of 0.1% sodium atoms to 1% acetylene, three different reaction levels were detected by ESR experiments. The first stage observed after co-condensation was the isolated matrix of sodium atoms. After irradiation, two subsequent stages were produced, which consisted of an acetylene radical anion and a vinylidene-sodium complex, respectively. A higher concentration of acetylene (~4%) gave acetylene dimers, which reacted with sodium atoms to form vinyl radicals.

In a classical paper, Skell and Girard reported[58] the preparation of nonsolvated Grignard reagents by co-condensing Mg atoms with alkyl halides (Table 2.1). The formation of RMgX reagents did not occur at 77 K (-196 °C). The authors proposed the formation of a black RX-Mg complex that rearranged to RMgX by warming.

Table 2.1

Unsolvated Grignard Reagents

Alkyl Halide	Yield (%)[a]
$CH_3CH_2CH_2I$	76
$(CH_3)_2CHBr$	55
$(CH_3)_3CBr$	5
C_6H_5Cl	58
$CH_2{=}CHBr$	78[b]

[a] Yields were determined by hydrolysis
[b] Products stable at -78 °C

Curiously, these nonsolvated Grignard reagents display unusual characteristics. Acetone was not added by these reagents, and crotonaldehyde gave 1,2-addition rather than 1,4 as usually. Probably, the anomalous behavior of these solvent-free Grignard reagents is due to their special physical nature. Skell and Girard proposed that Mg atoms are weakly bound to the halogen atom, so that metal atoms can be easily transferred to other alkyl halides. It should be pointed out that magnesium atoms were formed by thermal vaporization at high vacuum ($\sim 10^{-5}$ torr), which produces ground-state magnesium atoms (^1S). Alternatively, excited state magnesium atoms (^3P) are formed by arcing.

Likewise, ground-state magnesium atoms (^1S) were codeposited with a mixture of 2-butene and ammonia to produce *trans*-2-butene (87% yield) upon warming.[59] A proposed mechanism seems to involve the formation of a transient alkene radical anion, which reacts with ammonia to form 2-butene radical. Further reaction with magnesium atoms and proton abstraction from ammonia yields *trans*-2-butene.[43] It should be pointed out that the thermally stable (^1S) magnesium atoms were essential in this process. When a mixture of both reactants were co-condensed on a magnesium film, a very poor yield (less than 5%) of product was obtained. Also, in the presence of metastable, excited state magnesium atoms (^3P), *trans*-2-butene was produced in 28% yield. This was attributed to a possible competition between 2-butyne and ammonia toward the highly reactive excited magnesium atoms.

Similarly to organomagnesium compounds, the formation of calcium atoms constitutes an extremely promising and useful method for the preparation of highly reactive organocalcium compounds.[60,61] Calcium atoms have been scarcely employed in metal vapor chemistry. An interesting result (Scheme 2.1) is the oxidative addition of calcium atoms (strong base) to perfluoroolefins to give C-F insertion products.[62] Perfluoroalkanes were inert under the same reaction conditions.

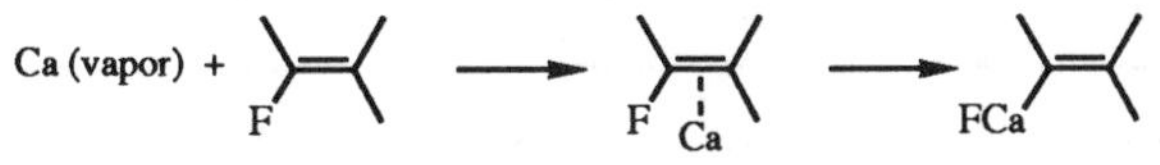

Ca (vapor) + F ⟶ F Ca ⟶ FCa

Scheme 2.1

C. TRANSITION METAL VAPORS

Transition metal atoms have been extensively employed because of their ability to form organometallic compounds with a wide variety of ligands.

Many important and useful organometallics contain inorganic ligands such as CO, NO, or phosphines. Condensation of metal vapors with these ligands at -196 °C constitutes a useful way of making zero-valent complexes. Thus, the normal routes to trifluorophosphine complexes utilize high pressures.[63] However, when metal vapors are condensed with trifluorophosphine in at least a 1:8 mole ratio, the corresponding complexes are obtained in high yields.[1,64] Table 2-2 summarizes the approximate yields of such experiments.

Table 2.2

Reactions of Transition Metal Vapors with PF_3

Metal	Product	Yield (%)[a]
Cr	$Cr(PF_3)_6$	65
Mn	—[b]	—
Fe	$Fe(PF_3)_5$	25
	$(PF_3)_3Fe(PF_2)_2Fe(PF_3)_3$	25
Co	$[Co(PF_3)_4]_x$[c]	50
Ni	$Ni(PF_3)_4$	>85
Cu	—[b]	—
Pd	$Pd(PF_3)_4$	70

[a] Yields based on the metal vaporized. [b] No volatile product. Decomposition on warming. [c] Polymeric product.

Carbon monoxide cannot be used directly in these low temperature condensations because of its high vapor pressure at -196 °C. However, the very important (and highly poisonous) reagent $Ni(CO)_4$ can be prepared employing an alternative route[1] which uses carbon dioxide as the precursor of the ligand CO. When nickel vapor is condensed with CO_2 at -196 °C, a solid is instantaneously formed which releases CO on warming above -150 °C. Nickel tetracarbonyl is obtained in about 10% yield.

A considerable progress in metal vapor techniques was introduced by Ozin and co-workers, who developed cryopumped electron guns and cryopumps in the extended temperature range from 300 to 10 K. This enabled access to ligands such as CO, CH_4, H_2, N_2, O_2, or C_2H_4 that are noncondensable at -196 °C, *i.e.*, liquid nitrogen temperature.[65,66]

Table 2.2 seems to reveal that PF_3 adds randomly to metal vapors. This tendency is also illustrated in the synthesis of PH_3 complexes. Thus, an attempt to form

$Ni(PH_3)_4$ by co-condensation of nickel vapor with phosphine at -196 °C was not successful.[1] Instead, the process was accompanied by hydrogen evolution and no volatile product was obtained. When a stronger ligand like PF_3 competes with PH_3, interesting species can be formed. Reaction of nickel vapor at -196 °C with an equimolar mixture of PF_3 and PH_3 affords several species, but only $Ni(PF_3)_4$, $Ni(PF_3)_2(PH_3)_2$, and $Ni(PF_3)_3PH_3$ are stable enough to be isolated.[64]

The well-established dehalogenating properties of copper atoms in organic synthesis can be also used at cryogenic temperatures. Timms studied the reaction between copper vapor and boron trichloride at -196 °C to give diboron tetrachloride.[1,67] Moreover the reaction can be performed on a large scale, and about 300 mmoles of copper can be evaporated in 50 min and condensed with the substrate to give about 10 g of diboron tetrachloride (eq. 2.1).[68]

$$2\,Cu\,(vapor) + 2\,BCl_3\,(g) \xrightarrow{-196\,^\circ C} 2\,CuCl + B_2Cl_4 \qquad (2.1)$$

Silver vapor also reacts with boron trichloride, but diboron tetrachloride is obtained in a poor yield. However, B_2Cl_4 was not formed from the reaction with nickel atoms under the same conditions.

Halogen atom abstraction also occurred with other inorganic compounds. Copper atoms reacted with $SiCl_4$ at -196 °C to give perchlorosilanes. Curiously, the success of this synthesis was dependent on the vaporization technique. When copper vapor was generated from copper heated in a crucible, reaction with $SiCl_4$ did not occur whereas the process took place with the copper vapor formed by electron bombardment vaporization. The reason is very likely the formation of excited copper atoms with the last method. Copper and silver vapors react rapidly with phosphorus trichloride at -196 °C to give P_2Cl_4, but the yields are small (10-15%) due to the coordination of dechlorinated species to the metal.[1]

Copper, silver, and gold atoms also react with simple alkyl halides at -196 °C. The process involves the formation of alkyl radicals which undergo homocoupling and disproportionation.[1,67] The reaction of copper vapor with ethyl bromide yields butane and very small amounts of ethylene and ethane. Silver vapor produced ethylene and ethane with a small amount of butane. Finally, gold vapor gave butane and an almost equimolar mixture of ethylene and ethane. Therefore, the reactions are scarcely useful and only copper atoms favor carbon-carbon coupling.

The importance of metal vapor chemistry arises from the reaction of transition metal vapors with organic substrates. In an attractive paper, Skell and co-workers reported that the co-condensation of iron atoms and 1,3-cyclohexadiene at 77 K followed by warming at 97 K for 1 h, afforded a mixture of cyclohexene (33%), benzene (36%), and cyclohexadiene (30%). A very small amount (~2%) of the complex benzene(1,3-cyclohexadiene)iron(0) was also produced.[69] The reaction is really a catalytic disproportionation of cyclohexadiene to cyclohexene and benzene, and a few mmoles of iron atoms converted 120 mmoles of the cyclohexadiene into the product mixture.

In an excellent review,[30] Klabunde has surveyed the reactivity of many types of organic molecules by using a variety of main-group and transition metal atoms. Some reactions deserve a special consideration with regard to improved preparations and anomalous reactivities.

Among a great number of chemical properties, the reactivity of a metal-substrate pair is markedly influenced by the acid-base properties and the π-complexation, that is, the availability of orbitals to establish effective π-bonds. It would be expected that transition metals with available d-orbitals should complex well with unsaturated molecules, which have nonbonding or π-electrons. Also, single electron-transfer processes (SET mechanism) must be particularly important when metal atoms interact with weak bonds.

Since these processes take place at very low temperatures, reactions possess very low activation energies. A competing process (with low activation energy) and an almost unavoidable problem of this technique, is the repolymerization of the vaporized species.

Thus in the presence of CO, metal atoms form metal carbonyl dimers in a low-temperature matrix.[70] Likewise, when π-electrons are not available in the organic substrate, metal atoms repolymerize even at 77 K. Molecules with σ-electrons and nonbonding electrons complex rarely with metal atoms. In fact, alkanes and perfluoroalkanes do not prevent metal atom recombination at low temperatures.[30]

Oxidative additions of carbon-halogen bonds to transition metal atoms represent a suitable way of access to the preparation of non-solvated organometallic compounds. The reaction with calcium atoms has been previously reported.[62]

Perfluoroalkylzinc halides have been prepared from the corresponding zinc atoms (Scheme 2.2).[71] Various fluorinated substrates were co-condensed with zinc atoms at liquid nitrogen temperatures. The resulting organometallic matrices decomposed rapidly upon warming (-80 °C). Thus, trifluoromethyl iodide (**1**) gave mainly hexafluoroethane and tetrafluoroethylene as volatile products. Furthermore, fluorine and iodine-containing polymer and fluoride and iodide ions were also detected. The presence of water had a scavenging effect and fluoroform was formed. The results of this conspicuous research indicated free-radical decompositions of fluoroorganic zinc compounds. It should be noted that only perfluoroalkyl iodides reacted with zinc vapor. Moreover, these nonsolvated organozinc reagents were unstable even at low temperatures, which contrasts with the corresponding solvated analogs.

$$\underset{\mathbf{1}}{CF_3I} + Zn\,(vapor) \xrightarrow{-196\,^\circ C} \underset{\mathbf{2}}{[CF_3\text{-}Zn\text{-}I]} \xrightarrow{-80\,^\circ C} \underset{\mathbf{3}}{CF_3\text{-}CF_3} + \underset{\mathbf{4}}{CF_2{=}CF_2} + \underset{\mathbf{5}}{CHF_3}$$

Scheme 2.2

Codeposition of silver vapor with fluoroalkyl iodides R_fI (*e.g.*, CF_3I) at -196°C followed by matrix warmup gave AgI and perfluoroalkylsilver compounds. These reactions were the first examples of the formation of organometallic compounds employing the metal atom deposition, and of the formation of primary perfluoroalkylsilver compounds. It should be mentioned that, contrary to expected results, high R_fI/Ag ratios favor the formation of R_fAg compounds. This point was confirmed by matrix dilation experiments.[72] Similarly, oxidative insertion of palladium atoms into carbon-fluor bonds has been also observed.[73]

Metal insertions have also been performed in the preparation of some RMX, ArMX, and RCOMX (M = Ni, Pd, Pt, Co, Fe, Mn, and others). Although in some cases reactions only afforded metal halides and products resulting from homocoupling and/or disproportionation processes, it is also possible to produce the desired reagents. Low temperatures and inert conditions were essential to stabilize the organometallic species. In general, after co-condensation of metal atoms and alkyl(aryl) or acyl halides, the matrix was warmed (usually at -78 °C), and then organometallic species were trapped with another substrate such as triethylphosphine.[30,31,73-75] The major portion of this research has been carried out by using fluorinated alkyl and aryl halides, since it is known that fluorinated systems form stronger carbon-metal bonds. Scheme 2.3 visualizes these processes for reactions of alkyl, aryl, or acyl halides with palladium or nickel followed by trapping of product.

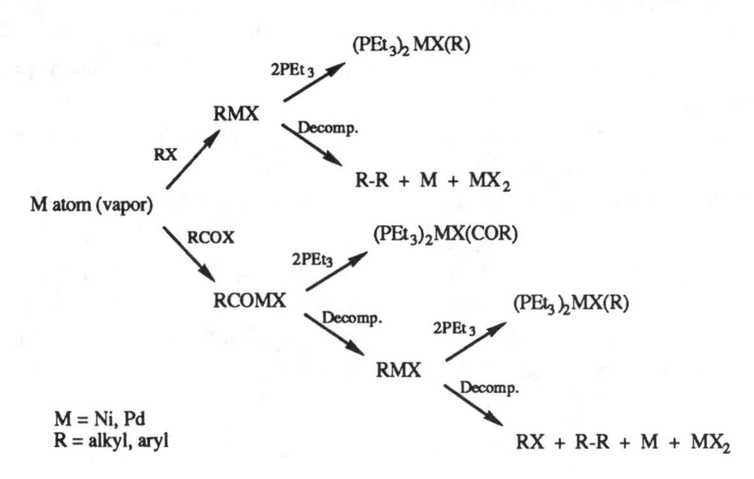

Scheme 2.3

Several and important conclusions can be obtained from these experiments: a) RMX species can be definitely formed at low temperatures, before phosphine trapping; b) aryl halides are better than alkyl halides in the efficiency of insertion and the concomitant stability of the intermediates; c) non-fluorinated alkyl halides only produced very unstable organometallic species. In fact, phosphine trapping was not successful, and decomposition products were formed at low temperatures. Thus, in the co-condensation of nickel atoms and methyl iodide, methane was detected even at -196 °C. Similarly, Pd atoms and ethyl iodide formed a wide variety of products, such as ethane, ethylene, butane, and 1-butene. Free radicals are apparently involved in these processes.[31,74] The situation is also applicable to non-fluorinated acyl chlorides, which released CO at very low temperatures and organometallic species could not be trapped. Curiously, the alkylpalladium chloride generated from benzyl chloride and Pd atoms was synthesized, and did not decompose upon warming to room temperature.[31] This result demonstrates the importance of aromatic π-systems in coordinating with the open sites on these organometallic intermediates. d) The efficiency of the insertion follows the order C-I > C-Br > C-Cl. Although a detailed mechanism for these processes has not been postulated, some stereochemical evidences were also confirmed. Trapping with phosphine always provides *trans* complexes. The *trans* geometry is sterically favored over *cis*, which would indicate that phosphine addition to RMX species must proceed in a stepwise and selective fashion. When the halide, phosphine, and metal atoms are condensed simultaneously, the formation of RMX species may not be the first step. Thus, metal-phosphine complexes could be formed and then react with the halide. Nevertheless this sequence must be discarded, since metal-phosphine combinations evolve hydrogen, which was not observed in simultaneous co-condensations.[75]

Olefins and unsaturated halides also react with metal atoms. Thus, nickel, platinum, and palladium atoms can be co-condensed with allyl chlorides or bromides to give the corresponding π-allyl dimers.[30]

Unsubstituted olefins react with a variety of metal atoms. The experiments have made possible to establish differences and similarities between the metals. Organometallic species could not be isolated, but their presence was evidenced by D_2O hydrolysis, in order to mark carbon-metal bonds. Thus, reaction of aluminum atoms with 1-propene afforded mainly σ-complexes of 1,2-dialuminoalkanes.[76] Furthermore, spectroscopic studies demonstrated that the dialumino-alkanes were formed even at -196 °C. Unlike aluminum atoms, nickel and platinum produced mainly π-complexes.[77-79] Thus, 1-propene and nickel atoms (vapor) were codeposited to give the corresponding olefin-Ni matrix. Upon D_2O hydrolysis, nondeuterated propene and monodeuteriopropene were detected (Scheme 2.4).[78]

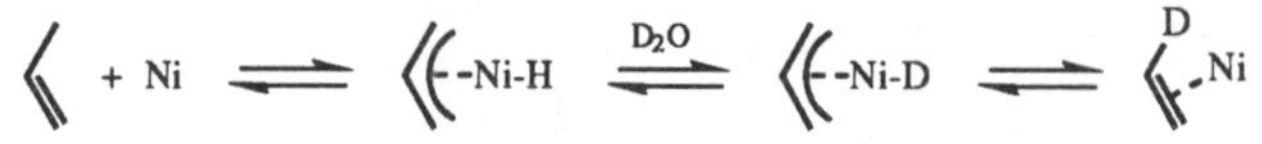

+ Ni ⇌ -Ni-H $\xrightarrow{D_2O}$ -Ni-D ⇌ D -Ni

Scheme 2.4

Although carbon-halogen insertions have been extensively studied, other attractive oxidative insertions have been envisaged (C-C, C-O, O-O, S-S, N-X, S-X, Si-X, etc.). Oxidative insertions of nickel and palladium atoms into carbon-carbon and carbon-oxygen bonds have been successfully reported.[75] Perfluoroolefins form complexes with these metallic atoms at low temperatures. Palladium vapor was co-condensed with perfluorobutane (**6**) and triethylphosphine to give the metal-alkene matrix. After warming to 25 °C, excess olefin and phosphine were evacuated (vacuum pumpoff). Conventional work-up afforded crystalline octafluoro-2-butene-bis(triethylphosphine)palladium (**7**) in moderate yield. The palladium-olefin intermediate is stable at -90 °C and can be trapped with triethylphosphine at that temperature to give (**7**) in 30% yield. On warming to -20 °C, extensive decomposition took place and palladium metal and olefin were isolated (Scheme 2.5).[75]

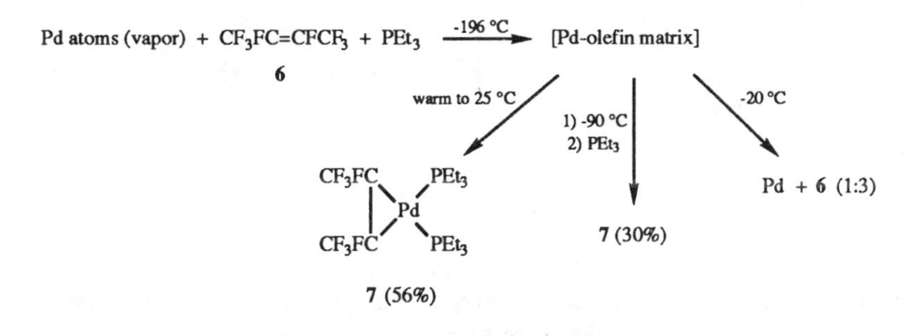

Scheme 2.5

The authors noted that this process should be considered an oxidative addition[80] of a C-C bond to a metal atom, since olefin and palladium become σ-bonded in the complexes.

Stable arene sandwich compounds can be easily prepared by condensing metal atoms with the appropriate arene ligand. In general, electron-withdrawing groups have a stabilizing effect.[31,78,81,82] Thus, bis(arene)chromium(0) complexes possessing the F- and CF_3-substituents are indefinitely stable to air at room temperature. Numerous π-arene complexes of transition metals with a wide variety of substituents are actually prepared by co-condensation techniques.[30,31,40,83,84]

Neutral sandwich complexes of annellated arenes such as compounds (**8**)-(**10**), are also accessible by this method. It should be pointed out that in the case of arenes highly annellated like anthracene, phenanthrene, triphenylene, and coronene, the metal atom coordinates preferentially to the ring with the highest index of local aromaticity, that is, the ring possessing the major number of Kekule structures among its canonical forms. In general, this is the terminal and least annellated ring.[84]

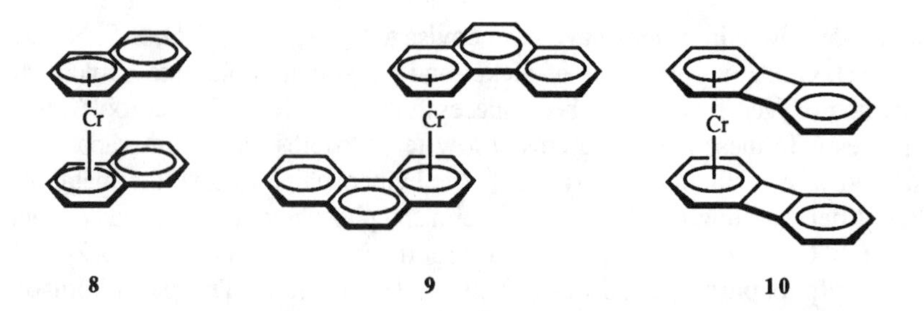

Skell and his team were able to synthesize 1,3-butadiene complexes of transition metals, such as Fe, Cr, Ni, Mo or W.[69,78,79,85] The products have a trigonal-prismatic coordination (Scheme 2.6).

$$\mathrm{M\,(vapor)} + 3\,\mathrm{C_4H_6\,(g)} \xrightarrow[\text{2) warm to 25 °C}]{\text{1) -196 °C}}$$ **11**

M = Cr, Mo, W

Scheme 2.6

Cyclopentadiene is also reactive with metal atoms, and the corresponding metallocenes can be produced with hydrogen (gas) as a by-product. In some cases, hydrogen transfer processes take place to give the complexes shown below (Scheme 2.7).[1,86-88]

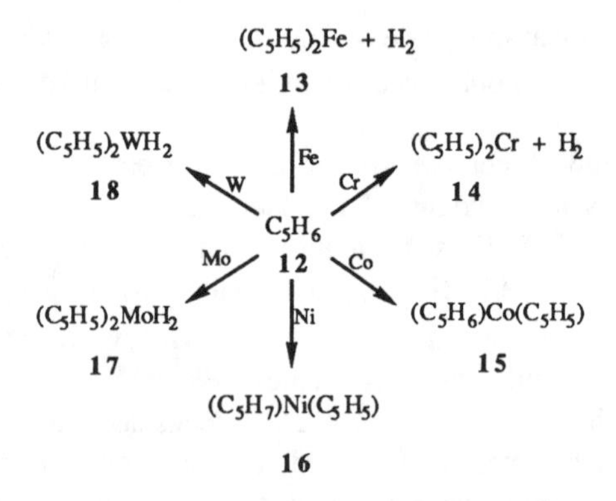

Scheme 2.7

The interesting ligand 1,5-cyclooctadiene (COD) may interact well with transition metals, and complexes with platinum[77] or iron[89,90] among others have been prepared. The latter decomposes above -20 °C and it is a versatile starting material for the preparation of other iron(0) complexes (Scheme 2.8).

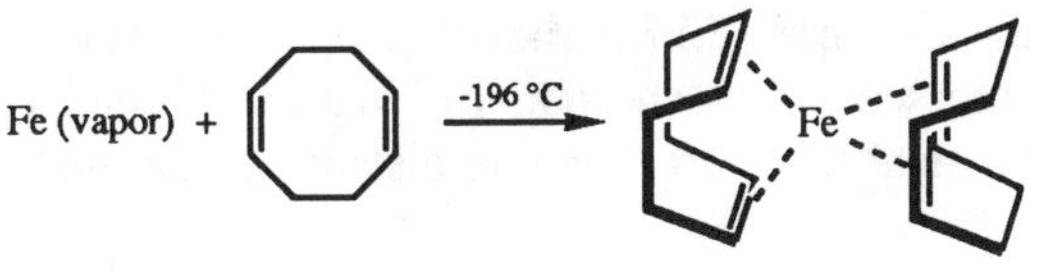

Fe (vapor) + [cyclooctene] $\xrightarrow{-196\,°C}$ **19**

Scheme 2.8

Transition metal atoms can interact with organic molecules at low temperatures without complexation. The process often involves a rearrangement of the organic substrate. In this sense, the use of metal vapors as chemical reagents constitutes an attractive approach. Thus, styrene can be polymerizated by chromium atoms.[91] The condensation of chromium atoms with benzyl sulfide (**20**) at -196 °C gave desulfurization products, like bibenzyl (**21**) and *trans*-stilbene (**22**). Apparently, metal complexes were not detected in this reaction (Scheme 2.9).[92]

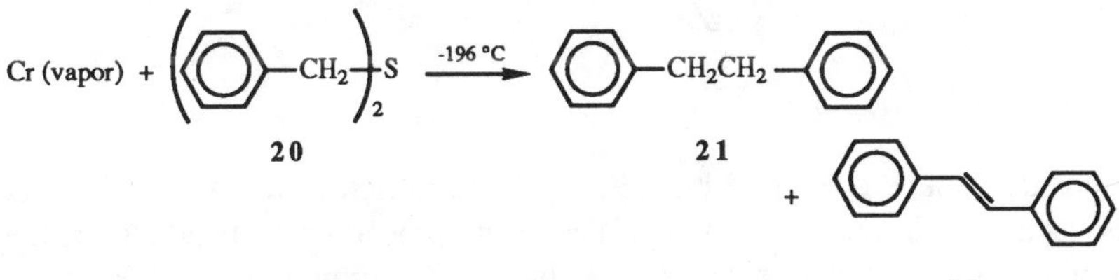

Cr (vapor) + $(C_6H_5CH_2)_2S$ (**20**) $\xrightarrow{-196\,°C}$ $C_6H_5CH_2CH_2C_6H_5$ (**21**) + **22**

Scheme 2.9

This result contrasts clearly with the reaction of chromium atoms with benzyl ether, in which coordination to one phenyl group was observed (Scheme 2.10).

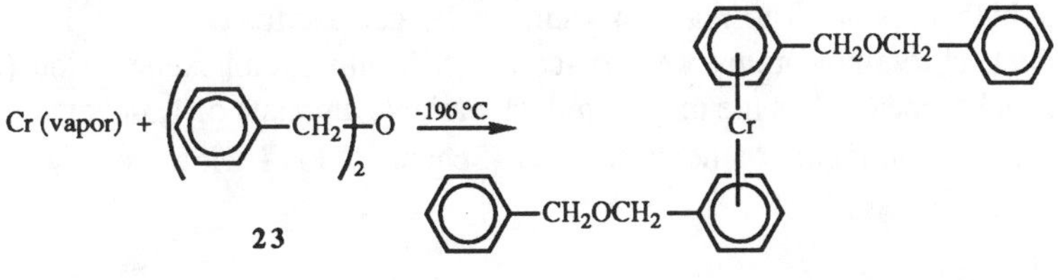

Cr (vapor) + $(C_6H_5CH_2)_2O$ (**23**) $\xrightarrow{-196\,°C}$ **24** ($C_6H_5CH_2OCH_2$–Ph–Cr–Ph–$CH_2OCH_2C_6H_5$)

Scheme 2.10

Nevertheless, it is also possible to coordinate metal atoms to both phenyl groups of the organic molecule. Thus, the interesting complexes **25** and **26** could be successfully obtained by reaction of chromium atoms at -196 °C with [2.2]-paracyclophane.[93]

The compound **26** (η^{12}-[2.2]-paracyclophane)chromium(0), constitutes an example of a compressed sandwich complex. In fact, the metal-ligand distance is considerably shorter than the corresponding distance in the bis(arene)chromium(0) complex (η^6-$C_6H_6)_2$Cr.

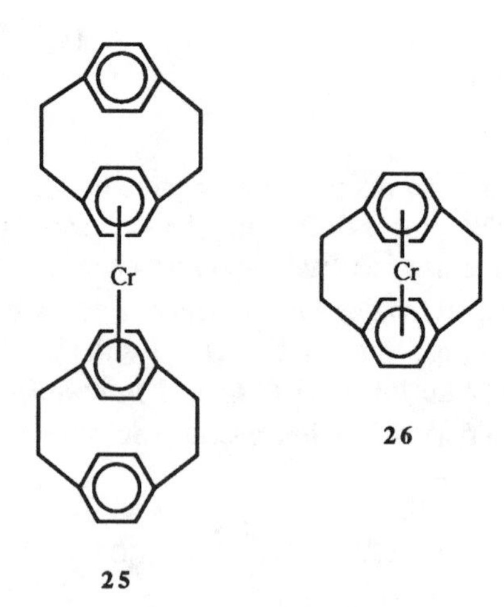

Epoxides can be converted into alkenes by transition metal atoms. The reaction represents a consequence of the relevant reducing properties of metal atoms. Several first-row transition metal atoms were examined, and chromium and vanadium atoms, particularly the former, appear as the most effective deoxygenating species for this purpose, abstracting over 2 equiv. of oxygen. Titanium gave lower yields, with epoxide polymerization in some cases, and nickel atoms were considerably less reactive.

Co-condensations were conducted with at least a tenfold excess of epoxide at 77 K under high vacuum. The reaction seems to be general for epoxides. In a typical experiment, chromium atoms were co-condensed with cyclohexene oxide (**27**) to afford cyclohexene (**28**) as the major product, variable amounts of 1,3-cyclohexadiene (**29**), and small amounts of benzene (**30**) (Scheme 2.11).[94] However, cyclohexane was not detected.

Cr (vapor) + **27** (O) —(-196 °C)→ **28**(86%) + **29**(11%) + **30**(3%) + [Cr=O]

Scheme 2.11

With *cis*-stilbene oxide double-bond migrations were not observed, but *cis-trans* isomerization occurred. Likewise, *cis*-4-decene underwent isomerization without olefin dismutation. Authors concluded that deoxygenation might be stereospecific. The presence of chromium atoms was essential for success. When deoxygenation of cyclohexene oxide was performed on a chromium surface, cyclohexene yields were decreased.

Chromium atoms also served as deoxygenating agents for heteroatom oxides. Thus, dimethyl sulfoxide gave dimethyl sulfide, triethylphosphine oxide afforded triethylphospine, and 2,6-dimethylpyridine *N*-oxide was deoxygenated to 2,6-dimethylpyridine without formation of π-complexes.[95] In contrast, Lagowski and co-workers have obtained bis(η^6-2,6-dimethylpyridine)chromium(0) from the condensation of chromium atoms with 2,6-dimethylpyridine.[96]

Organic isocyanates and isothiocyanates display some differences towards transition metal atoms.[95] Thus, isocyanates are not deoxygenated to the corresponding isocyanides by metal atoms. Instead, urea and amide derivatives are isolated from the reaction mixture. When isothiocyanates are co-condensed with chromium or vanadium atoms, isonitriles are the major products.

Interestingly, the co-condensation of nitrosobenzene with chromium atoms afforded azoxybenzene, azobenzene, and aniline (Scheme 2.12). The yields were dependent on the organic substrate/chromium ratio. With high ratios of nitrobenzene to metal, nitrosobenzene was the major product (~77%). It should be noted that reaction do not occur on preformed chromium surfaces.

$$\underset{\mathbf{31}}{PhNO_2} + \text{Cr (vapor)} \xrightarrow{-196\,^{\circ}\text{C}} \underset{\mathbf{32}}{PhNO} + \underset{\mathbf{33}}{PhNH_2} + \underset{\mathbf{34}}{PhN{=}N^{+}(O^{-})Ph} + \underset{\mathbf{35}\ \text{(cis + trans)}}{PhN{=}NPh}$$

Scheme 2.12

From these experiments it appears quite probable that chromium atoms are implicated as active species, since the alternative reaction on a preformed chromium surface provides little or no deoxygenation. However, the authors have observed that the origin of some products, particularly coupled products, is not completely unambiguous.[95] For example, the nitrosobenzene formed from nitrobenzene exists in solution in equilibrium with its dimer, so that the formation of azoxybenzene might occur by deoxygenation of the nitrosobenzene dimer or *via* nitrene or nitrenoid intermediates. Similarly, azobenzenes and arylamines can be commonly formed from decomposition of aryl nitrenes.

Very important results were found by Gladysz and his associates[95] in the co-condensations of metal atoms with ketones. Deoxygenation takes place in tandem with intermolecular reductive coupling to give pinacol and McMurry-type products (see Part Two). Furthermore, aldol products can also be obtained. In contrast with the deoxygenation of epoxides, oxygen abstraction from ketones was scarce (up to 0.05 equiv. of oxygen per metal atom). Experiments were conducted with cyclohexanone or cycloheptanone and chromium, cobalt, or nickel atoms (Scheme 2.13). After the usual matrix warm-up, volatile C_6 or C_7 products (cyclohexene, cycloheptane, cycloheptene, etc.) were not isolated, but a large amount of starting ketone was recovered. Products were obtained in poor yields, although chromium atoms were the most effective species in this regard. Nickel atoms gave only traces of products.

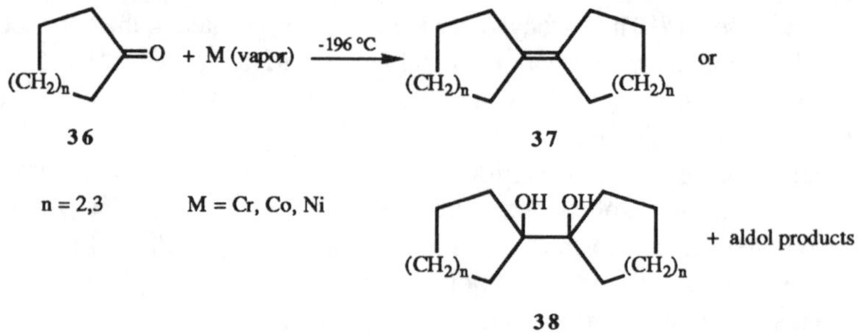

$(CH_2)_n$ =O + M (vapor) —(-196 °C)→ $(CH_2)_n$ $(CH_2)_n$ or

36 37

n = 2,3 M = Cr, Co, Ni

OH OH $(CH_2)_n$ $(CH_2)_n$ + aldol products

38

Scheme 2.13

An important transition-metal-atom-mediated organic transformation is the oligomerization of alkynes.[41,43] Trimerization and tetramerization of alkynes are usually performed with transition metal catalysts to give benzenes and/or cyclooctatetraenes.[98] The reaction has an additional importance for the preparation of very crowded molecules.

Skell reported that some acetylenes were trimerized to benzenes with chromium vapor, although bis(arene)chromium(0) complexes were not isolated.[78] In an attempt to prepare cyclophanes, Gladysz co-condensed cyclic diynes (cyclododeca-1,7-diyne and cyclotetradeca-1,8-diyne) with chromium atoms.[99] An organic trimer (**40**) was formed and only one triple bond of the diyne was involved in the trimerization (Scheme 2.14). No cyclophane was detected in the reaction mixture.

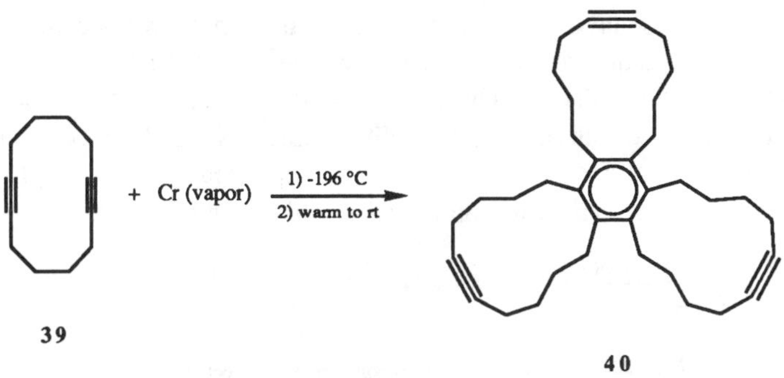

+ Cr (vapor) 1) -196 °C 2) warm to rt

39

40

Scheme 2.14

Co-condensations of chromium atoms with other acetylenes resulted in the formation of several chromium complexes.[43] From the reports by Skell[78] and Gladysz[99] it seems that one chromium atom can be coordinated to three alkynes at one time, so that cyclotrimerization is possible, but not the formation of bis(arene) complexes. However in a further study, it was reported that reaction of chromium atoms with diphenylacetylene gave bis(hexaphenyl-η^6-benzene)chromium.[100] This indicates that chromium atoms must coordinate to six alkynes. Although these reactions cannot be strictly compared, it is obvious that a detailed mechanism should be formulated for a better understanding of the process.

The important group VIII transition metals also react with alkynes to afford a variety of products.[43] Co-condensation of iron atoms with alkynes yields benzenes, cyclooctatetraenes, and metal-containing complexes.[101] Apparently, the formation of products is dependent on the nature of the starting alkyne. Thus, iron complexes were isolated from the reactions of 2-butyne, 2-pentyne, and methylphenylacetylene. However, propyne produced a complex polymer, but metal-containing products were not formed.

The iron complexes were originally formulated as Fe(alkyne)$_5$ on the basis of high-resolution mass spectrometry. A further analysis by X-ray crystallography revealed that these compounds were the unexpected bis(substituted-cyclopentadienide)iron complexes. Iron complexes have been also isolated from other alkynes. Reaction of iron atoms with trimethylsilylacetylene, followed by reaction with carbon monoxide provided the cluster $Fe_3(CO)_9$(alkyne). In addition, condensation of iron atoms with bis(trimethylsilyl)acetylene and subsequent reaction with CO gave $Fe(CO)_4$(alkyne).[43]

Nickel atoms exhibit a similar behavior to iron atoms in their reactions with alkynes. In a preliminary communication, Skell reported that alkynes, when co-condensed with nickel atoms, are oligomerized to trialkylbenzenes.[79] A more concise study on nickel-atom-alkyne condensations was reported later by Simons and Lagowski.[43,102] In all cases, nickel-alkyne polymers were the major products of these reactions, as well as an oligomeric mixture of benzenes, cyclooctatetraenes, and dimeric products. Again, propyne was an exception and the nickel-propyne polymer was the sole product isolated. Scheme 2.15 visualizes the condensation of nickel atoms with a variety of alkynes.[102] In addition, Lagowski and Simons reported that nickel-containing polymers are active, homogeneous catalysts for alkyne oligomerization and arene hydrogenation under mild conditions.[102,103]

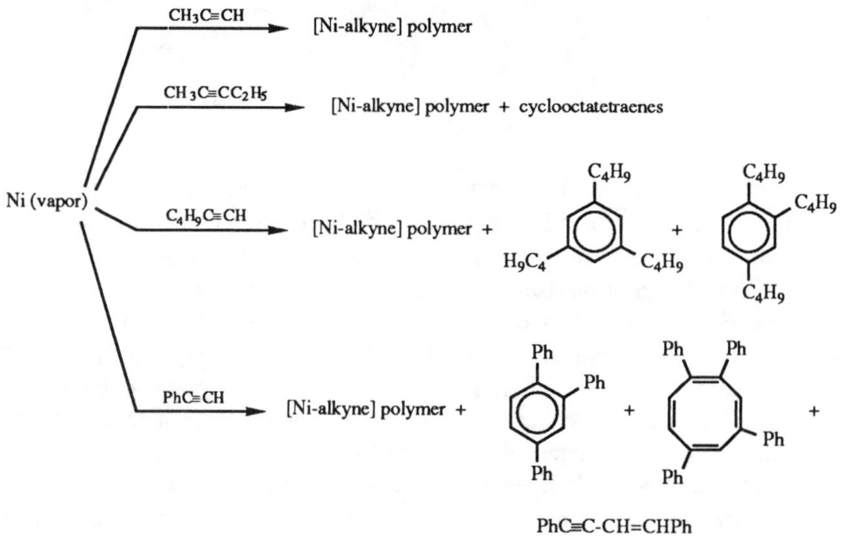

Ni (vapor) $\xrightarrow{CH_3C{\equiv}CH}$ [Ni-alkyne] polymer

Ni (vapor) $\xrightarrow{CH_3C{\equiv}CC_2H_5}$ [Ni-alkyne] polymer + cyclooctatetraenes

Ni (vapor) $\xrightarrow{C_4H_9C{\equiv}CH}$ [Ni-alkyne] polymer + 1,3,5-tri(C_4H_9)benzene + 1,2,4-tri(C_4H_9)benzene

Ni (vapor) $\xrightarrow{PhC{\equiv}CH}$ [Ni-alkyne] polymer + 1,2,4-triphenylbenzene + tetraphenylcyclooctatetraene + PhC≡C-CH=CHPh

Scheme 2.15

Klabunde and his associates studied the reaction of nickel, palladium, and platinum atoms with perfluoroalkynes, followed by reaction with CO, to yield metal carbonyl clusters.[104] Other substituted alkynes, such as dimethylacetylene dicarboxylate and silyl-substituted acetylenes were co-condensed with nickel atoms to give nickel-alkyne complexes with variable stoichiometries.[43]

D. METAL ATOM-SOLVENT CONDENSATIONS

An additional application of metal vapor techniques is the preparation of highly active metals from metal atom-solvent co-condensations. Although this process is frequently referred to in the literature as metal activation by solvents, it should be emphasized that the formation of metal slurries implies the condensation with metal vapors. The acronym SMAD (solvated metal atom dispersion) has been introduced to denote this procedure. By this process, metal atoms are solvated at low temperatures in an appropriate solvent, and upon warming metal clusters result.

These clusters incorporate carbonaceous fragments (C_1, C_2, C_3, etc., even carbon polymers) to form amorphous pseudoorganometallic powders.

When magnesium vapor was co-condensed with THF at 77 K followed by warming, a black Mg-THF slurry resulted.[105] This type of activated magnesium is extremely reactive for Grignard reagent preparations. Moreover, Mg-THF slurries are so finely divided that they can be handled by syringe. This form of magnesium was successful for preparation of hitherto inaccessible Grignard reagents. For example, cyclopropylmethylmagnesium bromide (**42**), as usually prepared in ether, rearranges to 3-butenylmagnesium bromide (**43**). This rearrangement occurs to only a slight extent when the Mg-THF slurry is used (Scheme 2.16).

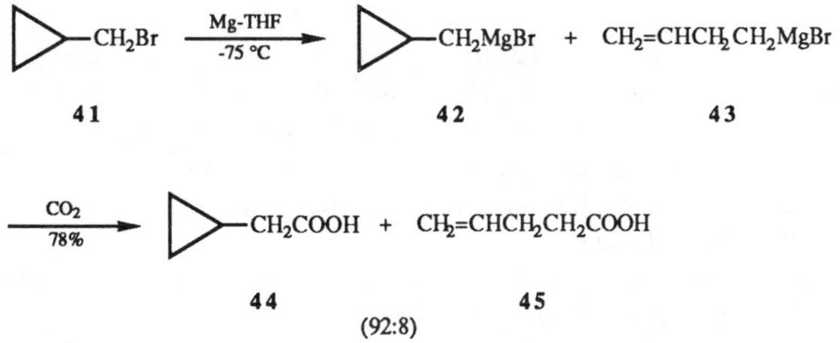

$$\triangleright\!-CH_2Br \xrightarrow[-75\ ^\circ C]{Mg\text{-}THF} \triangleright\!-CH_2MgBr + CH_2{=}CHCH_2CH_2MgBr$$

41 **42** **43**

$$\xrightarrow[78\%]{CO_2} \triangleright\!-CH_2COOH + CH_2{=}CHCH_2CH_2COOH$$

44 **45**

(92:8)

Scheme 2.16

A very interesting and detailed study on metal vapor-solvent co-condensations was reported by Klabunde and co-workers.[22] These authors studied the formation of solvated nickel atoms and their free clusters in organic media. Since solvents do not react with metal atoms at cryogenic temperatures, several discrete stages were distinguished during the formation of metal-solvent slurries. Thus, a weak σ- and/or π-coordination should be important in the first stage, which involves the formation of the metal-solvent colored complex at -196 °C.

Nickel vapors were allowed to react with three different solvents (hexane, toluene, and THF) to give very reactive metal slurries having high surface areas. Furthermore, these metal-solvent complexes can react with triethylphosphite to afford nickel clusters. Interestingly, metal slurries can be deposited on supports such as alumina, silica, molecular sieves, or activated carbon. Additionally the metal particles formed from metal slurries on warming can serve as active catalysts. Nickel-hexane powders are more active hydrogenation catalysts than Raney nickel. In contrast, Ni-THF powder is a poor and unreactive hydrogenation catalyst, but it is efficient for alkene disproportionation.

Klabunde reported[106] the preparation of highly active cadmium and zinc slurries by metal atom-solvent co-condensations, and the further reaction with alkyl bromides and iodides to give the corresponding organometallic species (Scheme 2.17). The method was also applied to the preparation of active slurries of Al, In, Sn, Pb, and Ni, which were allowed to react with alkyl and aryl halides.[107] It is noteworthy that powders prepared from these slurries by solvent evaporation can be stored for several months without appreciable loss in activity.

$$\text{M atoms (vapor)} + \text{Solvent (excess)} \xrightarrow[\text{2) warm}]{\text{1) -196 °C}} [\text{M-alkane slurry}] \xrightarrow[\text{reflux}]{\text{R-X}} R_2M + RX_2$$

M = Cd; R = C_2H_5; X = I
M = Zn; R = C_2H_5, n-C_3H_7, n-C_4H_9; X = Br
Solvent = THF, Diglyme, Dioxane, Hexane, Toluene

Scheme 2.17

The activated metal-solvent slurries were produced when metal atoms (vapor) and excess solvent were codeposited on the walls of a metal atom reactor at 77 K, followed by warming. The colored matrices initially formed turned black on melt down. Alkyl iodides were added before melting and the reaction mixture was initially warmed and finally refluxed. It should be noted that slurries are pure, clean, and finely divided powders (dispersed forms).

The procedure can be carried out in a multigram sequence (~9 g of cadmium with ~60 mL of solvent), and can be also applied to other metals in a wide range of polar and nonpolar solvents. Apparently the solvent plays a crucial role, since different metallic forms with different properties are obtained with different solvents (for the same metal). Furthermore, yields of organometallic species are highly dependent on solvents (Table 2.3). The data of this table show an important difference between zinc and cadmium atoms. Thus, the former produced dialkylzinc compounds in high to quantitative yields, except in hexane and toluene. Cadmium atoms afforded moderate to good yields of organometallic reagents.

E. OTHER APPLICATIONS

Besides metal clusters and organometallic compounds, metal vapors can now be routinely utilized in the preparation of new materials which have found important applications in polymer chemistry and catalysis.[42]

Some recent and innovating contributions should be mentioned, such as the formation of a) polymer-protected colloidal metals and metal oxides, b) carbon-attached metal clusters, c) metal cluster polymers, d) intrazeolite metal clusters, and e) bimetallic solvated metal atom dispersed (SMAD) catalysts.[42,44,108]

Polymer-protected colloids are easily accessible from the reaction of metal atoms with commercially available polymers. When poly(phenylmethyl-*co*-dimethyl)siloxane interacts with resistively generated iron atoms at ~220 K, the complex on the polymer decomposes and iron atoms recombine to form polymer-encapsulated colloidal iron. These metal clusters can be additionally oxidized in oxygen atmosphere at 250-300 K to form a polymer-protected iron oxide colloid.[109,110]

Table 2.3

Reactions of Zn and Cd Slurries from Metal Atom-Solvent Condensation[a]

Metal (mol)	R-X	R_2M (%)[b]	Solvent
Zn (0.127)	EtBr	100	Diglyme
Zn (0.107)	n-PrBr	100	Diglyme
Zn (0.077)	n-BuBr	100	Diglyme
Zn (0.149)	n-PrBr	93.5	Dioxane
Zn (0.357)	n-PrBr	89.0	THF
Zn (0.203)	n-PrBr	57.7	Hexane
Zn (0.126)	n-PrBr	28.5	Toluene
Cd (0.083)	EtI	82.7	Diglyme
Cd (0.026)	EtI	74.0	Dioxane
Cd (0.088)	EtI	61.8	Hexane
Cd (0.069)	EtI	61.7	THF
Cd (0.115)	EtI	54.5	Toluene

[a]Reactions were performed at reflux temperature and at reduced pressure.
[b]Yields based on RX as limiting reagent.
(Reprinted with permission from Murdock, T. O.; Klabunde, K. J. *J. Org. Chem.* **1976**, *41*, 1076. Copyright 1976 American Chemical Society).

With a related methodology, metal-carbon composites are formed. Metal atoms (Ag, Pd, or Pt) are reacted with a carrier molecule (usually a polyether, an oligo- or polyolefin, and an ethereal or aromatic solvent), in the presence of carbon powders. The process involves the formation of a stabilized colloidal metal that interacts with carbon powder at low temperatures. When this system is warmed to room temperature, metal atoms rapidly agglomerate to form a carbon-supported metal colloid. This constitutes an interesting example of cluster growth with a determined particle size on the surface and within the carbon support. With this strategy, the authors were able to prepare Ag/C, Pd/C, and Pt/C clusters, which were incorporated into porous oxygen electrodes. A further voltammetric study of these systems revealed comparable results to those of commercial Pt/C electrodes.[42,65] Carbon-assembled metal clusters are expected to find applications in the preparation of carbon and modified electrodes which represent a state-of-the-art field of modern electrochemistry.

Organometallic and metal clusters can be also stabilized as polymer-supported clusters.[42] Arene functionalized oligo- and polyethers, siloxanes, or silastyrenes are codeposited with metal atoms to form a polymer-attached mononuclear sandwich complex, which must act as the nucleation center for the growth of clusters. Mixtures of metal atoms can be also employed to form polymer-stabilized intermetal clusters. Schemes 2.18 and 2.19 illustrate such processes.

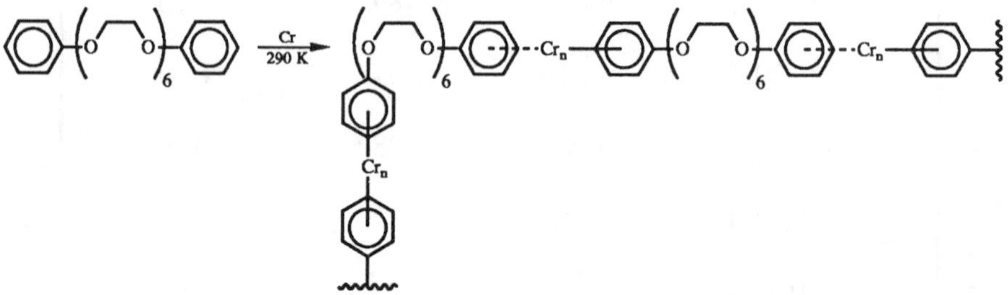

Cr, 290 K; Cr_n

Scheme 2.18

(Reprinted with Permission from Ozin, G. A. *CHEMTECH* **1985**, *15*, 488. Copyright 1985 American Chemical Society)

Solvated metal atoms are also a convenient source of very interesting supported metal particle catalysts denoted as solvated metal atom dispersed (SMAD) catalysts. By using the metal vapor SMAD method, Klabunde and his associates prepared successfully naked metal clusters which were stabilized on supports such as magnesium oxide, silica, or alumina.[111,112]

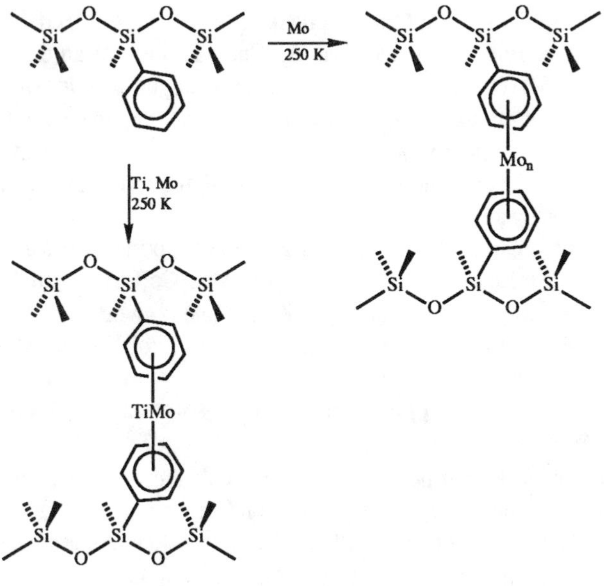

Scheme 2.19

(Reprinted with Permission from Ozin, G. A. *CHEMTECH* **1985**, *15*, 488. Copyright 1985 American Chemical Society)

The supported catalysts can be generated from the reaction of metal atoms with a solvent at cryogenic temperatures. The solvated atom solutions are exposed to high surface area supports. On warming, the metal atom co-condensates decompose and metal atoms are deposited on the support surface (Scheme 2.20). The method can be applied to aromatic, aliphatic, and ether solvents using a wide range of metals.

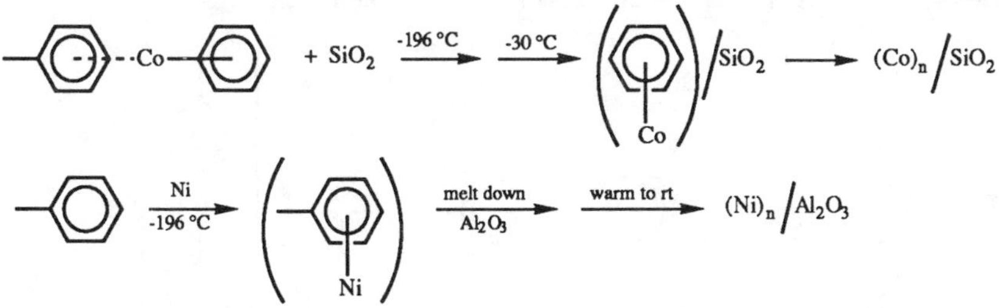

Scheme 2.20

In recent years numerous examples of syntheses and reactivities of catalysts prepared by SMAD procedure have been reported.[44,46,111,112] Also, the method was effective for the preparation of unsupported and supported bimetallic catalysts, such as Fe-Mn, Fe-Co, Ni-Mn, and Co-Mn clusters.[113-115] These mixed catalysts exhibit relevant catalytic activities and selectivities. Thus, addition of an equimolar amount of Mn-Co/SiO_2 SMAD catalyst[114] increased the catalytic activity of hydrogenation and hydrogenolysis processes. Further data demonstrated that the Mn addition affected the activities (but not selectivities) of Co to a large extent. Curiously, Mn itself is catalytically inactive. The authors have proposed that Mn favors dispersion and affects Co catalysis by an electronic effect.

The structures of both the silica- and alumina-supported Co-Mn catalysts have been reported.[115] Surface species were found to be $Co(OH)_2$, MnO, and Mn_2O_3. Since Mn is more oxophilic than cobalt, Klabunde and co-workers concluded that Mn atoms interact first with surface OH groups. Now, cobalt can be deposited on manganese atoms that act as nucleation sites. This suggests the formation of a layering system, which is supported by X-ray absorption spectroscopy (EXAFS) experiments.[114]

Metal cluster-oxide composites generated with the SMAD technique are also useful photoheterogeneous catalysts. Timms and Francis dispersed toluene-solvated iron atoms on TiO_2 powders to form Fe/TiO_2 composites. Aqueous suspensions of this system catalyzed the photoconversion of nitrogen to ammonia.[116] Recently, other metal cluster-oxide semiconductor materials have shown promising applications in light-driven processes.[117]

Finally, another recent application of metal vapor chemistry is the formation of intrazeolite metal clusters.[42,108,118] Typical examples of zeolite-entrapped metal clusters are iron and cobalt. The process involves the formation of bis(toluene)iron(0) or bis(toluene)-cobalt(0) from the corresponding metal vapors. The complexes are then inserted into a zeolite at ~170 K (below the decomposition temperatures of the complexes) to give a zeolite-encapsulated metal cluster. This is filtered, washed with toluene at 170 K, and warmed to room temperature. Then, the zeolite-metal cluster decomposes to give toluene and naked metal atoms. With this procedure, small metal clusters are deposited in the supercages of zeolite (Scheme 2.21).

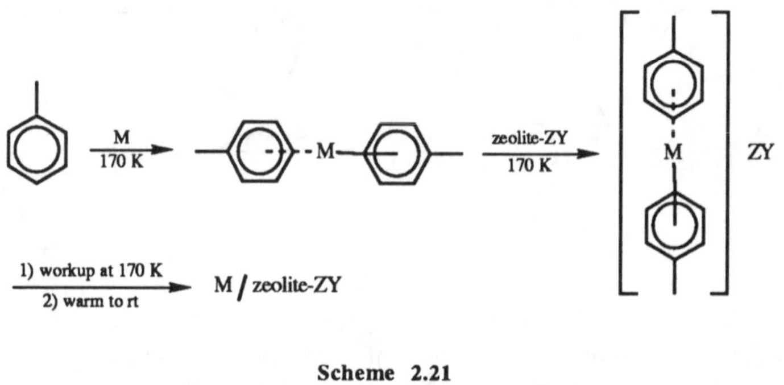

M 170 K
zeolite-ZY 170 K
M ZY
1) workup at 170 K 2) warm to rt M / zeolite-ZY

Scheme 2.21

In conclusion, the chemistry of highly reactive atomic elements enjoys increasing interest due to the formation of species possessing unusual properties and reactivities. Reactions of metal and non-metal atoms/clusters with some inorganic and organic substrates at low temperatures under matrix isolation conditions are currently reported.

Metal atom chemistry provides alternative methods in the synthesis of new organometallic compounds and novel materials. Active metal powders generated in this way are particularly useful in catalysis and for use in low-temperature organometallic synthesis schemes.

Although the metal vapor technique is a fascinating branch of chemistry and a nontraditional research area, several inherent advantages and limitations must be noted. Thus, a) bulk metals as starting materials are cheap and accessible chemical reagents; b) metal slurries and metal clusters formed at low temperatures followed by warming, show unusual reactivities in some cases compared to similar activated metals; c) for a number of fundamental organometallics such as some bis(arene)metal compounds, co-condensation techniques represent the exclusive means of access; d) the technique requires special equipment, but vapor synthesis reactors are now commercially available, which enable large scale preparations (up to kilograms) to be made.

In contrast, cryogenic temperatures are sometimes undesirable for many organic processes requiring increased reaction rates. Likewise, some nonsolvated organometallic species are reactive and unstable. Repolymerization or aggregation of metal atoms is otherwise a serious complication, even at low temperatures. This aspect is particularly important when the appropriate ligands are molecules with σ- and nonbonding electrons. Other drawbacks of this method are the frequent low yields and the large amount of cooling agent required.

In any event, metal vapor chemistry will be doubtless a future research area, and many exciting processes await systematic study. In fact, the topic of chemical vapor deposition (CVD) is currently indexed in the Chemical Abstracts Selects series.

References

1. For a preliminary and excellent review on this matter: Timms, P. L. *Adv. Inorg. Chem. Radiochem.* **1972**, *14*, 121.
2. Rice, F. O.; Freamo, M. *J. Am. Chem. Soc.* **1951**, *73*, 5529.
3. Bass, A. M.; Broida, H. P. *Formation and Trapping of Free Radicals*; Academic Press: New York, 1960.
4. Skell, P. S.; Westcott, L. D. *J. Am. Chem. Soc.* **1963**, *85*, 1023.
5. Jolly, W. L. *Adv. Chem. Ser.* **1969**, *80*, 156.
6. Siegel, B. *J. Chem. Educ.* **1961**, *38*, 496.

7. a) Honig, R. E. *The Characterization of High Temperature Vapors*; Margrave, J. L., Ed.; Wiley: New York, 1967; p 475. b) Brewer, L.; Rosenblatt, G. M. *Adv. High Temp. Chem.* **1969**, *2*, 1.
8. *Handbook of Chemistry and Physics*; Weast, R. C., Ed.; CRC Press Inc.: Boca Raton, 1979; p D-62.
9. Skell, P. S.; Engel, R. R. *J. Am. Chem. Soc.* **1966**, *88*, 3749.
10. Skell, P. S.; Engel, R. R. *J. Am. Chem. Soc.* **1967**, *89*, 2912.
11. Skell, P. S.; Westcott, L. D.; Golstein, J. P.; Engel, R. R. *J. Am. Chem. Soc.* **1965**, *87*, 2829.
12. Siegel, B. *Quart. Rev. Chem. Soc.* **1965**, *19*, 77.
13. Kant, A. *J. Chem. Phys.* **1964**, *41*, 1872.
14. Timms, P. L. *J. Am. Chem. Soc.* **1967**, *89*, 1629.
15. *Handbook of Chemistry and Physics*; Weast, R. C., Ed.; CRC Press Inc.: Boca Raton, 1979; p F-220.
16. Gaydon, A. G. *Dissociation Energies and Spectra of Diatomic Molecules*; Chapman and Hall: London, 1968.
17. Skell, P. S., Engel, R. R. *J. Am. Chem. Soc.* **1965**, *87*, 1135.
18. Brewer, L.; Chang, C.; King, B. *Inorg. Chem.* **1970**, *9*, 814.
19. Davis, S. C.; Klabunde, K. J. *Chem. Rev.* **1982**, *82*, 153.
20. Davis, S. C.; Klabunde, K. J. *J. Am. Chem. Soc.* **1978**, *100*, 5973.
21. Davis, S. C.; Severson, K. J.; Klabunde, K. J. *J. Am. Chem. Soc.* **1981**, *103*, 3024.
22. Klabunde, K. J.; Efner, H. F.; Murdock, T. O.; Ropple, R. *J. Am. Chem. Soc.* **1976**, *98*, 1021.
23. Meyer, B. *Low Temperature Spectroscopy, Optical Properties of Molecules in Matrices, Mixed Crystals, and Organic Glasses*; American Elsevier: New York, 1970.
24. Bassler, J. M.; Timms, P. L.; Margrave, J. L. *Inorg. Chem.* **1966**, *5*, 729.
25. Klabunde, K. J.; Jeong, G. H. *J. Am. Chem. Soc.* **1986**, *108*, 7103.
26. Klabunde, K. J.; Tanaka, Y. *J. Am. Chem. Soc.* **1983**, *105*, 3544.
27. Parnis, J. M.; Ozin, G. A. *J. Am. Chem. Soc.* **1986**, *108*, 1699.
28. Jeong, G. H.; Klabunde, K. J.; Pan, O.-G.; Paul, G. C.; Shevlin, P. B. *J. Am. Chem. Soc.* **1989**, *111*, 8784.
29. Jeong, G. H.; Boucher, R.; Klabunde, K. J. *J. Am. Chem. Soc.* **1990**, *112*, 3332.
30. Klabunde, K. J. *Acc. Chem. Res.* **1975**, *8*, 393.
31. Klabunde, K. J. *Angew. Chem. Int. Ed. Engl.* **1975**, *14*, 287.
32. Klabunde, K. J. *New Synth. Methods* **1975**, *3*, 135.
33. Klabunde, K. J.; Timms, P. L.; Skell, P. S.; Ittel, S. *Inorg. Synth.* **1979**, *19*, 59. A good description of metal vapor techniques in general.
34. Klabunde, K. J.; Murdock, T. O. *Chem. Technol.* **1975**, *5*, 624.

35. Moskovits, M.; Ozin, G. A. *Cryochemistry*; Wiley: New York, 1976.
36. Klabunde, K. J. *Chemistry of Free Atoms and Particles*; Academic Press: New York, 1980.
37. Breckenridge, W. H.; Umemoto, H. *Adv. Chem. Phys.* **1982**, *50*, 325.
38. Timms, P. L.; Turney, T. W. *Adv. Organomet. Chem.* **1977**, *15*, 53.
39. Blackborrow, J. R.; Young, D. *Metal Vapor Synthesis in Organometallic Chemistry*; Springer-Verlag: Berlin, 1979.
40. Green, M. L. H. *J. Organomet. Chem.* **1980**, *200*, 119.
41. Power, W. J.; Ozin, G. A. *Adv. Inorg. Chem. Radiochem.* **1980**, *23*, 79.
42. Ozin, G. A. *CHEMTECH* **1985**, *15*, 488.
43. Zoellner, R. W.; Klabunde, K. J. *Chem. Rev.* **1984**, *84*, 545.
44. Imizu, Y.; Klabunde, K. J. *Catalysis of Organic Reactions*; Augustine, R. L., Ed.; Marcel Dekker: New York, 1985; p 225.
45. Klabunde, K. J.; Davis, S. C.; Hattori, H.; Tanaka, Y. *J. Catal.* **1978**, *54*, 254.
46. Matsuo, K.; Klabunde, K. J. *J. Catal.* **1982**, *73*, 216.
47. Klabunde, K. J.; Ralston, D. H.; Zoellner, R. W.; Hattori, H.; Tanaka, Y. *J. Catal.* **1978**, *55*, 213.
48. Horn, E.; Polanyi, M.; Style, D. W. G. *Trans. Faraday Soc.* **1934**, *30*, 189.
49. Warhurst, E. *Quart. Rev. (London)* **1951**, *5*, 44.
50. Skell, P. S.; Petersen, R. J. *J. Am. Chem. Soc.* **1964**, *86*, 2530.
51. Doerr, R. G.; Skell, P. S. *J. Am. Chem. Soc.* **1967**, *89*, 3062.
52. Doerr, R. G.; Skell, P. S. *J. Am. Chem. Soc.* **1967**, *89*, 4684.
53. Skell, P. S.; Goldstein, E. J.; Petersen, R. J.; Tingey, G. L. *Chem. Ber.* **1967**, *100*, 1442.
54. Skell, P. S.; Doerr, R. G. *J. Am. Chem. Soc.* **1967**, *89*, 4688 and references cited therein.
55. Mile, B. *Angew. Chem. Int. Ed. Engl.* **1968**, *7*, 507.
56. Gusel'nikov, L. E.; Polyakov, Yu. P. *Synthesis of Hydrocarbons and Silahydrocarbons Using Alkali Metal Vapors*; Harwood Academic Publishers: New York, 1989; pp 3-35.
57. Kasai, P. H. *J. Phys. Chem.* **1982**, *86*, 4092.
58. Skell, P. S.; Girard, J. E. *J. Am. Chem. Soc.* **1972**, *94*, 5518.
59. Skell, P. S. *Plenary Main Sect. Lect. Int. Congr. Pure Appl. Chem., 23rd, 1971* **1971**, *4*, 215.
60. Ioffe, S. T.; Nesmeyanov, A. N. *The Organic Compounds of Magnesium, Beryllium, Calcium, Strontium, and Barium*; North-Holland: Amsterdam, 1967.
61. a) Lindsell, W. E. *Comprehensive Organometallic Chemistry*; Wilkinson, G.; Stone, F. G. A.; Abel, E. W., Eds.; Pergamon Press: Oxford, 1982; Vol 1, Chapter 4, p 223. b) Wakefield, B. J. *ibidem*; Vol 7, Chapter 44.

62. Klabunde, K. J.; Low, J. Y. F.; Key, M. S. *J. Fluorine Chem.* **1972**, *2*, 207.
63. Kruck, T. *Angew. Chem. Int. Ed. Engl.* **1967**, *6*, 53.
64. Timms, P. L. *J. Chem. Soc. A* **1970**, 2526.
65. Ozin, G. A.; Andrews, M. P.; Nazar, L. F.; Huber, H. X.; Francis, C. G. *Coord. Chem. Rev.* **1983**, *48*, 203.
66. Ozin, G. A.; Molnar, K. US Patent 588 616, 1984.
67. Timms, P. L. *Chem. Commun.* **1968**, 1525.
68. Timms, P. L. *J. Chem. Soc. A* **1972**, 830.
69. Williams-Smith, D. L.; Wolf, L. R.; Skell, P. S. *J. Am. Chem. Soc.* **1972**, *94*, 4042.
70. Hanlan, L. A.; Ozin, G. A. *J. Am. Chem. Soc.* **1974**, *96*, 6324.
71. Klabunde, K. J.; Key, M. S.; Low, J. Y. F. *J. Am. Chem. Soc.* **1972**, *94*, 999.
72. Klabunde, K. J. *J. Fluorine Chem.* **1976**, *7*, 95.
73. Klabunde, K. J.; Low, J. Y. F. *J. Organomet. Chem.* **1973**, *51*, 33.
74. Klabunde, K. J.; Low, J. Y. F. *J. Am. Chem. Soc.* **1974**, *96*, 7674.
75. Klabunde, K.J.; Low, J. Y. F.; Efner, H. F. *J. Am. Chem. Soc.* **1974**, *96*, 1984.
76. Skell, P. S.; Wolf, L. R. *J. Am. Chem. Soc.* **1972**, *94*, 7919.
77. Skell, P. S.; Havel, J. J. *J. Am. Chem. Soc.* **1971**, *93*, 6687.
78. Skell, P. S.; Williams-Smith, D. L.; McGlinchey, M. J. *J. Am. Chem. Soc.* **1973**, *95*, 3337.
79. Skell, P. S.; Havel, J. J.; Williams-Smith, D. L.; McGlinchey, J. J. *J. Chem. Soc. Chem. Commun.* **1972**, 1098.
80. For concepts of oxidative addition and reductive elimination in alkane reactions on metallic catalysts, see: Garin, F.; Maire, G. *Acc. Chem. Res.* **1989**, *22*, 100.
81. Klabunde, K. J.; Efner, H. F. *Inorg. Chem.* **1975**, *14*, 789.
82. Klabunde, K. J.; Efner, H. F. *J. Fluorine Chem.* **1974**, *4*, 115.
83. Gastinger, R. G.; Klabunde, K. J. *Transition Met. Chem.* **1979**, *4*, 1.
84. Elschenbroich, Ch.; Salzer, A. *Organometallics*; VCH Publishers: Weinheim, 1989; p 344-345 and references cited therein.
85. Skell, P. S.; Van Dam, E. M.; Silvon, M. P. *J. Am. Chem. Soc.* **1974**, *96*, 626.
86. Van Dam, E. M.; Brent, W. N.; Silvon, M. P.; Skell, P. S. *J. Am. Chem. Soc.* **1975**, *97*, 465.
87. Timms, P. L. *Chem. Commun.* **1969**, 1033.
88. D'Amiello, M. J.; Barefield, E. K. *J. Organomet. Chem.* **1974**, *76*, C50.
89. Mackenzie, R.; Timms, P. L. *J. Chem. Soc. Chem. Commun.* **1974**, 650.
90. Timms, P. L. *Angew. Chem. Int. Ed. Engl.* **1975**, *14*, 273.

91. Blackborow, J. R.; Grubbs, R.; Miyashita, A.; Scrivanti, A. *J. Organomet. Chem.* **1976**, *120*, C49.
92. Gladysz, J. A.; Fulcher, J. G.; Bocarsly, A. B. *Tetrahedron Lett.* **1978**, 1725.
93. Zenneck, U.; Elschenbroich, Ch.; Möckel, R. *Angew. Chem. Int. Ed. Engl.* **1978**, *17*, 531.
94. Gladysz, J. A.; Fulcher, J. G.; Togashi, S. *J. Org. Chem.* **1976**, *41*, 3647.
95. Togashi, S.; Fulcher, J. G.; Cho, B. R.; Hasegawa, M.; Gladysz, J. A. *J. Org. Chem.* **1980**, *45*, 3044 and references cited therein.
96. Simons, L. H.; Riley, P. E.; Davis, R. E.; Lagowski, J. J. *J. Am. Chem. Soc.* **1976**, *98*, 1044.
97. Gladysz, J. A.; Fulcher, J. G.; Togashi, S. *Tetrahedron Lett.* **1977**, 521.
98. March, J. *Advanced Organic Chemistry*; Wiley: New York, 1985; p 774 and references cited therein.
99. Gladysz, J. A.; Fulcher, J. G.; Lee, S. J.; Bocarsly, A. B. *Tetrahedron Lett.* **1977**, 3421.
100. Yur'eva, L. P.; Zaitseva, N. N.; Zakurin, N. V.; Vasil'kov, A. Yu.; Vasyukova, N. I. *J. Organomet. Chem.* **1983**, *247*, 287.
101. Simons, L. H.; Lagowski, J. J. *J. Organomet. Chem.* **1983**, *249*, 195.
102. Simons, L. H.; Lagowski, J. J. *Fundam. Res. Homogeneous Catal.* **1978**, *2*, 73.
103. Simons, L. H.; Lagowski, J. J. *J. Org. Chem.* **1978**, *43*, 3247.
104. Klabunde, K. J.; Groshens, T.; Brezinski, M.; Kennelly, W. *J. Am. Chem. Soc.* **1978**, *100*, 4437.
105. Klabunde, K. J.; Efner, H. F.; Satek, L.; Donley, W. *J. Organomet. Chem.* **1974**, *71*, 309.
106. Murdock, T. O.; Klabunde, K. J. *J. Org. Chem.* **1976**, *41*, 1076.
107. Klabunde, K. J.; Murdock, T. O. *J. Org. Chem.* **1979**, *44*, 3901 and references cited therein.
108. Gallezot, P. *Metal Clusters*; Moskovits, M., Ed.; Wiley: New York, 1986; p 219 and references cited therein.
109. Francis, C. G.; Ozin, G. A.; Huber, H. X. *J. Am. Chem. Soc.* **1979**, *101*, 6250.
110. Ozin, G. A.; Francis, C. G.; Huber, H. X.; Nazar, L.; Andrews, M. P. *J. Am. Chem. Soc.* **1981**, *103*, 2453.
111. Klabunde, K. J.; Tanaka, Y. *J. Mol. Catal.* **1983**, *21*, 57.
112. Matsuo, K.; Klabunde, K. J. *J. Org. Chem.* **1982**, *47*, 843.
113. Klabunde, K. J.; Imizu, Y. *J. Am. Chem. Soc.* **1984**, *106*, 2721.
114. Tan, B. J.; Klabunde, K. J.; Tanaka, T.; Kanai, H.; Yoshida, S. *J. Am. Chem. Soc.* **1988**, *110*, 5951.

115. Tan, B. J.; Klabunde, K. J.; Sherwood, P. M. A. *J. Am. Chem. Soc.* **1991**, *113*, 855.
116. Francis, C. G.; Timms, P. L. *J. Chem. Soc. Chem. Commun.* **1977**, 460.
117. Nishimoto, S.; Ohtani, B.; Yoshikawa, T.; Kagiya, T. *J. Am. Chem. Soc.* **1983**, *105*, 7180 and references cited therein.
118. Ozin, G. A.; Gil, C. *Chem. Rev.* **1989**, *89*, 1749 and references cited therein.

III. DEPASSIVATING METHODS

Many metals have an intrinsic tendency to undergo passivation processes, so that several metal activation procedures have been extensively used.[1] In general, they consist of two strategies: a) the removal of the deactivating metal oxide layer from the metal surface by chemical or mechanical means, and b) achieving a fine distribution of the metal in the reaction medium.

Solid metals may be commercially available in the form of foil, granules, pellets, powders, sponges, turnings, or wires.[2] Metal powders are generally more reactive than bulk materials, although they are also more hazardous. For a wide range of synthetic purposes, metals can be employed without further activation or purification. Moreover, pure metals can be currently prepared by electrolysis.

However, some organic substrates and processes require a very reactive form of the metal to ensure appreciable yields of products and/or practical reaction rates. Several methods of metal activation have been developed:

a) Mechanical activation of the metal surface, by etching, polishing, rolling, etc.

b) Simple washing with dilute aqueous solutions and/or organic solvents.

c) Addition of reagents, which act as catalysts or activators to the reaction mixture.

d) Formation of supported metallic catalysts, with the aid of alumina, molecular sieves, or silica.

e) Reduction of metal salts by hydrogen or hydride donors.

f) Reduction of metal halides with an alkali metal.

g) Formation of alloys or intermetallic couples.

h) Metal vapor deposition (MVD) techniques, including metal vapor-solvent condensations (see Chapter II).

i) Activation by sonication.

j) Formation of metal graphites.

k) Decomposition of organometallics.

These activation methods have different limitations and advantages and provide different degrees of activation. Thus, metal activation by using reagents or solvents is extremely simple and can be performed in conventional glassware. In contrast, metal vaporizations require special equipment, and low yields are frequently obtained.

However, metal halide-alkali metal reductions, metal vapors, metal graphites, and sonication, yield the most active forms of metals.

A. ACTIVATION BY REAGENTS AND SOLVENTS

Despite their comparatively limited effects, numerous depassivating methods employing reagents and solvents are still generally used. As previously indicated, many common metals are easily activated by simple washing of the commercial material. Thus zinc, for the preparation of organozinc reagents, is activated by washing it rapidly with dilute sodium hydroxide solution, water, dilute acetic acid, water, ethanol, acetone, and ether.[3] Then, it is dried in vacuum at 100 °C for 2 h. A modified procedure[4-6] is to wash zinc with dilute hydrochloric acid, with water until neutral, and finally with acetone and ether. Likewise, zinc dust is also activated by Cava's method,[7] by stirring it with saturated ammonium chloride solution followed by decantation and successive washings with water, ethanol, ether, and DMF. The advantage of these procedures is that they can be easily applied, even in large scale preparations.

Iodine constitutes one of the earliest catalysts used in the activation of metals.[8,9] Particularly, iodine-activated magnesium has been widely employed for the preparation of versatile Grignard reagents.[10-13] This chemically activated magnesium can be stored under nitrogen and reactivated by simple heating before use.

Preparation of Iodine-Activated Magnesium.[10]

To dry benzene (100 mL) are added under nitrogen magnesium (5 g, 205.8 mmol), iodine (2.5 g, 9.8 mmol), and dropwise addition of diethyl ether (5 mL). The reaction mixture is stirred until the iodine color disappears completely. The solvent is distilled and the black residue is heated at 150 °C for 5 min. The activated magnesium is cooled under nitrogen and transferred to a dry vessel.

Probably, the activation involves the formation of magnesium(I) iodide that is more reactive than magnesium metal and it is regenerated during the reaction, so that a catalytic amount of iodine is usually added.[13] Iodine is also utilized as activator in the preparation of organozinc reagents.[14-16]

In some cases, iodine is both activator and reagent. This feature is illustrated in the preparation of unstable metal iodides, which act as the effective reagents. The very promising organocerium reagents can be obtained by transmetallation of alkyllithiums with cerium(III) iodide. The latter is prepared *in situ* by reaction of cerium(0) with iodine.[17]

Reactive alkyl halides have been utilized as entrainers to provide a continuous activation of the metal.[13,18] This chemical cleansing of the metal surface forces the substrate to react subsequently. Catalytic amounts of reactive alkyl halides are commonly employed to initiate the reaction of the formation of Grignard reagents, although stoichiometric quantities are added with inert substrates. Methyl iodide, ethyl bromide, and especially 1,2-dibromoethane[19,20] are utilized as common entrainers.

A wide variety of inorganic and organic substrates, such as metal halides,[21-29] chlorotrimethylsilane,[30-33] ammonia,[34,35] or amines,[36,37] have been also reported as metal activators. Metals can be also activated by deposition on supports like alumina,[38-40] molecular sieves,[24,25] or silica[41,42] to form mild and selective reagents.

It is known that some metal-mediated reactions should be conducted on certain solvents. For example, direct reaction of cadmium metal with alkyl halides required the presence of very polar solvents, such as DMSO, DMF, or HMPA.[43] Similarly, sodium, potassium, and lithium dissolve in HMPA to give blue solutions up to 1M which are stable for several hours and then change to red. This aprotic solvent has high solvent power for organometallic compounds, and with it both organolithium and Grignard reagents can be obtained in good yields.[44,45] A wide range of organic substrates undergo reductive processes when treated with potassium in HMPA.[44-50]

By adding powdered sodium hydride[51-54] or sodium[53] to excess DMSO, the highly reactive reagent sodium methyl sulfinylmethylide (dimsylsodium or Corey's reagent)[54] is obtained.

Organic acid or anhydrides are also utilized as the appropriate medium for many reductive processes.[55-57] Despite their limitations and unpredictable results, some metal-solvent combinations are particularly useful. Thus, iron in glacial acetic acid is one of the earliest reducing methods in synthesis. Owsley and Bloomfield[56] made a systematic study of such a reagent and found that it was a useful alternative to catalytic hydrogenation, and was simpler to carry out than reductions with other metals. Although either iron powder or iron wire can be employed, the former gives faster and cleaner reactions.

However, exceptional reactivity of the chemically activated metals has been demonstrated by some reactions, which are now described.

B. RIEKE METALS

Highly reactive metal powders can be prepared by the reduction of the metal halide with an alkali metal (lithium, sodium, or potassium, with the latter being preferred) in an ethereal or hydrocarbon solvent. The so-called Rieke's method[58-60] has become an useful procedure in metal activation (eqs. 3.1-3.7).

This technique has also allowed several organic reactions, which hitherto generally required long reaction times and/or high-boiling solvents, to proceed efficiently under milder conditions. These reductions are usually carried out under an argon atmosphere, and freshly distilled, dry solvents must be added to the flask with a syringe. In general, metal powders will not spontaneously ignite if removed from the reaction mixture wet with solvent. In any event, authors have recommended that metal powders be transferred under inert conditions. In fact, some dry metal powders such as magnesium or aluminum are pyrophoric when exposed to the air. Use of sodium-potassium alloy as the reducing agent is not recommended, and can lead to explosions.

$$MX_n + n\,K \xrightarrow[X = Cl, Br, I]{} M^* + n\,KX \qquad (3.1)$$

$$ZnCl_2 + 2\,K \xrightarrow[\text{reflux, 2.5 h}]{\text{THF}} Zn^* + 2\,KCl \qquad (3.2)$$

$$TlCl + K \xrightarrow[\text{reflux, 24 h}]{\text{xylene}} Tl^* + KCl \qquad (3.3)$$

$$TiCl_3 + 3\,K \xrightarrow[\text{reflux, 45 min}]{\text{THF}} Ti^* + 3\,KCl \qquad (3.4)$$

$$MgCl_2 + KI + 2\,K \xrightarrow[\text{reflux, 3 h}]{\text{THF}} Mg^* + 2\,KCl \qquad (3.5)$$

$$NiI_2 + PEt_3 + 2\,K \xrightarrow[\text{reflux, 20 h}]{\text{THF}} Ni^* + 2\,KI \qquad (3.6)$$

$$NiI_2 + PPh_3 + 2\,K \xrightarrow[\text{reflux, 2 h}]{\text{THF}} Ni^* + 2\,KI \qquad (3.7)$$

The powders can be immediately used to prepare the organometallic reagent. The alkali salts generated in this process are not removed, and the organic substrate is added directly to the suspension of powdered metal. If necessary, the hydrocarbon or ethereal solvent can be removed and replaced by the desired solvent in the synthesis. In general, this manipulation does not affect the reactivity of metal powders.

A great number of highly reactive metal powders can be generated by this outstanding procedure.[58-68] Rieke's method is particularly useful for some metals possessing an intrinsic inactivity. Indium metal, for instance, reacts with organic halides only under extreme conditions,[69-71] whereas potassium reduction of indium(III) chloride gives active indium powder, which reacts with alkyl and aryl halides at 80-150 °C to afford quantitative yields of organoindium reagents.[59]

It is noteworthy that this reduction method is highly dependent on the solvent, anion, reducing agent, etc. among other factors.[58-60] Thus, activated zinc powder is formed by reduction of anhydrous zinc chloride or bromide with K or Li in an ethereal solvent, mostly THF or glyme. THF has been found to be the best solvent in which the reduction of zinc halide proceeds easily. However, the yields of the Reformatsky reactions are generally improved in diethyl ether, and the use of this solvent in the activation process complicates handling. Fortunately, as zinc can be activated in THF using the standard procedure, this solvent is stripped off and dry ether is then added.

Likewise, finely divided black aluminum, indium, and thallium powders of exceptional reactivity are formed by reduction of the corresponding metal chlorides with potassium in refluxing xylene. In these cases, ethereal solvents do not work since they undergo extensive reductive cleavage. Similarly, highly reactive black chromium powder is generated by reduction of $CrCl_3.3THF$ in benzene, due to the high reactivity of chromium(II) salts toward organic ethers.

Rieke and co-workers have also observed that in some cases the addition of simple alkali salts or Lewis bases before potassium reduction, provides metal powders of even higher reactivity.[63,72,73] The reactivity of chromium powders can be increased by performing the reduction in the presence of KI. This effect is particularly important in the case of activated magnesium. When KI is added prior to the reduction of anhydrous $MgCl_2$ with potassium in refluxing THF, a more reactive black magnesium powder is obtained (eq. 3.5). Apparently, only KI and NaI among several salts, had an activating effect. Scanning electron microscope techniques have shown that this magnesium is a sponge-like material with a much smaller particle size.

The important transition metals Ni, Pd, and Pt can be activated with potassium in THF. However, they exhibit a limited reactivity toward oxidative addition. Reduction in the presence of a triaryl- or trialkylphosphine gives rise to a highly reactive black powder. Reaction times are dependent on the phosphine. When triethylphosphine is added to nickel(II) iodide in THF, the highly soluble diiodobis(triethylphosphine)nickel(II) complex was formed. Potassium pieces were added and the reaction mixture was refluxed for 20 h (eq. 3.6). In the presence of triphenylphosphine, potassium reduction was accomplished within 2 h (eq. 3.7).

Preparation of Rieke Zinc. Typical Procedure.[68]

A suspension of freshly cut potassium (2.0 g, 51.3 mmol) (**CAUTION**: Care must be taken upon cutting potassium to be sure that there is no evidence of peroxides on the metal surface, which may react with the storage oil violently), and fused zinc chloride (3.5 g, 25.6 mmol) in anhydrous THF (80 mL) is heated with stirring under nitrogen. The reaction is very exothermic and should be heated very slowly at first. When the potassium melts, the heating bath is removed immediately. After the initial vigorous reaction is over, the reaction mixture is refluxed for 2.5 h under nitrogen to obtain a deep black zinc powder.

C. METAL ANTHRACENE/NAPHTHALENIDE

Another important reducing method that appears to be complementary in a variety of ways to the metal halide-potassium reduction involves the reduction of metal salts with metal naphthalenides or metal-anthracene complexes.

Naphthalene does not react with sodium in ether; however, sodium dissolves in a solution of naphthalene in an ethereal solvent to give a deep green solution.[74] A further study indicated that the metal dissolves by transferring an electron to napththalene to give an anion radical.[75] Naphthalene-sodium is easily prepared by stirring clean pieces of sodium with a slight excess of naphthalene in THF under argon or nitrogen at room temperature.[76] Similarly, lithium naphthalenide is also prepared by stirring lithium (1 mole) with 0.3-0.5 mole of naphthalene in dry THF under inert conditions.[77-79]

Naphthalene-magnesium has been, however, utilized scarcely. This green complex probably has the composition $MgC_{10}H_8$. It is prepared by stirring a 1:1 mixture of magnesium powder and naphthalene in liquid ammonia. Addition of dry ethanol discharges the color and further workup provides 1,4-dihydro-naphthalene.[80,81]

These metal naphthalenides are useful reagents in the preparation of activated metals. Thus, Mg and Zn have been prepared in a very active and dispersed form by reduction of $MgCl_2$ or $ZnCl_2$ with sodium naphthalenide in THF. The reaction is almost instantaneous at 20 °C as evidenced by the discharge of the green color of the reagent.[82]

However, Rieke and his associates have accomplished the most important contributions of this methodology for the preparation of highly reactive metal powders.[83-101] Metal halides can be reduced easily with lithium (2.0-2.5 equiv.) and a catalytic amount (0.1-0.25 equiv.) of an electron carrier (usually naphthalene) to speed the reduction. The mixture is stirred at room temperature under argon for 6-15 h. With regard to the solvent in which the reduction is carried out, 1,2-dimethoxyethane (DME or glyme) and THF are definitely the best solvents. If necessary a hydrocarbon solvent like toluene can be also employed in some cases. Moreover, naphthalene is easily removed by rinsing prior to a further reaction.

This reducing method is very simple, has an inherent safeness, and can be routinely employed when an active form of metal is desired. A wide variety of active metals are readily prepared by this outstanding procedure, and particularly the important organomagnesium, organocopper, organozinc, and organonickel reagents have been generated from a highly reactive form of the corresponding metal. It should be mentioned that some reactions are impossible to perform with organometallics derived from ordinary metal powders. In general, the reductions are readily carried out from metal halides under mild conditions (eqs. 3.8 and 3.9).

$$\underset{(1\ \text{equiv})}{NiI_2} + \underset{(2.3\ \text{equiv})}{2\ Li} \xrightarrow[\text{rt, Ar, 15 h}]{\text{naphthalene (0.1 equiv), glyme}} Ni^* + 2\ LiI \quad (3.8)$$

$$\underset{(1.2\ \text{equiv})}{ZnCl_2} + \underset{(2.5\ \text{equiv})}{2\ Li} \xrightarrow[\text{rt, Ar, 15 h}]{\text{naphthalene (0.25 equiv), glyme}} Zn^* + 2\ LiCl \quad (3.9)$$

However, other important modifications have been introduced to improve the reactivity or product yields. Thus, a more reactive magnesium powder[87] can be produced by using lithium-naphthalene reduction in the presence of alumina, although this procedure is not always reproducible.

Cadmium powders[90] are prepared according to three general methods:

a) Reduction of cadmium chloride with lithium naphthalenide in glyme or THF at room temperature for 6-12 h. under argon.

b) A more highly reactive cadmium is obtained by this procedure. Lithium naphthalenide is prepared by sonicating lithium, naphthalene, and TMEDA in toluene for 8-12 h. under argon. The deep purple solution is then transferred to cadmium chloride.

c) The well-known cadmium-lithium alloy, Cd_3Li can be prepared as follows: lithium, cadmium chloride, and a catalytic amount of naphthalene are mixed in glyme or THF and the mixture refluxed for 3-4 h.

The methods for the preparation of highly reactive copper powder have been extensively studied.[83,88,93-96]

a) Initially, copper metal was produced by reducing CuI with a stoichiometric amount of potassium in a 10 mol % solution of naphthalene in DME under argon for 8 h. The method affords a grey-black granular solid. Stirring, however, favors the sintering (particle aggregation), which results in loss of reactivity.

b) To a 1 equiv. of a preformed solution of lithium naphthalenide in DME was added CuCl. The reduction requires only 0.5 h. to give a more reactive copper, and sintering is considerably minimized.

c), d) Methods utilize the reduction of a soluble copper salt by a solution of 1 equiv. of preformed lithium naphthalenide, in order to increase the reactivity and particle size. Method c) uses $CuCl.S(CH_3)_2$ and method d) $CuI.PEt_3$ or $CuI.PBu_3$. These salts may be dissolved in ethereal solvents such as THF or DME. These methods provide the most reactive copper, which can react with the less reactive systems such as vinyl bromides and alkyl halides. Most alkyl halides undergo self-coupling in less than 1 minute after addition of this zerovalent copper. In contrast, methods a) and b) work well with reactive organic halides such as perfluoroaryl iodides, allyl bromide, and benzyl bromide.

Although the physical nature of metal powders formed by this procedure has been studied in most cases, the exact structure of other metals remains unknown. Thus, highly reactive calcium has been recently prepared[101] by the reduction of $CaBr_2$ or CaI_2 in THF with a preformed lithium biphenylide under argon at room temperature. The colored calcium species generated is relatively soluble in THF, however, the reactive calcium complex formed from lithium naphthalenide is insoluble in THF and gives an active black solid. Upon hydrolysis this material releases naphthalene and THF. It seems, therefore, plausible that it has a structure of calcium-naphthalene-THF complex similarly to the well-known magnesium-anthracene.3THF complex, which will be discussed later. It is noteworthy that highly reactive barium has been similarly prepared by the reduction of barium iodide with lithium biphenylide.[102]

Reduction of Nickel(II) Iodide by Lithium Naphthalenide.[86]

Nickel(II) iodide (3.82 g, 12.2 mmol), Li (0.2 g, 28.1 mmol), and naphthalene (0.16 g, 1.25 mmol) are suspended in anhydrous 1,2-dimethoxyethane (DME or glyme; 30 mL), and stirring at room temperature under argon is continued for 12 h to ensure complete reduction. The nickel slurry is allowed to settle, and the solvent is removed via syringe. The nickel powder is rinsed twice with DME (2x20 mL) to remove the naphthalene.

The reaction of Mg with anthracene in THF to form a Mg-anthracene complex was discovered by Ramsden[103] in 1965. Additionally this complex is obtained by reaction of $MgBr_2$ with sodium anthracenide; or with anthracene, excess magnesium bromide, and Mg(0) in THF in the absence of air and moisture. It can be isolated as clear orange needles with the formula magnesium-anthracene complexed with 3 THF ($MgC_{14}H_{10}.3THF$).[104,105] It is hydrolyzed by water to give 9,10-dihydroanthracene quantitatively, and is therefore a magnesium anthracene dianion (Scheme 3.1).

Mg + 1 ⇌ (THF, 20-60 °C) 2 (Mg. 3 THF)

Scheme 3.1

Complexed THF molecules can be replaced by other mono-, bi-, or tridentate ligands such as 1,4-dioxane, DME, bis(alkyl)amines, and 1,4,7-triazacyclononane, to give penta-coordinate complexes[106,107] $MgC_{11}H_{10}.L_n$.

The formation of Mg-anthracene complex depends on the magnesium surface area and on the anthracene concentration. Since the conversion of (1) into (2) is a temperature-dependent reversible equilibrium, kinetic measurements have been achieved; (**2**) is favored at low temperatures. By decreasing the content of anthracene and/or THF, the equilibrium can be shifted to the left, which is an outstanding method for the preparation of a finely dispersed magnesium.

Magnesium-anthracene complexes exhibit a low thermal stability. The molecular structure of the related magnesium 1,4-dimethylanthracene.3THF has been studied.[108] The magnesium atom is pentacoordinate and displays long Mg-C bonds (2.32 Å). The complex has therefore a tendency to undergo homolytic Mg-C bond cleavage, so that it can be easily transformed to radical anion species acting as a single-electron donor. Also, the anthracene ring is folded along the C_9-C_{10} line, and the central ring has lost the aromaticity. Bogdanovic has conspicuously reviewed the properties and chemistry of these Mg-anthracene complexes.[109,110]

An important application of magnesium-anthracene complex is the catalytic activation of magnesium for the synthesis of magnesium hydrides. Hydrogenation of magnesium powder is accomplished by reacting Mg-anthracene with chromium(III) chloride or titanium(IV) chloride in the presence of excess Mg (Scheme 3.2).

$$Mg + H_2\ (1\text{-}80\ bar) \xrightarrow[THF,\ 20\text{-}65\ °C]{Mg\text{-}anthracene\ [MX_n]} MgH_2^*$$

$MX_n = CrCl_3, TiCl_4, FeCl_2$

Scheme 3.2

The hydrogenation of magnesium can be improved by adding $MgCl_2$ as a cocatalyst, which reduces the time of the process from 16-20 h to 1-2 h. Important reactions of activated MgH_2 have been discovered and studied, particularly for the formation of organometallic and intermetallic compounds[110] and for reversible chemical storage.[111] The latter constitutes an ongoing research activity as a medium for hydrogen and high-temperature heat storage. Many of these processes are extremely difficult or impossible to carry out with the ordinary magnesium hydride. Unfortunately, activated MgH_2 is very pyrophoric, so that air-stable nickel-doped magnesium powders are recommended for energy storage.

The formation of a highly reactive magnesium powder from Mg-anthracene complex can be performed either in a solvent or in the solid state. The complex decomposes at room temperature in solvents such as toluene or ether via $MgC_{14}H_{10}$.2THF as intermediate. Magnesium powder is also generated by heating solid $MgC_{14}H_{10}$.3THF at 200 °C under high vacuum to remove THF and anthracene. The resulting magnesium is a black pyrophoric powder with a high surface area.

Similarly, dehydrogenation of activated magnesium hydride at >250 °C under vacuum or at >300 °C under normal pressure also yields activated magnesium, although with a minor surface area (Scheme 3.3). This active magnesium is also less reactive than magnesium generated from Mg-anthracene. In any event, it should be pointed out that this magnesium displays a reactivity comparable to that of Rieke-magnesium or magnesium vapor. Indeed, Grignard reagents, cleavage of THF, or rapid reaction with hydrogen are easily accessible with this magnesium metal.

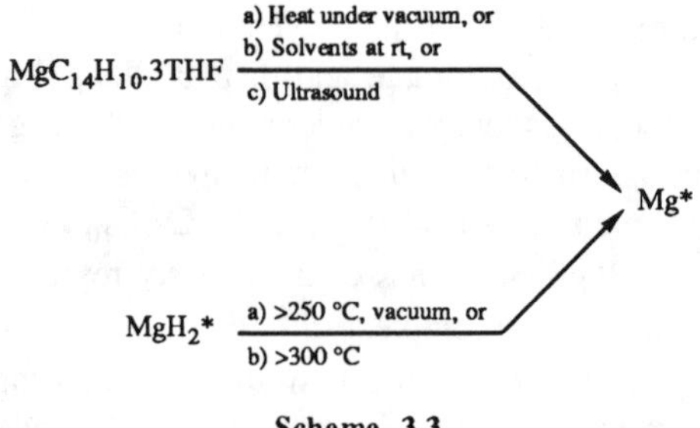

Scheme 3.3

Another important application of $MgC_{14}H_{10}$.3THF is the preparation of finely divided metal powders. The complex is utilized as a reducing agent of metal salts.[112] The surface area is similar to those obtained by other classical methods, such as Rieke reduction or metal vaporization (Scheme 3.4).

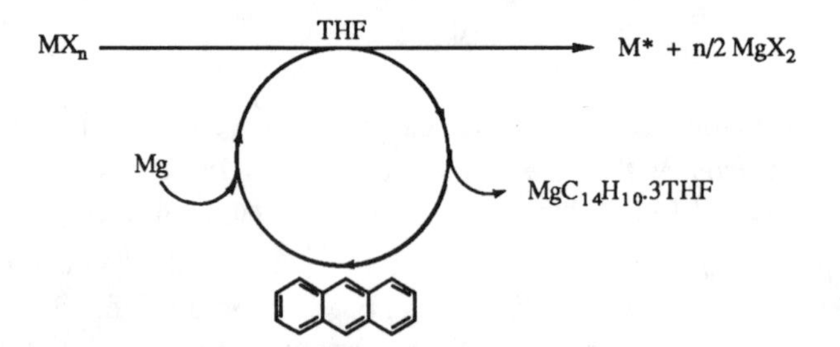

MX_n = $FeCl_2$, $CoCl_2$, $NiCl_2$, $PdCl_2$, $PtCl_2$, CuCl, CuBr, $ZnCl_2$, ...

Scheme 3.4

Sonication has a beneficial effect on the preparation of this magnesium electron-transfer complex. Magnesium powder in THF reacts smoothly with anthracene in a cleaning bath of low intensity ultrasound.[113] The magnesium produced in this way is an excellent reducing agent for metal salts and when the reduction is performed in the presence of Lewis base ligands it is a useful route to organo-transition metal complexes.

The magnesium-anthracene complex is also a convenient route to allyl Grignard reagents at temperatures as low as -78 or -35 °C in high yields.[114,115] This eliminates the coupling of allyl magnesium halides with the starting alkyl halide which is a common side-reaction with conventional methods. The complex is particularly useful for conversion of benzylic chlorides or bromides into the Grignard reagents in yields generally of >92% by reaction in THF at 20 °C. Even benzylic di-Grignards can be prepared in high yield.[116]

Interestingly propargyl chloride does not react with ordinary magnesium (or with other active forms), however, in the presence of magnesium-anthracene complex, magnesium powder reacts with this organic halide in THF to give allenylmagnesium chloride (Scheme 3.5).[110]

$$HC{\equiv}CCH_2Cl + \text{Mg powder} \xrightarrow[\text{10 mol \% of } MgC_{14}H_{10}\cdot 3THF]{\text{THF, 20 °C}} H_2C{=}C{=}CHMgCl$$

Scheme 3.5

D. OTHER REDUCING METHODS

In addition to the well-established hydrogenation methods on supported catalysts[117,118] and reductions by dissolving metals,[119] active catalysts can be prepared by reduction of metal salts by hydrogen or hydride donors (see also Chapter I).

Treatment of platinum metal salts with sodium borohydride solutions results in the formation of finely divided black precipitates, which serve as effective catalysts both for hydrolysis[120] and hydrogenation[121] processes. Additionally, the reduction of noble metal salts with a solution of sodium borohydride in ethanol, in the presence of decolorizing carbon produces an active carbon-supported metal catalyst.[122,123] Rhodium, palladium, and platinum showed the highest activities for the hydrogenation, whereas ruthenium, osmium, and iridium were less effective.

The method can be extended to other transition metals, particularly nickel, which gives two types of finely divided materials. Reduction of nickel salts in aqueous solutions with sodium borohydride affords a granular material, denoted as P-1 nickel, which has a catalytic activity comparable to that of Raney nickel. Treatment of ethanolic nickel(II) acetate with sodium borohydride produces a nearly colloidal black material, designated as P-2 nickel.[124,125] This consists probably of an amorphous mixture of nickel and boron atoms.[124] This simple catalyst exhibits a great selectivity, and one double bond can be selectively reduced in the presence of others. Reduction of alkynes produces pure *cis*-alkenes, especially in the presence of 1,2-diaminoethane,[126] and conjugated dienes can be semihydrogenated. In addition, hydrogenation is usually performed without hydrogenolysis.[125]

The intermediacy of nickel(0) species is also invoked in other catalytic reactions. Thus, a nickel catalyst for the arylation and vinylation of lithium ester enolates is generated by the addition of *n*-butyllithium to nickel(II) bromide in THF at low temperatures.[127] This catalyst gradually loses activity if allowed to reach room temperature. The absence of coordinating ligands plays a crucial role for success, and thus the addition of triaryl- or trialkylphosphines produces totally inactive material.

An efficient nickel catalyst is generated *in situ* from anhydrous nickel salt and triphenylphosphine in the presence of a reducing metal (Zn, Mg, or Mn).[128] Excess reducing metal was essential. When nickel chloride was treated with zinc in the presence of triphenylphosphine followed by filtration to remove excess zinc, the resulting zero-valent nickel complex gave low yields of coupled products. These solutions were reactivated when transferred to a flask containing zinc. The catalyst can be generated from several nickel salts, but nickel(II) chloride and bromide were most effective. Likewise, triarylphosphines were the best ligands, whereas trialkylphosphines gave slower reactions in low yields, and bidentate arylphosphines were ineffective in coupling reactions. In addition, nickel(II) chloride can be reduced by sodium, and the corresponding red solutions can be employed in reductive processes. Coupling reactions, however, were scarce with this system.

Decomposition of organometallics in the vapor phase constitutes another old procedure of metal activation.[129-131] Metallic nickel can be formed by thermal decomposition of nickel complexes such as nickel oxalate,[132] nickel formate,[133] nickel di-*tert*-butoxide,[134] or $Li_2[Ni(i\text{-}PrO)_4]$-LiBr-3THF.[135] Recently, a detailed study of the thermal decomposition of nickel diisopropoxide in 2-propanol has been reported.[136]

Anhydrous nickel chloride reacts with an alkali metal isopropoxide in dry 2-propanol to give the corresponding nickel diisopropoxide. This can be decomposed to nickel metal, hydrogen, and acetone. The rate of decomposition is highly dependent on the alkali metal derivative. In general, the catalyst is prepared from lithium (60 °C) or potassium isopropoxide (100 °C) which ensure a complete reaction, whereas when sodium isopropoxide (100 °C) is used 85% conversion is obtained (Scheme 3.6).

$$2\,Li(^{i}PrO) + NiCl_2 \longrightarrow 2\,LiCl + Ni(^{i}PrO)_2$$

$$Ni(^{i}PrO)_2 \xrightarrow[2\,h]{60\,°C} Ni^* + H_2 + 2\,Me_2C{=}O$$

Scheme 3.6

The nickel generated was initially a black colloid which aggregates on heating to give a precipitate. Although, the particle size of nickel clusters was not reported, a further study by X-ray diffraction analysis identified the catalyst as metallic nickel.

Preparation of Activated Nickel from Nickel Diisopropoxide.[136]

A 250-mL, two-necked, round-bottomed flask, equipped with a magnetic stirring bar, a reflux condenser, and a tube connected to a source of argon, is charged with lithium (0.14 g, 20 mmol) and dry 2-propanol (40 mL). To this solution is added anhydrous nickel chloride (1.3 g, 10 mmol) and the mixture is vigorously stirred at 100 °C (external oil bath) for 2 h. A black colloidal nickel is formed, which can be employed in situ for transfer hydrogenation reactions.

An outstanding reduction procedure has been recently developed by Bönnemann and co-workers.[137] In this method, finely divided metal and alloy powders are prepared by reducing metal salts in organic phases. The salts are kept in solution as hydroorganoborates, using organoboron complexing agents.

Table 3.1

Preparation of Metal Powders by Reduction of Metal Salts in THF

Metal Salt	Reducing Agent	Time (h)	T(°C)	Metal Content (%)
$Fe(OEt)_2$	$NaBEt_3H$	16	67	96.8
$Co(OH)_2$	$NaBEt_3H$	2	23	94.5
$Co(CN)_2$	$NaBEt_3H$	16	67	96.5
$Ni(OH)_2$	$NaBEt_3H$	2	23	94.7
CuCN	$LiBEt_3H$	2	23	97.3
$Pd(CN)_2$	$NaBEt_3H$	16	67	95.5
$CrCl_3$	$NaBEt_3H$	2	23	93.3
$MnCl_2$	$LiBEt_3H$	1	23	94.0
$COCl_2$	LiH + 10% BEt_3	16	67	95.8
$NiCl_2$	$NaBEt_3H$	16	67	96.9
$ZnCl_2$	$LiBEt_3H$	12	67	97.8
$RhCl_3$	$LiBEt_3H$	2	23	96.1
$OsCl_3$	$NaBEt_3H$	2	23	95.8
$IrCl_3$	$KBPr_3H$	16	67	94.7
$PtCl_2$	LiH + 10% BEt_3	5	67	98.8

(Reprinted with Permission from Bönnemann, H.; Brijoux, W.; Joussen,T. *Angew. Chem. Int. Ed. Engl.* **1990**, *29*, 273. Copyright 1990 VCH).

Metal powders of groups 6-12 and 14 of the periodic table have been prepared by reducing their metal salts with hydroorganoborates in THF, diglyme, or hydrocarbon solvents (Scheme 3.7). Since the organoboron complexes formed by reaction of hydroxides, alkoxides, cyanides, cyanates, and thiocyanates are highly soluble in organic solvents, the metal powders can be easily isolated by simple filtration. Moreover, the organoboron complexing agent can be essentially recovered by treatment with acid, which constitutes an additional advantage of this procedure.

$$n\,MX_m + m\,M'(BR_3H)_n \xrightarrow{THF} n\,M^* + m\,M'(BR_3X)_n + 1/2\,nm\,H_2$$

M = metal powder
M' = alkali or alkaline-earth metal
R = C_1-C_6-alkyl
X = OH, OR, CN, OCN, SCN

Scheme 3.7

In contrast, metal halides do not form stable complexes, although in many cases they remain with the trialkylborane in the solution of the organic solvent (eq. 3.10). The reduction of metal halides can also be carried out with a catalytic amount of trialkylborane, since the organoboron complexing agents liberated react *in situ* with the corresponding metal hydrides (eq. 3.11) (Table 3.1).

$$n\,MX_m + m\,M'(BR_3H)_n \xrightarrow{THF} n\,M^* + m\,M'X_n + m\,BR_3 + 1/2\,nm\,H_2 \quad (3.10)$$

$$n\,MX_m + m\,M'H_n \xrightarrow{BR_3,\ THF} n\,M^* + m\,M'X_n + 1/2\,nm\,H_2 \quad (3.11)$$

Table 3.2

Preparation of Alloy Powders by Co-Reduction of Metal Salts in THF

Metal Salt	Reducing Agent	Time (h)	T (°C)	Metal	Content (%)
$Co(OH)_2$ $Ni(OH)_2$	$NaBEt_3H$	7	67	Co Ni	48.3 45.9
$FeCl_3$ $CoCl_2$	Li + 10% BEt_3	6	67	Fe Co	47.0 47.1
$CoCl_2$ $PtCl_2$	$LiBEt_3H$	7	67	Co Pt	21.6 76.3
$RhCl_3$ $IrCl_3$	$LiBEt_3H$	5	67	Rh Ir	33.5 62.5
$PdCl_2$ $PtCl_2$	$LiBEt_3H$	5	67	Pd Pt	33.6 63.4
$PtCl_2$ $IrCl_3$	$NaBEt_3H$	12	67	Pt Ir	50.2 48.7
$CuCl_2$ $SnCl_2$	$LiBEt_3H$	4	67	Cu Sn	49.6 47.6
$FeCl_3$ $CoCl_2$ $NiCl_2$	$LiBEt_3H$	1.5	23	Fe Co Ni	30.1 31.4 30.9

(Reprinted with Permission from Bönnemann, H.; Brijoux, W.; Joussen,T. *Angew. Chem. Int. Ed. Engl.* **1990**, *29*, 273. Copyright 1990 VCH).

Analogously, alloys of two or more metals were prepared by simple co-reduction of salts of different metals (Table 3.2). A further study by X-ray diffraction has revealed that metal powders generated are microcrystalline to amorphous, but alloy powders are practically single phase, nearly amorphous solids. The authors have noted that this feature is particularly important in catalysis, since alloys can be deposited on support reagents. It should be pointed out that metal powders and alloys prepared in this way are essentially boron-free materials as revealed by the boron content in the product.

Preparation of an Iron-Cobalt Alloy.[137]

A solution of $FeCl_3$ (9.1 g, 56 mmol) and $CoCl_2$ (3.1 g, 24 mmol) in THF (2.5 L) was added dropwise to a stirred solution of 1.7M $LiBEt_3H$ (255 mmol) in THF at 23 °C for 5 h. The reaction mixture was stirred overnight, the iron-cobalt alloy powder was separated from the clear solution and washed with THF (2x200 mL). The product was then stirred with ethanol (300 mL) and subsequently with an ethanol-THF mixture (1:1, 400 mL) until the evolution of gas ceased. The product was washed again with THF (2x200 mL) and dried under high vacuum to yield the pure alloy powder (5.0 g).

E. INTERMETALLIC COUPLES

An intermetallic couple is basically a homogeneous dispersion of a metal or metal salt in a finely divided metal powder. The latter can be activated prior to the couple formation. The importance of these systems arises from some experimental observations in synthesis. It is well known that zinc-copper couples react with alkyl halides under milder conditions and more efficiently than zinc alone.[138-143] Addition of 10 to 20% of copper powder to zinc was originally reported to give increased yields of Reformatsky reactions.[144,145] The incorporation of 1-2% sodium in lithium increases the yield of certain organolithium reagents.[146] Likewise, the reactivity of magnesium toward alkyl bromides is increased by alloying with lithium and decreased by alloying with more electronegative metals.[18]

Among a great number of intermetallic systems and alloys, Zn-Cu and Zn-Ag couples have found a wide applicability in many organic transformations, and they will be discussed particularly in this Chapter. These couples play a crucial role in two important zinc-mediated processes, Reformatsky and Simmons-Smith reactions, as well as in other reductive transformations.

The zinc-copper couple has been prepared using several procedures:

a) Zinc activated with hydrochloric acid and an aqueous solution of copper(II) sulfate.[147,148] This couple offers the advantages of ease of preparation and of good reproducibility. It can be easily stored under vacuum and is then ready for use.

Probably, this couple has been the most widely used in many syntheses.

b) Reduction of a mixture of zinc powder and copper(II) oxide at 500 °C in an atmosphere of hydrogen.[149-151] The so-called Simmons-Smith couple was initially utilized in cyclopropanation studies. The main advantages are reproducibility and good storage properties. In fact, the couple is active even in the presence of oxygen and atmospheric moisture. In contrast, the disadvantage is the tedious preparation compared with other couples.

c) Zinc dust and a hot solution of copper(II) acetate monohydrate in acetic acid.[152] This couple has been also extensively employed, its preparation is very simple and is highly active.

d) A suspension of zinc dust and copper(I) chloride in dry ether is heated at reflux under nitrogen.[153] This couple uses equimolar amounts of both zinc and copper(I) halide. Its preparation is very rapid and the couple is highly active in synthesis.

Preparation of Zinc-Copper Couple.[148,151]

A mixture of zinc dust (62.0 g, 0.95 g-atom) and 3% hydrochloric acid (50 mL) is stirred vigorously for 1 minute. The solution is decanted and the zinc powder is washed successively with more 3% hydrochloric acid solution (3x50 mL), distilled water (5x100 mL), 2% aqueous copper(II) sulfate solution (2x100 mL), distilled water (5x100 mL), absolute ethanol (4x100 mL), and ether (5x100 mL). The couple is filtered, washed with ether, covered with a rubber dam, and dried by suction to room temperature. Finally, the Zn-Cu couple is dried *in vacuo* over P_2O_5.

A comparative analysis of these couples to effect cyclopropanations has been accomplished by Simmons *et al.*,[151] including detailed experimental procedures. In some cases, the reaction is very dependent on the source and particular batch of zinc-copper couple. This is due probably to some unknown contaminant or passivation processes on the couple surface. On the other hand, the Zn-Cu couple is relatively insensitive towards poisoning by impurities in the solvents or starting materials. When solvents of reagent-grade purity are used without purification, some reactions (*e.g.* Reformatsky reactions) have to be initiated but good yields are also obtained.

A safe and feasible preparation under inert conditions was recently reported by McMurry and Rico.[154] The couple was prepared from zinc dust and $CuSO_4.5H_2O$ in water purging with argon. The black couple is stirred with continued argon purging, filtered under argon, washed with argon-purged water, dry acetone, and ether. The slurry is dried in vacuo and can be stored under argon. This couple was utilized in conjunction with $TiCl_3$ to generate a low-valent titanium reagent. It should be mentioned that the $TiCl_3$/Zn-Cu system gave the best yields in titanium-induced pinacol coupling reactions.

Considerable information on the use and preparations of Zn-Cu couples was obtained through the Simmons-Smith reaction.[151] However, when this reaction was applied to α,β-unsaturated ketones, enamines, enol ethers, and others, poor results were obtained. Conia and collaborators[155] utilized a zinc-silver couple in place of the zinc-copper couple with substituted olefins and improved yields of cyclopropanes resulted.

Preparation of Zinc-Silver Couple.[151,155]

Granular zinc (17.0 g, 0.26 g-atom) is added to a hot stirred solution of silver acetate (100 mg) in acetic acid (100 mL). The mixture is stirred for 30 seconds, then the supernatant liquid is decanted, and the zinc-silver couple is washed with ether (5x100 mL). Dry ether is poured onto the product and two or three small pieces of silver wool are added.

A modified preparation of the couple was described by Clark and Heathcock.[156] Zinc dust is stirred with 10% hydrochloric acid solution and then washed with acetone and ether. A suspension of silver acetate in boiling acetic acid is added. The solvent is removed and the black zinc-silver couple is successively washed with acetic acid, ether, and methanol.

Although zinc-silver couple was initially reported to be unsufficiently reactive towards diidomethane,[152] this couple is more reactive than Zn-Cu couple in many reductive processes, which will be discussed in further Chapters.

F. ULTRASONIC IRRADIATION

The use of ultrasonic waves in organic chemistry has led to important findings and improvements in several synthetic processes. As evidence for the growing and varied usefulness of sonochemistry, some extended review articles[157-164] and books[165-167] have been recently published. These accounts also discuss the theoretical principles and quantitative approaches of sonochemistry, which will not be considered for the purpose of this Chapter. However, some qualitative and technical notes are briefly outlined.

1. Sonochemistry. Basic Principles

Ultrasound is sound pitched above 16 kHz, beyond the normal range of human hearing. The chemical effects of ultrasound derive from the creation, expansion, and collapse of small bubbles that result when a liquid is irradiated by an ultrasonic beam. This phenomenon, called cavitation, generates high temperatures and pressures in certain spots within the liquid. Nevertheless, the surrounding liquid cools these hot spots quickly, so that a cavity of a few microns disappears in less than a millionth of a second. The temperature of cavitation has originated a great number of theoretical hydrodynamic models and tedious analyses. The temperature predictions have varied over a wide range, from 1000 K to more than 10000 K. Recently, Suslick and

Flint[168] have developed a suitable scheme for temperature measurement by using medium resolution sonoluminescence spectra. Cavitation temperatures lie apparently in the range from 4500 to 5500 K.

It should be noted that acoustic radiation is a mechanical, nonquantified energy, which is transformed in part, into thermal energy. In contrast with a photochemical process, this energy is not absorbed by the molecules so that the effect of frequency cannot be easily evaluated. Since the process of cavitation can be produced in a broad range of frequencies, a sonochemical reaction occurs at a frequency capable of inducing cavitation. Unfortunately, frequency effects can affect extensively the process and many reactions are not reproducible. In this sense, technical descriptions of the sonochemical apparatus such as dimensions, frequency, and intensity of ultrasonic radiation are necessary and frequently reported. These features can be particularly important with heterogeneous mixtures, such as solids in suspension, metal powders, or reactions on solid surfaces.

A great variety of ultrasonic generators for chemical laboratories are commercially available. With solids, cleaning baths and probe disruptors are preferred. In a cleaning bath it is quite important to locate the reaction mixture precisely, since ultrasonic waves propagate linearly.

Special attention should be paid to solvents such as hydrocarbons, since they can be cleaved by pyrolysis under the influence of ultrasound. Particularly, the volatility, viscosity, surface tension, and other hydrodynamic parameters are considered to choose the appropriate solvent system.[164]

2. Reactions on Metal Surfaces

Reactions involving metals can be grouped in two types: a) reactions in which the metal is a reagent and which proceed with consumption of the metal, and b) reactions in which the metal serves as a catalyst.

In these heterogeneous reactions the enhancements of reactivity are mostly due to the well-known cleaning effect of the ultrasonic waves. Sonication creates and exposes new clean surfaces to the reagents. Likewise, finely divided materials are originated under the effect of ultrasound, which increases the effective surface area for reaction and minimizes the particle size. In this sense, sonication is comparable to chemical activation by reagents or solvents which act as entrainers to remove surface contaminants and promote metal activity.

However, the cleaning effect of ultrasound is not sufficient to explain the enhanced reactivity. Many chemical processes are absolutely impeded in the absence of sonication, and some ultrasonically induced reactions take place through a different pathway. Absorption-desorption phenomena are thought to occur, and sonication must sweep reactive intermediates or products. Apparently this effect is not so effective under normal or even vigorous stirring.

The benefit of ultrasound in particle size reduction and simultaneous surface activation can be applied to generate extremely fine emulsions from mixtures of immiscible solvents, by increasing the interfacial contact area. Interestingly, heterogeneous reactions conducted under phase-transfer catalysis can now be performed using sonication. However, sonication favors particle fragmentation up to a limit beyond which ultrasound has no effect.

One important application of the use of ultrasound on metallic catalysts is the sonochemical hydrogenation.[169] Ultrasonic waves play a crucial role in the activation of platinum, palladium, and rhodium metals. Nevertheless, the results of this ultrasonic activation measured in various processes could not be easily rationalized. Palladium and platinum catalysts were prepared by reduction of aqueous solutions of metal salts with formaldehyde under sonication. These systems showed enhanced activity in the decomposition of hydrogen peroxide, the oxidation of ethanol, and the hydrogenation of 1-hexene. Curiously, the catalytic activity of platinum catalysts increased with increasing frequency of ultrasound employed in their production, whereas palladium catalysts had the opposite effect. More systematic studies have been achieved on sonochemical hydrogenations and hydrogenolysis in the presence of Raney-nickel.[164,170,171] In addition, high intensity ultrasound favors the formation of layered ionic solids[172] and enhances intercalation reactions.[173]

3. Reactive Metal Powders

Ultrasonic waves have proved to be an extremely useful means of producing highly active metal powders. In general, ultrasound produces fine dispersions due to cavitation phenomena, which are also responsible for erosion of the metal surface. Initial studies of the erosion of the metal revealed a sequence opposite to that of the hardness.[174] Several metals were dispersed in liquids in an ultrasonic field. The loss of weight by erosion at 400 kHz in water increases in the order Cu, Al, Ni, Sn, Pt, Au, Cd, Zn, Mg, Bi, Pb, and in paraffin oil in the order Pb, Cd, Bi. Other metals were not dispersed in the mineral oil. More recent experiments indicated, however, that ultrasonic irradiation of metals increases the hardness.[175] Since the hardening is produced by dislocations and deformations of the crystal lattice, which can serve as active sites, it seems to be consistent with improved activation.

In a pioneering paper, Fry reported the formation of ultrasonically dispersed mercury emulsions in acetic acid.[176] These emulsions were useful for the reduction of α,α'-dibromo-ketones.

Colloidal alkali metals are readily prepared by sonication in a cleaning bath (35 kHz). Potassium can be transformed into a blue colloidal suspension in a few minutes in toluene or xylene. However, sodium forms dispersions in xylene but not in toluene. This solvent effect is due to the powerful cavitational energy in a higher boiling solvent. In other solvents with a weak cavitation like THF, alkali metals give no dispersion.[177]

Ultrasonically dispersed potassium is particularly useful in Thorpe-Ziegler and Dieckmann-type cyclizations, as well as in the formation of ylide reagents in Wittig and Wittig-Horner reactions.[177] Reductions of sulfones have been also observed with this sonicated potassium.[178,179] These reactions will be treated in later Chapters.

Interestingly, alkali metals react with aromatic rings under sonication and milder conditions to give important aromatic radical anions.[180,181] Lithium naphthalenide, sodium biphenylide, sodium naphthalenide, or sodium-anthracene are prepared in this way in THF or DME. If necessary, the ethereal solvent can be replaced by benzene or toluene in the presence of TMEDA.[182,183] Similarly, magnesium-anthracene complex is also produced from magnesium powder and anthracene under low intensity ultrasound.[113]

A finely divided copper powder is obtained when commercial copper-bronze is subjected to ultrasound (20 kHz) in DMF at 60 °C. The method is also convenient to remove metal salts from the copper surface.[184] Moreover, this active copper was very efficient in Ullmann coupling reactions.[185]

More importantly, reduction of metal halides with lithium in THF at room temperature under sonication in a cleaning bath (50 kHz) affords active metal powders.[186] These were as reactive as those of the Rieke method. Highly reactive metal powders were easily obtained in less than 40 min by ultrasonic irradiation, whereas the comparable Rieke powders required several hours under more drastic conditions. It should be emphasized, however, that metals having a high hardness or those with a strong passivation cannot be directly activated by sonication. In fact, many ultrasound-promoted reactions most likely occur on the metal surface, so that with a weak cavitational energy the reaction can be uneffective. In contrast, other metals with appropriate solvents give rise to fine dispersions or emulsions and sonication accelerates the process.

An active nickel powder[187] can be prepared by sonication in octane at 0 °C. This material was highly effective in the hydrogenation of alkenes, while normal nickel powder is practically inactive as a hydrogenation catalyst. In addition, a catalytically active nickel is prepared by ultrasonic reduction of nickel(II) chloride with zinc dust. The latter is also activated, and in an aqueous medium, zinc reduces the water to give hydrogen.[188] This system constitutes a new sonochemical hydrogenation method of high selectivity. In fact, conjugated double bonds can be selectively hydrogenated in the presence of carbonyl groups.[188,189]

Important findings and studies on the sonochemical activation of metal powders and ultrasonically-generated metals from organometallics have been accomplished by Suslick and his associates.[190-197] Thus, ultrasonic irradiation of iron pentacarbonyl, $Fe(CO)_5$, produces nearly pure amorphous iron.[197] This material can be generated by sonolysis with a high-intensity ultrasonic probe (20 kHz, 100 W/cm^2) from pure $Fe(CO)_5$ or 4 M solutions in decane at 0 °C for 3 h under argon. Elemental analysis revealed the powder to be 96% iron by weight, with traces of carbon and oxygen. The iron powder was examined by several techniques, including scanning and transmission electron microscopy, differential scanning calorimetry (DSC), X-ray powder diffraction, and electron-beam microdiffraction. All the tests confirmed the amorphous character of the powder. DSC exhibits a large exothermic transition at 308 °C, indicating that the amorphous iron crystallizes. Heated to 350 °C, the iron powder sinters and becomes metallic in color, which is indicative of crystallization. This amorphous iron powder catalyzes the Fischer-Tropsch process (hydrogenation of carbon monoxide) at 200 °C, and is 10 times more active than the corresponding commercial powder. Likewise, amorphous iron powder catalyzes the hydrogenolysis-dehydrogenation of saturated hydrocarbons. As result of sintering and crystallization, catalytic activity decreases above 300 °C.

4. Organometallic Compounds

Ultrasonic waves have been intensively applied in the preparation of organometallic compounds. By using sonochemical techniques, syntheses are generally conducted under milder conditions, in shorter times, and often with greater yields.

As early as 1950, Renaud[198] obtained improved yields of organometallic compounds by direct metallation of alkyl halides with Li, Mg, and Al in undried diethyl ether under sonication (960 kHz, 2W). The method, however, was unsuccessful for Be, Ca, Zn, and Hg.

This exciting finding was not followed by other results on the same area until thirty years later, when Luche and Damanio[199] prepared organolithium and Grignard reagents by sonication in a simple laboratory cleaning bath (50 kHz, 60 W). Surprisingly, organometallic reagents were prepared without previous activation of the metal and in commercial undried diethyl ether. Primary organolithiums (*n*-propyl, *n*-butyl, and phenyllithium) were obtained in good yields by reaction of the corresponding alkyl bromide with Li wire or Li-2% Na sand (Scheme 3.8). The reactions were almost instantaneous. Secondary and tertiary alkyl bromides required sonication for longer times (>1 h).

$$\text{RBr} \xrightarrow[\text{Et}_2\text{O, rt,)))}]{\text{Li or Li-2\%Na (4 equiv)}} \text{RLi}$$

R	Yield (%)
Pr	90
Bu	61
Ph	95

Scheme 3.8

Induction times were studied in the formation of Grignard reagents with or without sonication and with various grades of diethyl ether (even 50% saturated with water). Although the results cannot be strictly compared in all cases, the initiation and the rate are much more rapid with sonication. It is noteworthy that previous sonication of magnesium in ether had no effect on the induction period. This suggests that ultrasound removes adsorbed water from the metal surface besides a simple cleaning effect.[200]

Functionalized organolithium[201] and Grignard[202,203] reagents have been readily prepared by sonication, including α-lithium or magnesium reagents.[204] Organolithium compounds can be also prepared by hydrogen-metal exchange. These reactions often involve the use of butyllithiums to effect the metallation. Thus, sonication allows a rapid and easier preparation of LDA in high yield (eq. 3.12).[205,206]

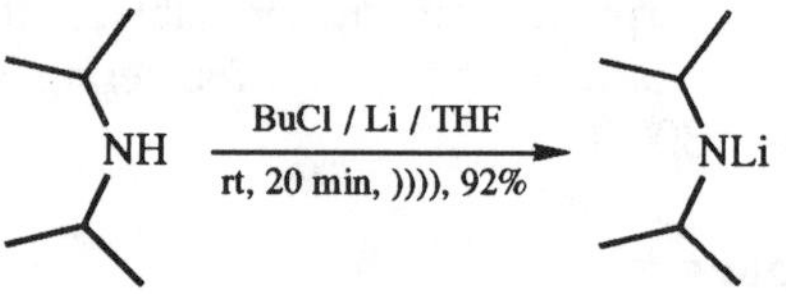

$$(i\text{-}Pr)_2NH \xrightarrow[\text{rt, 20 min,)))), 92\%}]{\text{BuCl / Li / THF}} (i\text{-}Pr)_2NLi \qquad (3.12)$$

Ultrasound also facilitates the formation of other useful organometallics from halides. Thus, alkyl aluminum halides are readily obtained in this way from chlorides or bromides and aluminum metal which is previously irradiated ultrasonically. The yields of isobutenyl aluminum sesquichloride (Scheme 3.9, R = $CH_2C(CH_3)CH_2$) are higher in dioxane than in diethyl ether, probably due to a better cavitation in the higher boiling dioxane, although yields remain unaffected by changes in ultrasound frequency.[207]

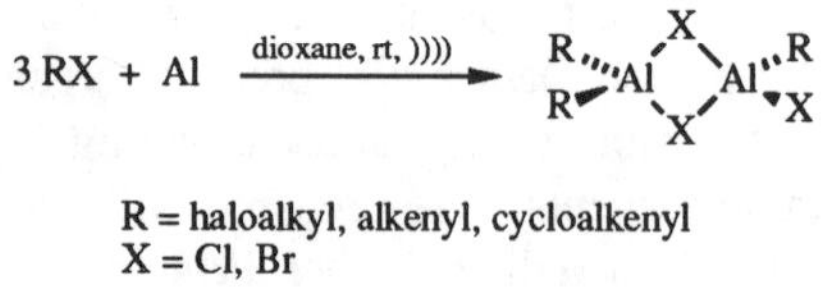

$$3\,RX + Al \xrightarrow{\text{dioxane, rt,))))}} R_2Al(\mu\text{-}X)_2AlRX$$

R = haloalkyl, alkenyl, cycloalkenyl
X = Cl, Br

Scheme 3.9

Similarly, bromoethane reacts with Al in THF to give ethyl aluminum sesquibromide with ultrasound at room temperature in only 19 min. Curiously, no reaction was detected under mechanical agitation.[208] The authors also reported the preparation of methyl aluminum sesquiiodide by sonication at room temperature in 2 h. Further reaction with triethyl aluminum gave trimethyl aluminum (eq. 3.13).[209] The nonultrasonic process required 6.5 h at 100 °C.

$$CH_3I + Al \xrightarrow[96\%]{\text{rt, 2 h,))))}} (CH_3)_3Al_2I_3 \xrightarrow[\text{))))}]{Et_3Al,\ 30\ min} (CH_3)_3Al \qquad (3.13)$$

Allylic organozinc[210] and organopalladium[211] reagents can be also prepared by a direct sonochemical metallation (eqs. 3.14 and 3.15).

$$\text{CH}_2\text{=C(COOEt)CH}_2\text{Br} + \text{HC}\equiv\text{C-CH}_2\text{OSi(CH}_3)_3 \xrightarrow[45\text{-}50\ ^\circ\text{C},\ 68\%]{\text{Zn, THF,))))}} \text{CH}_2\text{=C(COOEt)CH}_2\text{C(=CH}_2\text{)CH}_2\text{OSi(CH}_3)_3 \quad (3.14)$$

$$\text{CH}_2\text{=CH-CH}_2\text{Br} + \text{Pd (black)} \xrightarrow[4\ \text{h},\ 85\%]{\text{DMF,))))}, 55\text{-}60\ ^\circ\text{C}} [(\eta^3\text{-C}_3\text{H}_5)\text{PdBr}]_2 \quad (3.15)$$

However, unlike alkali and alkali-earth metals, most transition metals and other elements are not easily metallated by sonication, which is consistent with their inherent hardness. A more reliable route involves transmetallation with organolithium or Grignard reagents. Thus, diarylzincs can be obtained using an aryl halide, zinc bromide, and lithium wire in THF or diethyl ether in a cleaning bath (50 kHz) (eq. 3.16).[212]

$$\text{H}_3\text{C-C}_6\text{H}_4\text{-Br} \xrightarrow[\text{rt, 30 min,))))}]{\text{Li / ZnBr}_2\text{ / Et}_2\text{O}} (\text{H}_3\text{C-C}_6\text{H}_4)_2\text{Zn} \quad (3.16)$$

In contrast, dialkylzincs require a more energetic probe (a sonic horn).[213] Toluene containing a little THF was the best solvent in this reaction, which demonstrates again the better cavitation in the higher boiling toluene. With sonication, dialkylzinc reagents are obtained in quantitative yields in 20-30 min, whereas the nonultrasonic process requires 2 h and products are obtained in only 75% yield.

Similarly, organocopper reagents are generated by transmetallation of copper(I) iodide with alkyl- or aryllithiums in THF-ether (1:1) using a simple cleaning bath (50 kHz, 90 W).[214]

Trialkylboranes can be prepared in high yields by sonication *via* the formation of *in situ* generated Grignard reagents followed by transmetallation with boron trifluoride etherate (Scheme 3.10).[215]

$$3\ \text{RX} \xrightarrow[\text{15-30 min,))))}]{\text{Mg, BF}_3\text{.OEt}_2\text{, Et}_2\text{O}} \text{R}_3\text{B}$$

R	Yield (%)
$CH_3CH_2CH_2$	100
C_6H_5	97
$C_6H_5CH_2$	99
$CH_2=CHCH_2$	94

Scheme 3.10

With this method, symmetrical trialkylboranes such as *n*-Pr_3B, are obtained with a purity >99% whereas conventional hydroboration produces 93% purity. The sonochemical strategy is also useful for the synthesis of hindered boranes, and thus trinaphthylborane is obtained in 93% yield in only 15 min. The silent process requires 24 h and gives 91% yield. Triethylborane was prepared by sonication of bromoethane with aluminum and triethylborate (eq. 3.17).[208] Curiously, these sonochemical hydroborations which have a weak cavitation energy, were carried out in diethyl ether, or in the last case in bromoethane itself.

$$3\ CH_3CH_2Br \xrightarrow[\text{10-20 min,)))), 90\%}]{\text{Al, B(OEt)}_3} (CH_3CH_2)_3B \qquad (3.17)$$

It is noteworthy that low intensity ultrasound is beneficial in the preparation of other useful trialkylborane reagents. Brown and Racherla have reported[216] the sonochemical hydroboration of alkenes with various boron hydrides. Reactions are generally complete in only 1 hour, whereas classical hydroboration requires 24 h.

By successive sonochemical transmetallations in diethyl ether, organoaluminum and then organozinc reagents were obtained. The yields of triethylaluminum were unaffected by changes in ultrasound frequency in the range 43-80 kHz (Scheme 3.11).[217]

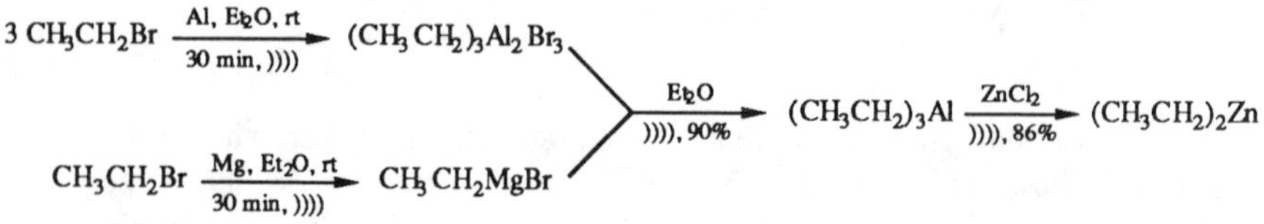

$$3\ CH_3CH_2Br \xrightarrow[\text{30 min,))))}]{\text{Al, Et}_2\text{O, rt}} (CH_3CH_2)_3Al_2Br_3$$

$$CH_3CH_2Br \xrightarrow[\text{30 min,))))}]{\text{Mg, Et}_2\text{O, rt}} CH_3CH_2MgBr$$

$$(CH_3CH_2)_3Al_2Br_3 + CH_3CH_2MgBr \xrightarrow[\text{)))), 90\%}]{\text{Et}_2\text{O}} (CH_3CH_2)_3Al \xrightarrow[\text{)))), 86\%}]{\text{ZnCl}_2} (CH_3CH_2)_2Zn$$

Scheme 3.11

Organosilanes and organostannanes can be prepared from trialkylchlorosilanes or stannanes with lithium in THF under low intensity ultrasound.[218-223] Platinum-catalyzed hydrosilylation of alkenes is performed at only 30 °C in an ultrasonic cleaning bath. This temperature is the lowest reported temperature for Pt-C catalysis.[224,225] The same authors also reported nickel-catalyzed hydrosilylation of alkenes (Scheme 3.12).[226] Activated nickel was prepared by lithium reduction of nickel (II) iodide under sonication.

$$CH_2{=}CH(CH_2)_3CH_3 + RCl_2SiH \xrightarrow[\text{THF, 20-25 °C}]{\text{NiI}_2\text{, Li,))))}} RCl_2Si(CH_2)_5CH_3$$

$R = Cl$ (94%)
$R = CH_3$ (92%)

Scheme 3.12

Sonication facilitates electroreduction of selenium or tellurium to their anions, which react further with electrophiles to give organoselenium or organotellurium compounds (Scheme 3.13).[227]

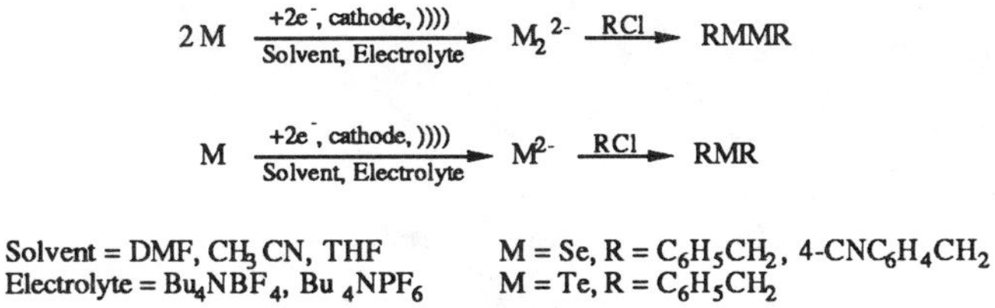

$$2M \xrightarrow[\text{Solvent, Electrolyte}]{+2e^-,\ \text{cathode},\ ))))} M_2^{2-} \xrightarrow{RCl} RMMR$$

$$M \xrightarrow[\text{Solvent, Electrolyte}]{+2e^-,\ \text{cathode},\ ))))} M^{2-} \xrightarrow{RCl} RMR$$

Solvent = DMF, CH_3CN, THF
Electrolyte = Bu_4NBF_4, Bu_4NPF_6

M = Se, R = $C_6H_5CH_2$, 4-$CNC_6H_4CH_2$
M = Te, R = $C_6H_5CH_2$

Scheme 3.13

Many interesting organometallic compounds are produced by ligand displacement reactions which are facilitated and accelerated by ultrasound. Suslick *et al.* have studied the sonolysis of iron pentacarbonyl in hydrocarbon solvents.[190,191] Sonication of this metal carbonyl produces finely divided iron and $Fe_3(CO)_{12}$. The formation of this cluster is favored in solvents having high vapor pressures such as heptane in which the yield is >82%, whereas in low vapor pressure solvents such as decalin a yield of only 5% is obtained. These results indicate that cavitation energies are inversely proportional to solvent vapor pressure. Clusterification is a process of lower activation energy, and it is then favored in lower boiling point solvents which have a weak cavitation.

More importantly, sonochemical results are quite different to photo- and thermolytic processes. Thermolysis of iron pentacarbonyl above 100 °C affords finely divided iron, and photolysis with ultraviolet light gives $Fe_2(CO)_9$ (Scheme 3.14).

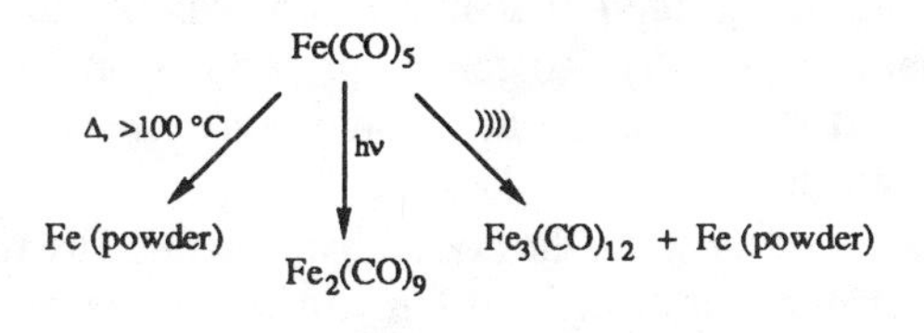

Scheme 3.14

In the presence of phosphines or alkenes, the coordinatively unsaturated intermediate species can add these ligands to give new organometallics. Finely divided iron can be also produced, but only in low vapor pressure solvents. With alkenes, isomerization takes place and the rate of this process is considerably higher than in the corresponding nonultrasonic reaction (Scheme 3.15).

Similar sonochemical substitutions have been also observed[192,228,229] with other metal carbonyls such as $Mn_2(CO)_{10}$, $Re_2(CO)_{10}$, $Ru(CO)_5$, $Cr(CO)_6$, $Mo(CO)_6$, or $W(CO)_6$.

$$Fe(CO)_5 \xrightarrow{)))))} Fe(CO)_{5-n} + n\,CO \xrightarrow{L} Fe(CO)_3L_2 + Fe(CO)_4L + Fe(CO)_2L_3 + Fe\ (powder)$$

$L = PPh_3$, alkene

$$\xrightarrow[)))))]{Fe(CO)_5\ (cat)}$$

Scheme 3.15

Ultrasonically generated active metals are also useful for the synthesis of organometallics. Ruthenium chloride is sonochemically reduced with zinc dust in methanol in the presence of 1,5-cyclooctadiene using a cleaning bath (50 kHz) to give (η^6-1,3,5-cyclooctatriene)(η^2-1,5-cyclooctadiene)ruthenium in 93% yield (eq. 3.18).[230] The nonultrasonic method involves a two-step procedure and less than 35% yield of product was obtained.

$$RuCl_3.3H_2O + Zn \xrightarrow[2)\ \Delta,\ 70\ ^\circ C]{1)\ MeOH,\ COD,\ )))),\ rt} \quad Ru \qquad (3.18)$$

G. METAL GRAPHITES

1. Introduction

The structure and composition of some supports can be modified to afford new compounds. This may be done by two general procedures: a) intercalation, in which atoms or ions are inserted into the structure, and b) ion exchange in which ions in the structure are replaced with other external ions.

Both processes can occur when the crystalline structure has an inherent openness which permits the incorporation or diffusion of atoms or ions. Among the compounds possessing this structural characteristic, graphite can intercalate a wide variety of ions and molecules such as alkali metals, halide ions, metal halides, metal oxides, ammonia, amines, or oxysalts, between the carbon layers to give graphite intercalation compounds (GIC).[231-240] Moreover, these intercalation compounds have been extensively utilized as catalysts and as reagents in numerous reactions.[240-246]

2. Structural Considerations

Unlike alumina or silica, graphite presents a lamellar structure. Similarly, a layer or lamellar disposition also exists in other polymorphic compounds. In the β-alumina structure open layers are found and sodium ions may be replaced by other cations. The zeolite molecular sieves possess three-dimensional, alumino-silicate framework structures containing networks of cages and tunnels. These contain hydrated cations which can be exchanged by other cations. Likewise, the disulfides of transition metals of groups 4, 5, and 6 have layered structures and can form intercalation compounds.

Graphite has a planar, hexagonal structure in which carbon layers are separated so that the carbon coordination number is only three. The layers are normally stacked following a pattern ABAB. However, this sequence differs somewhat from a general hexagonal stacking. Thus some carbon atoms are directly superimposable over other carbon atoms in the layer below, but others are over the space in the middle of the rings. Another usual sequence is ABCABC stacking (rhombohedral graphite) in which every third layer of carbon atoms is superimposable. The carbon atoms are sp^2 -hybridized, and the additional p-orbitals with a single electron form an infinite, delocalized π-system. This structure resembles an aromatic macromolecule, and graphite is generally considered as an extended benzene molecule. The carbon-carbon bond distance in graphite, 1.41 Å, is similar to that in benzene and is intermediate between single and double bond distance. In contrast to the strong carbon-carbon bond within the graphite layers, the adjacent layers are held by weak van der Waals bonds with an interlayer spacing of 3.35 Å (Figure 3.1). This favors the intercalation phenomenon and thus foreign atoms can intercalate between the carbon layers.

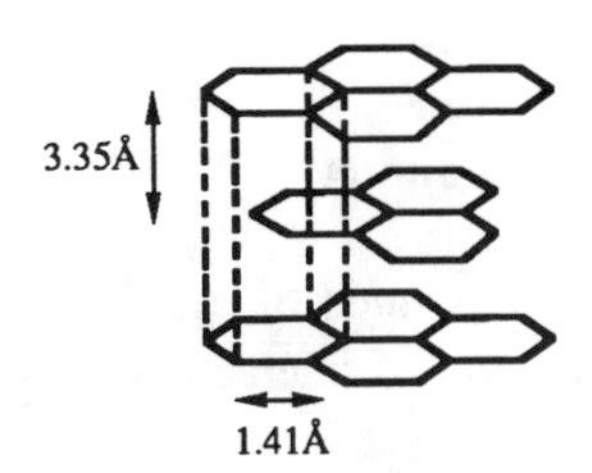

Figure 3.1. The hexagonal structure of graphite showing the two-layer stacking sequence

Strictly speaking, intercalation is the insertion of ions, atoms, or molecules into the interlayer spaces, while retaining the lamellar structure and bonding network of the host. However, other structural characteristics such as stacking order, bond distances, or bond directions may be altered.

The structures of the intercalation compounds remain unknown in most cases. Their formulas cannot be deduced from usual valences or oxidation states. In addition, they present nonstoichiometric compositions and are often characterized on the basis of their structural or crystallographic data. Hence, GIC have been precisely defined as topochemical compounds.[240]

An important feature of intercalation compounds is the stage, which is defined as the ratio of host layers to guest layers. A first-stage intercalation compound is one in which every interlayer is filled, and is the most concentrated. Second- and third-stage compounds are also frequent, and other well-defined stages can be obtained from 1 to 12. Figure 3.2 shows the nonclassical view of staging proposed by Herold[247] indicating the sequences of carbon layers and intercalants. This model explains stage evolution and intercalation-deintercalation phenomena.[248]

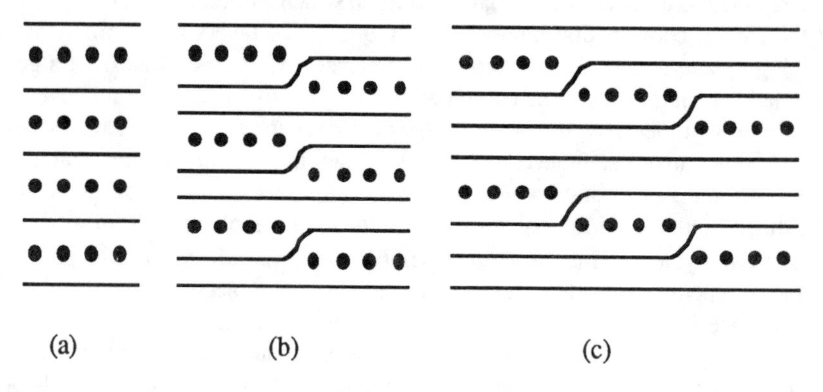

Figure 3.2. The nonclassical view of staging proposed by Herold (ref.: 247). (a) First-stage, (b) second-stage, and (c) third-stage compounds. (——) Carbon layers, (●) intercalant.

3. Graphite Intercalation Compounds

Structurally, intercalation compounds of graphite can be grouped in three categories,[238] according to the strength of interaction between intercalating species and graphite.

a) Covalent compounds, which result from the attack of strong oxidizing agents such as nitric acid, fluorine, or permanganate on graphite. Then, the aromatic planarity of the graphite is destroyed and a buckled sp^3-hybridized layer is formed.

b) Lamellar compounds, generated by the attack of moderately strong oxidants or reductants. The aromaticity of graphite is maintained, and more importantly the conductivity is enhanced.

c) Residue compounds, arising from the decomposition of lamellar compounds by thermolysis or *in vacuo* treatment.

Additionally, graphite can form surface compounds with the foreign species. In surface compounds, the atoms or molecules are absorbed on the surface of the graphite. These compounds are easily obtained from other GIC and are versatile and extremely useful reagents in synthesis.

Most of these intercalation reactions are reversible, when the structure and planarity of the carbon layers are essentially unaffected by intercalation. Thus, potassium-graphite forms on exposure of graphite to molten (or vapor) potassium, and the metal may be subsequently removed under vacuum.

According to the nature of intercalants and the reversibility of the process, a chemical classification of GIC has been also suggested.[245]

a) The intercalant participates with the graphite in an electron exchange process, and the intercalation is formally reversible. Alkali metals or bromine are representative examples of such species (eqs. 3.19 and 3.20).

$$\text{Graphite} + K \longrightarrow C_8K \xrightarrow[\text{vacuum}]{\text{partial}} C_{24}K \longrightarrow C_{36}K \longrightarrow C_{48}K \longrightarrow C_{60}K \quad (3.19)$$

$$\text{Graphite} + Br_2 \longrightarrow C_8Br \quad (3.20)$$

The electronic structure of graphite is modified by intercalation of potassium atoms since partial electron transfer from potassium takes place, and the resulting polar structure may be represented as $C_8^-K^+$. Thus, in C_8K, the graphite behaves as an electron acceptor. In contrast, in C_8Br, graphite acts as an electron donor to the halogen and the ionic formula $C_8^+Br^-$ may be attributed to graphite bromide.

b) The intercalation occurs with an irreversible loss of part of the intercalating agent. Hence the reaction of graphite with sulfuric acid to give the so-called "blue graphite" (eq. 3.21).[249]

$$\text{Graphite (anode)} + H_2SO_4 \xrightarrow{\text{electrolysis}} \underset{\text{(Blue graphite)}}{C_{24}^+(HSO_4)^-.2H_2SO_4} + H_2 \quad (3.21)$$

c) The intercalating agent forms one or more true covalent bonds with the graphite. Examples are provided by the formation of graphite oxide or graphite fluorides (eqs. 3.22 and 3.23).[238,245]

$$\text{Graphite} + HF/F_2 \xrightarrow{25\ ^\circ C} \underset{\text{(black graphite fluoride)}}{C_{3.6}F \text{ to } C_4F} \quad (3.22)$$

$$\text{Graphite} + HF/F_2 \xrightarrow{450 \text{ to } 630\ ^\circ C} \underset{\text{(white graphite fluoride)}}{CF_{0.7} \text{ to } CF} \quad (3.23)$$

Among the lamellar derivatives of graphite, alkali-graphite intercalation compounds have been widely studied[238,240] and applied in organic synthesis.[240,244-246] In these compounds, alkali metals are sandwiched between pairs of carbon rings. If all such sites were occupied, the stoichiometry C_2M would result, but in the usual formula C_8M only one quarter is occupied, in an ordered fashion. Figure 3.3 shows this disposition in which the graphite layers are superposed in projection and the alkali atoms are intercalated between the carbon rings. This arrangement of a C_8M

network is typical of many donor and acceptor compounds of graphite.

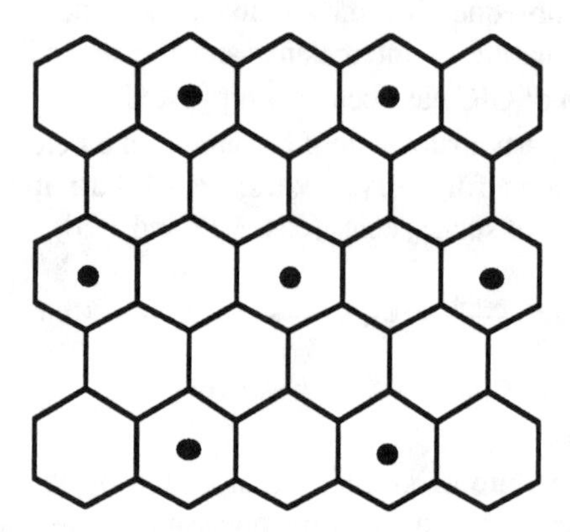

Figure 3.3. Typical structure of potassium-graphite (C_8K). Carbon layers are superposed in projection, and potassium atoms are sandwiched between pairs of carbon rings.

Potassium-graphite having the composition C_8K is the most commonly used alkali-graphite intercalation compound. In this substance all carbon layers are separated by a layer of potassium atoms. Several stages are also possible and the well-defined intercalation compounds $C_{12n}M$ (stage n >1) can be formed. Thus, in $C_{24}K$, $C_{36}K$, $C_{48}K$, and $C_{60}K$ the alkali metal is intercalated in each second, third, fourth, and fifth interlayer spacing, respectively.

A detailed crystallographic study of these compounds is extremely difficult and has been the subject of a comprehensive literature.[238,240] The crystallographic structures depend on temperature, pressure, stoichiometry of reagents, and other preparative conditions. Furthermore, complex processes can occur during the formation of GIC, such as intercalation and deintercalation reactions, stage changes, and the reversibility of the electron transfer process between host and guest. Early studies reported organized structures for GIC of various stages, particularly C_8M, $C_{12n}M$, and $C_{6n}Li$. However, only first-stages C_8M and C_6Li have stoichiometric structures.[240]

Special attention should be paid to the relevant physical properties of these compounds. In general, the intercalation of alkali metals does not affect the relative disposition or orientation of layers, so that the mechanical properties are essentially preserved. The compounds exhibit metallic characteristics and have been denoted as synthetic metals.[250] The colors of the intercalation compounds are a good guideline in their preparation. The first-stage C_8M compounds are bronze in reflection, whereas $C_{24}M$ (the second-stage compounds) are blue in reflection. The bronze C_8K can react with hydrogen to form a blue, nonstoichiometric, second-stage compound.[251] Interestingly, C_8K reacts with aromatic molecules in THF to give blue solutions of the corresponding radical anions.[252,253] More importantly, the electrical conductivity is greatly increased[254] and first-stage compounds display superconductivity.[255]

The preparation of graphite-alkali metal compounds was first reported by Weintraub[256] in the beginning of this century, and later by Fredenhagen.[257,258] Graphite intercalation compounds with compositions C_8M, $C_{24}M$, $C_{36}M$, $C_{48}M$, and $C_{60}M$ (M = K, Rb, Cs), are readily prepared from the molten alkali metals (or their vapors) with graphite in an evacuated system or under an inert atmosphere. Although several commercial graphites have proved to be equally satisfactory in the preparation of potassium-graphite,[244] a finely divided and degassed graphite powder reacts with striking readiness.

These alkali metal-graphites are extremely reactive, pyrophoric, ignite in air, and are very sensitive to moisture, reacting explosively with water.[231,244] The compounds, however, are stable in dry ethereal solvents (THF, DME, or diethyl ether) or pentane under argon for at least 24 h at room temperature.

Several experimental procedures have been utilized to synthesize the alkali metal-graphites,[233,235,238,240] including the use of alkali anion solutions, electrolysis of fused melts, or exchange from metal radical anions, among others. For the synthesis of stoichiometric compounds, the isobaric two-bulb procedure developed by Herold[240,259-261] is commonly preferred. However, the most easy and convenient method to prepare these compounds suitable for organic synthesis, is the direct melting of the alkali metal over graphite under argon.[262,263] Furthermore, the stoichiometry can be adjusted to give intercalated compounds of different stage.[263]

Preparation of Potassium-Graphite.[244,262,263]

Fresh, clean pieces of potassium (5.0 g, 0.13 mol) are added to finely divided graphite (12.5 g, 1.04 mol) with stirring for 20-30 min under argon at 150-160 °C. The reaction is very exothermic, and the potassium is added at a slow enough rate to ensure complete reaction before addition of more metal. When the potassium melts, the mixture is vigorously stirred for 10-15 min to afford C_8K as a fine bronze-colored powder.

Additionally, C_8K, $C_{24}K$, C_8Rb, and C_8Cs can also be generated at room temperature using metal transfer reactions in hydrocarbon solvents.[264,265] Graphite-alkali metal compounds are catalytically formed from graphite and alkali metals in an apolar medium.[265] The catalysts are monoolefintris(trialkylphosphine)cobalt(0) complexes, $[Co(olefin)(PR_3)_3]$, which act as metal carriers and are reversibly reduced. The reduced form, $M[Co(olefin)(PR_3)_3]_2$, transports the alkali metal to the graphite wherein the cobalt(0) complex is regenerated (Scheme 3.16).

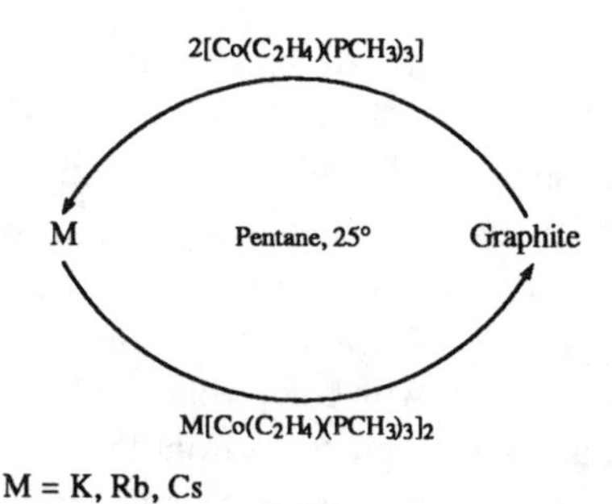

M = K, Rb, Cs

Scheme 3.16

In pentane, intercalation compounds free of solvent are formed, whereas in benzene first- and second-stage solvated compounds, with the compositions $C_{24}(C_6H_6)_xK$ and $C_{48}(C_6H_6)_xK$ were obtained. Interestingly, after longer reaction times the bronze C_8K is transformed into the blue $C_{24}K$ (eq. 3.24). Thus, the cobalt complex appears to catalyze the equilibrium between intercalates of different staging.

$$\text{Graphite} + \text{K} \xrightarrow[\text{Pentane, 25 °C, 1 day}]{10\%\ [\text{Co(C}_2\text{H}_4)(\text{PCH}_3)_3]} \text{C}_8\text{K} \xrightarrow[\text{Pentane, 25 °C, 4 days}]{10\%\ [\text{Co(C}_2\text{H}_4)(\text{PCH}_3)_3]} \text{C}_{24}\text{K} \qquad (3.24)$$

The intercalation compounds prepared in this way are very reactive, probably due to a greater disorder in the crystalline structure. This procedure can be also applied in the preparation of lithium-graphites,[264,266,267] although lithium metal is very sensitive to surface passivation. This difficulty can be overcome by using pure lithium powder. Curiously, this preparation is more rapid for lithium than for other alkali metals. This can be attributed to the higher mobility of lithium atoms in the intercalate rather than to rapid transfer reactions. In addition, high stages of sodium-graphite have been also reported.[264]

Unlike potassium, rubidium, and cesium; lithium and sodium atoms display an unusual behavior toward graphite. At first glance, this may be due to a lesser stability of lithium and sodium intercalates, at least in their concentrated stages.[241]

Lithium reacts under extreme conditions of temperature or pressure to give compounds with the formula $C_{6n}Li$ in most cases.[238,240] Moreover, the lithium carbide Li_2C_2 can be isolated in the final reaction mixtures, whereas K, Rb, or Cs do not form directly stable metal carbides.

In contrast, sodium has a scarce tendency to intercalate, but is easily adsorbed on the surface of supports such as charcoal or alumina to give highly reactive metals.[268] These forms have also an important synthetic value. However, sodium can react with crystalline graphite to give $C_{64}Na$.[269-271] The reaction requires an atmosphere of pure helium, since in the presence of nitrogen and hydrogen, graphite reacts with Na to give quaternary third-stage intercalates.[272]

The behavior of alkali metal amalgams toward graphite is also quite different. The reactivity is a balance between the affinity of the alkali metal for the graphite, which increases from Na to Cs, and for the mercury which decreases in the same order. Thus, sodium amalgam is very stable and does not react with graphite to give sodium intercalates. Potassium and rubidium amalgams display a similar reactivity giving ternary first-stage systems C_4HgK and C_4HgRb, and ternary second-stage compounds C_8HgK and C_8HgRb, respectively.[273] Finally, cesium amalgam affords pure cesium intercalation compounds.

Besides alkali metals, other elements can form similar lamellar compounds.[238,240] Alkali-earth metals such as calcium, barium, or strontium in vapor phase react with graphite to give yellow first-stage compounds.[274,275] Reactions with metal vapors are better than the direct heating of mixtures of metal powders and graphite. However, calcium intercalates are generated at about 450 °C, and higher temperatures result in the formation of calcium carbide.[275] Lanthanides also form intercalation compounds with graphite.[276-278] The processes either use a combination of graphite with the metal powder or with the vapor; the latter is preferred with the more volatile elements such as samarium, europium, thulium, and ytterbium. Interestingly, europium intercalates have been prepared from europium(III) chloride.[279,280] It should be pointed out, in connection with these findings, that Lalancette *et al.*[237] reported the intercalation of several metal chlorides in graphite from carbon tetrachloride solutions. Intercalation occurs by heating at reflux a mixture of metal halide and graphite in an atmosphere of chlorine. In the absence of this gas, no intercalation or poor results occur. The authors suggested that chlorine promotes the opening of the lattice to allow the intercalation. Furthermore, the intercalation proceeds with metal halides having low solubilities in CCl_4, whereas very soluble or insoluble salts led to no intercalation. The process was relatively effective for $AlCl_3$, $FeCl_3$, $NiCl_2$, and $CuCl_2$. Structural data on composition or staging were not reported, but the analyses of intercalates revealed the presence of different amounts of metal and chlorine.

4. Metal Dispersions on Graphite Surfaces

It has been previously noted that GIC can also react with metal salts to give new graphite species. Most of these compounds are now regarded as surface rather than intercalation compounds, and the process constitutes an excellent method for preparing highly active metals, dispersed on the graphite surface.[240,246]

Dispersed compounds are obtained by the reduction of anhydrous metal halides with potassium-graphite (C_8K) in dry ethereal solvents (THF or DME) under argon. A wide range of dispersed metals such as Mg, Ti, Mn, Fe, Co, Ni, Pd, Zn, Cu, or Sn can be prepared. Examples are provided by the equations 3.25-3.28.

$$2\,C_8K + ZnCl_2 \xrightarrow[\text{reflux, Ar}]{\text{THF}} C_{16}Zn + 2\,KCl \quad (3.25)$$

$$3\,C_8K + FeCl_3 \xrightarrow[\text{reflux, Ar}]{\text{THF}} C_{24}Fe + 3\,KCl \quad (3.26)$$

$$2\,C_8K + PdCl_2 \xrightarrow[\text{100°C, Ar}]{\text{DME}} C_{16}Pd + 2\,KCl \quad (3.27)$$

$$2\,C_8K + MgI_2 \xrightarrow[\text{reflux, Ar}]{\text{THF or ether}} C_{16}Mg + 2\,KI \quad (3.28)$$

The resulting metal-graphites have been widely used in organic synthesis due to their extreme reactivity associated with high surface areas and particle reduction.

Preliminary studies on zinc-graphite[281-283] reported the existence of lamellar intercalation compounds with different staging. Further results by the same authors suggested a dispersion of zinc on the graphite and only very weak reflections (X-ray diffraction lines) can be attributed to intercalated species.[284,285] Similarly, there have been claims of the existence of weak or very weak reflections assigned to metal lamellar systems in the case of magnesium,[282] titanium,[246,283,286] iron,[287,288] cobalt,[246] copper,[246] and tin.[246,286] However, opposite results indicate that all of the metal is essentially dispersed on the graphite surface.[240,289-293]

In general, metal-graphites undergo ready oxidation by air, and must be prepared and utilized *in situ* under argon. Palladium-graphite[294,295] and a less active form of nickel-graphite (Ni-Gr 2)[296] can be stored for longer periods.

Some preparations of these compounds deserve a special comment. Thus, a highly dispersed nickel on graphite (Ni-Gr 1)[297] is obtained by reduction of $NiBr_2.2DME$ dissolved in THF-HMPA (15:1) with C_8K (2 equiv) at 20 °C under argon. This material is slowly oxidized by air to give a less active metal (Ni-Gr 2),[296] which can be stored for several months.

Titanium-graphite, a versatile reagent in carbonyl coupling reactions, can be prepared according to several procedures.[240] The resulting reagents exhibit different reactivities, and it is quite difficult to ascertain the identity of species. Highly dispersed titanium on graphite is generated by reduction of titanium(IV) chloride with C_8K in benzene.[246,283] It has been suggested that titanium divalent species are involved in this reduction, and the reagent tends to favor pinacol products. In any event, the situation is not absolutely clear. Aromatic carbonyls produce alkenes (McMurry coupling), whereas aliphatic ones yield pinacols. This dichotomy in titanium-mediated coupling will be treated later (see Chapter V).

In contrast, the reduction of titanium(III) chloride with C_8K in THF affords a titanium-graphite,[286] probably consisting of Ti(0) species. The reductive coupling of ketones with this reagent produces alkenes (McMurry coupling) in good yields. More importantly reduction of $TiCl_3$ with C_8K in only 30 min provides a reagent, which promotes pinacol reaction. However, the reagent generated on prolonged reduction favors the formation of alkenes.[298-301]

Besides potassium-graphite, titanium chlorides also form low-valent titanium graphites by reaction with zinc/silver-graphite[290] (Zn/Ag-C_8), which can be prepared by reaction of C_8K with $ZnCl_2$/AgOAc in THF.

In addition, another titanium-graphite can be formed by reduction of titanium(IV) isopropoxide with C_8K in THF.[281,283] In this case, the formation of first- and fourth-stage intercalation compounds has been suggested.

Preparation of Zinc/Silver-Graphite.[290]

Potassium-graphite, prepared as described above from potassium (0.6 g, 15.4 mmol) and graphite (1.5 g, 125 mmol), is suspended in dry THF (15 mL) and then a mixture of freshly fused zinc chloride (1.0 g, 7.4 mmol) and silver acetate (0.1 g, 0.6 mmol) is added in portions. The exothermic reaction is refluxed for 30 min until complete reduction, to yield the zinc/silver-graphite which should be used immediately.

5. Other Metallic Graphite Compounds

As previously indicated, graphite-metal compounds are readily accessible either by direct interaction of graphite with metals or with metal halides. Under suitable conditions similar metal-graphite derivatives can be formed, regardless of whether they are true intercalates or highly dispersed metals on graphite surfaces. Anyway, the formation of complete lamellar metal-graphites appears to be highly dependent on the experimental conditions,[238,240,302] and the factors influencing the intercalation process have been surveyed.[303]

Graphite-metal halide compounds can be prepared following a number of general, synthetic techniques, which have been developed for other graphite compounds.[238]

a) By direct heating of the graphite with the metal halide in a sealed tube in the presence of halogen (chlorine or bromine). Additionally, compounds can be generated by direct combination of graphite and metal, or graphite and metal oxide, both in the presence of halogen.

b) Following the isobaric two-bulb procedure,[259-261] which ensures stoichiometric preparations. Here, an evacuated and sealed system consists of two interconnected bulbs, the first one contains the graphite and the second bulb is filled with the intercalant. This is vaporized and forced to react with the graphite in the first bulb. The graphite is generally maintained at a higher temperature thant the intercalant to avoid condensation of an excess on the graphite.

c) Intercalation from solutions in nonaqueous solvents.[304] The process can be limited because of partial leaching out of metal halide, so that some halides are intercalated only in a narrow range of solubility.[237] In some cases, the presence of chlorine is essential for success,[238] particularly for the intercalation of the less reacitve fluorides such as TiF_4, TaF_4, or NbF_5.[305,306]

d) Intercalation through a complexation reaction in the vapor phase.[304] In the presence of $AlCl_3$ at 500 °C, the nonvolatile cobalt(II) chloride can be intercalated, since its volatility is markedly increased. The resulting graphite-$CoCl_2$ compound contains only traces of $AlCl_3$.

Structural information on graphite-metal halides has been mostly limited to the determination of stage and crystallographic distances, although detailed X-ray diffraction studies have been performed with some transition metal halides.[238] In general, the structures of the free halides are largely maintained upon intercalation.

The reduction products of these systems have attracted considerable attention as potential catalysts or electrode material in high-energy-density batteries. Graphite-$FeCl_3$ can be reduced at 350 °C in hydrogen or nitrogen to afford very likely a nonintercalate.[238,307] When the heating is performed at higher temperatures, further reduction to α-Fe takes place,[308] although incomplete reduction occurs even up to 1000 °C.[309] Other reducing agents, such as lithium biphenylide or Na-liquid ammonia lead to iron metal, but π-complexes can also be formed. Additionally, other methods introduce iron clusters into the layered structure of graphite.[238] This iron in graphite layers is more resistant to oxidation than free iron.[310] Interestingly, the reduction of some graphite-metal halides can lead either to metals or sandwich-type π-complexes.[234,309,311]

Metal halide GIC have been employed with great benefit in numerous organic transformations,[240,245,312] such as halogenations, isomerizations, halogen-exchange reactions, reductions, or Friedel-Crafts alkylations, among others. These graphite supported reagents are more resistant and easier to handle than neat halides. Many reactions are effected very rapidly, under milder conditions, with good yields, and selectivity, which can differ from reactions with neat halides. In addition, the stereoselectivity can also be markedly affected. Thus, Kagan *et al.*[313] developed a graphite supported chiral rhodium catalyst, which was applied in asymmetric hydrogenation and asymmetric hydrosilylation reactions. The absolute configuration of products was opposite to that observed with the catalyst in homogeneous reactions.

A number of graphite-metal oxides has been also prepared.[238,302] Among them, the graphite-chromium trioxide intercalation compound emerges as a potential cathode in batteries,[314-316] and as an oxidizing agent. However the true intercalated oxide, formulated as $C_{13.6}CrO_3$ does not oxidize alcohols, but a mixture of graphite-Cr_2O_5-Cr_3O_8 does.[242,245,315,317,318] The latter is not a true intercalate. Similar compounds of iron, nickel, cobalt, or platinum, denoted graphimets, cannot be considered as GIC. Pt-Graphimet appears to be graphite-supported platinum.[292] Ni-Graphimet[234] is probably not a lamellar compound and is quite similar, from a morphological viewpoint, to nickel-graphite generated by reduction of nickel(II) chloride with potassium-graphite.[319,320]

In conclusion, graphite-metal compounds are new, versatile, and very promising reagents or catalysts in synthesis. They constitute a good example of how reactive reagents may have a chemistry modified by intercalation in graphite. Their unique properties have been presented concisely and their effect on numerous reactions will be outlined in part two of this book.

Despite the intrinsic advantages of metal-graphites, some limitations must be also considered. They are extremely reactive and are easily decomposed by traces of air or moisture. Eventually, reactive intermediates or reaction products may be intercalated within the metal-graphites. Finally, the disadvantages associated with the heterogeneous character of the reagents, particularly the reproducibility, should be kept in mind.

References

1. Klabunde, K. J.; Murdock, T. O. *J. Org. Chem.* **1979**, *44*, 3901 and references cited therein.
2. For a good report on commercial samples of metals: Solomon, S.; Bates, D. J. *J. Chem. Educ.* **1991**, *68*, 991.
3. Hauser, C. R.; Breslow, D. R. *Org. Synth., Coll. Vol. III* **1955**, 408.
4. Newman, M. S.; Arens, F. J. *J. Am. Chem. Soc.* **1955**, *77*, 946.
5. Goasdoue, N.; Gaudemar, M. *J. Organomet. Chem.* **1972**, *39*, 17.
6. Tsuda, K.; Ohki, E.; Nozoe, S. *J. Org. Chem.* **1963**, *28*, 783.
7. Kerdesky, F. A. J.; Ardecky, R. J.; Lakshmikanthan, M. V.; Cava, M. P.; *J. Am Chem. Soc.* **1981**, *103*, 1992.
8. Gilman, H.; Peterson, J. M.; Schulze, F. *Recl. Trav. Chim. Pays-Bas* **1928**, *47*, 19.
9. Gilman, H.; Heck, L. L. *Bull. Soc. Chim. Fr.* **1929**, 250.
10. Gilman, H.; Kirby, R. H. *Recl. Trav. Chim. Pays-Bas* **1935**, *54*, 577.
11. Summerbell, R. K.; Umhoefer, R. R. *J. Am. Chem. Soc.* **1939**, *61*, 3016.
12. Summerbell, R. K.; Hyde, D. K. A. *J. Org. Chem.* **1960**, *25*, 1809.
13. Lai, Y.-H. *Synthesis* **1981**, 585 and references cited therein.
14. Gaudemar, M. *Organomet. Chem. Rev. A* **1972**, *8*, 183.
15. Rathke, M. W. *Org. React.* **1975**, *22*, 423.
16. Nützel, K. *Houben Weyl-Methoden der organischen Chemie*; Georg Thieme Verlag: Stuttgart, 1973; Vol. XIII/2a, p 805.
17. Imamoto, T.; Kusumoto, T.; Yokoyama, M. *J. Chem. Soc. Chem. Commun.* **1982**, 1042.
18. Hill, C. L.; Sande, J. B. V.; Whitesides, G. M. *J. Org. Chem.* **1980**, *45*, 1020 and references therein.

19. Grignard, V. *C. R. Hebd. Acad. Sci. Ser. C* **1934**, *198*, 625.
20. Pearson, D. E.; Cowan, D.; Beckler, J. D. *J. Org. Chem.* **1959**, *24*, 504.
21. Kuwajima, I.; Nakamura, E.; Hashimoto, K. *Org. Synth.* **1983**, *61*, 122.
22. Masuoka, K.; Hashimoto, S.; Kitagawa, Y.; Yamamoto, H.; Nozaki, H. *Bull. Chem. Soc. Jpn.* **1980**, *53*, 3301.
23. Gaudemar-Bardone, F.; Gaudemar, M.; Mladenova, M. *Synthesis* **1987**, 1130.
24. Kuroboshi, M.; Ishihara, T. *Tetrahedron Lett.* **1987**, *28*, 6481.
25. Ishihara, T.; Kuroboshi, M. *Chem. Lett.* **1987**, 1145.
26. Sato, F.; Akiyama, T.; Iida, K.; Sato, M. *Synthesis* **1982**, 1025.
27. Onaka, M.; Matsuoka, Y.; Mukaiyama, T. *Chem. Lett.* **1981**, 531.
28. Toda, F.; Tanaka, K.; Tange, H. *J. Chem. Soc. Perkin Trans. 1* **1989**, 1555.
29. Tanaka, K.; Kishigami, S.; Toda, F. *J. Org. Chem.* **1990**, *55*, 2981.
30. Gawronski, J. K. *Tetrahedron Lett.* **1984**, *25*, 2605.
31. Picotin, G.; Miginiac, P. *J. Org. Chem.* **1987**, *52*, 4796.
32. Corey, E. J.; Pyne, S. G. *Tetrahedron Lett.* **1983**, *24*, 2821.
33. Vankar, Y. D.; Arya, P. S.; Rao, C. T. *Synth. Commun.* **1983**, *13*, 869.
34. Chapman, J. H.; Elks, J.; Phillipps, G. H.; Wyman, L. J. *J. Chem. Soc.* **1956**, 4344.
35. Mills, J. S.; Ringold, H. J.; Djerassi, C. *J. Am. Chem. Soc.* **1958**, *80*, 6118.
36. Leonard, N. J.; Steinhardt, C. K.; Lee, C. *J. Org. Chem.* **1962**, *27*, 4027.
37. Ashby, E. C.; Yu, S. H.; Beach, R. G. *J. Am. Chem. Soc.* **1970**, *92*, 433.
38. Pines, H.; Haag, W. O. *J. Org. Chem.* **1958**, *23*, 328.
39. Shabtai, J.; Gil-Av, E. *J. Org. Chem.* **1963**, *28*, 2893.
40. Savoia, D.; Tagliavini, E.; Trombini, C.; Umani-Ronchi, A. *J. Org. Chem.* **1980**, *45*, 3227.
41. Bonnet, M.; Geneste, P.; Rodriguez, M. *J. Org. Chem.* **1980**, *45*, 40.
42. Imamura, H.; Konishi, T.; Sakata, Y.; Tsuchiya, S. *J. Chem. Soc. Chem. Commun.* **1991**, 1527.
43. Gaudemar, M. *C. R. Hebd. Acad. Sci. Ser. C* **1969**, *268*, 1439.
44. Fraenkel, G.; Ellis, S. H.; Dix, D. T. *J. Am. Chem. Soc.* **1965**, *87*, 1406.
45. Normant, H. *Bull. Soc. Chim. Fr.* **1968**, 791.
46. Cuvigny, T.; Larchevêque, M.; Normant, H. *C. R. Hebd. Acad. Sci. Ser. C* **1972**, *274*, 797.
47. Cuvigny, T.; Larchevêque, M.; Normant, H. *Bull. Soc. Chim. Fr.* **1973**, 1174.
48. Larchevêque, M.; Cuvigny, T. *Bull. Soc. Chim. Fr.* **1973**, 1445.

49. Cuvigny, T.; Larchevêque, M. *J. Organomet. Chem.* **1974**, *64*, 315.
50. Chelucci, G. *Synthesis* **1991**, 474.
51. Corey, E. J.; Chaykovski, M. *J. Am. Chem. Soc.* **1962**, *84*, 867.
52. Sjöberg, K. *Tetrahedron Lett.* **1966**, 6383.
53. Kojima, T.; Fujisawa, T. *Chem. Lett.* **1978**, 1425.
54. Gololobov, Yu, G.; Nesmeyanov, A. N.; Lysenko, V. P.; Boldeskul, I. E. *Tetrahedron* **1987**, *43*, 2609. A review of dimethylsulfoxonium methylide (Corey's reagent).
55. Read, G.; Ruiz, V. M. *J. Chem. Soc. Perkin Trans. 1* **1973**, 1223.
56. Owsley, D. C.; Bloomfield, J. J. *Synthesis* **1977**, 118.
57. Pfaltz, A.; Anwar, S. *Tetrahedron Lett.* **1984**, *25*, 2977.
58. Rieke, R. D. *Top. Curr. Chem.* **1975**, *59*, 1.
59. Rieke, R. D. *Acc. Chem. Res.* **1977**, *10*, 301.
60. Rieke, R. D. *Science* **1989**, *246*, 1260.
61. Rieke, R. D.; Hudnall, P. M. *J. Am. Chem. Soc.* **1972**, *94*, 7178.
62. Rieke, R. D.; Uhm, S.; Hudnall, P. M. *J. Chem. Soc. Chem. Commun.* **1973**, 269.
63. Rieke, R. D.; Bales, S. E. *J. Am. Chem. Soc.* **1974**, *96*, 1775.
64. Rieke, R. D.; Chao, L. *Synth. React. Inorg. Met.-Org. Chem.* **1974**, *4*, 101.
65. Chao, L.; Rieke, R. D. *J. Organomet. Chem.* **1974**, *67*, C64.
66. Chao, L.; Rieke, R. D. *Synth. React. Inorg. Met.-Org. Chem.* **1975**, *5*, 165.
67. Chao, L.; Rieke, R. D. *J. Org. Chem.* **1975**, *40*, 2253.
68. Rieke, R. D.; Uhm, S. J. *Synthesis* **1975**, 452.
69. Spencer, J. F.; Wallace, M. L. *J. Chem. Soc.* **1908**, *93*, 1827.
70. Deacon, G. B.; Parrot, J. C. *Aust. J. Chem.* **1971**, *24*, 1771.
71. Gynane, M. J. S.; Waterworth, L. G.; Worrall, I. J. *J. Organomet. Chem.* **1972**, *40*, C9.
72. Rieke, R. D.; Bales, S. E. *J. Chem. Soc. Chem. Commun.* **1973**, 879.
73. Rieke, R. D.; Wolf, W. J.; Kujundzic, N.; Kavaliunas, A. V. *J. Am. Chem. Soc.* **1977**, *99*, 4159.
74. Scott, N. D.; Walker, J. F.; Hansley, V. L. *J. Am. Chem. Soc.* **1936**, *58*, 2442.
75. Paul, D. E.; Lipkin, D.; Weissman, S. I. *J. Am. Chem. Soc.* **1956**, *78*, 116.
76. Closson, W. D.; Wriede, P.; Bank, S. *J. Am. Chem. Soc.* **1966**, *88*, 1581.
77. Suga, K.; Watanabe, S.; Torii, I. *Chem. Ind.* **1967**, 360.
78. Suga, K.; Watanabe, S.; Pan, T. P. *Aust. J. Chem.* **1968**, *21*, 2341.
79. Suga, K.; Watanabe, S.; Kamma, K. *Can. J. Chem.* **1967**, *45*, 933.
80. Ivanov, Ch.; Markov, P. *Naturwissenschaften* **1963**, *50*, 688.

81. Markov, P.; Lasarov, D.; Ivanov, Ch. *Ann.* **1967**, *704*, 126.
82. Arnold, R. T.; Kulenovic, S. T. *Synth. Commun.* **1977**, *7*, 223.
83. Rieke, R. D.; Rhyne, L. D. *J. Org. Chem.* **1979**, *44*, 3445.
84. Rieke, R. D.; Li, P. Z.-J.; Burns, T. P.; Uhm, S. J. *J. Org. Chem.* **1981**, *46*, 4323.
85. Inaba, S.; Rieke, R. D. *Synthesis* **1984**, 844.
86. Matsumoto, H.; Inaba, S.; Rieke, R. D. *J. Org. Chem.* **1983**, *48*, 840.
87. Burns, T. P.; Rieke, R. D. *J. Org. Chem.* **1987**, *52*, 3674.
88. Ebert, G. W.; Rieke, R. D. *J. Org. Chem.* **1984**, *49*, 5280.
89. Inaba, S.; Matsumoto, H.; Rieke, R. D. *J. Org. Chem.* **1984**, *49*, 2093.
90. Burkhardt, E. R.; Rieke, R. D. *J. Org. Chem.* **1985**, *50*, 416.
91. Inaba, S.; Rieke, R. D. *J. Org. Chem.* **1985**, *50*, 1373.
92. Inaba, S.; Wehmeyer, R. M.; Forkner, M. W.; Rieke, R. D. *J. Org. Chem.* **1988**, *53*, 339.
93. Wu, T.-C.; Rieke, R. D. *J. Org. Chem.* **1988**, *53*, 2381.
94. Wu, T.-C.; Wehmeyer, R. M.; Rieke, R. D. *J. Org. Chem.* **1987**, *52*, 5057.
95. Wehmeyer, R. M.; Rieke, R. D. *J. Org. Chem.* **1987**, *52*, 5056.
96. Ebert, G. W.; Rieke, R. D. *J. Org. Chem.* **1988**, *53*, 4482.
97. Xiong, H.; Rieke, R. D. *J. Org. Chem.* **1989**, *54*, 3247.
98. Rieke, R. D.; Daruwala, K. P.; Forkner, M. W. *J. Org. Chem.* **1989**, *54*, 21.
99. O'Brien, R. A.; Rieke, R. D. *J. Org. Chem.* **1990**, *55*, 788.
100. Ginah, F. O.; Donovan, Jr., T. A.; Suchan, S. D.; Pfennig, D. R.; Ebert, G. W. *J. Org. Chem.* **1990**, *55*, 584.
101. Wu, T.-C.; Xiong, H.; Rieke, R. D. *J. Org. Chem.* **1990**, *55*, 5045.
102. Yanagisawa, A.; Habaue, S.; Yamamoto, H. *J. Am. Chem. Soc.* **1991**, *113*, 8955.
103. Ramsden, H. E.; U.S. Patent 3 354 190, 1967; *Chem. Abstr.* **1968**, *68*, 114744.
104. Freeman, P. K.; Hutchinson, L. L. *J. Org. Chem.* **1983**, *48*, 879.
105. Bogdanovic, B.; Liao, S.; Mynott, R.; Schlichte, K.; Westeppe, U. *Chem. Ber.* **1984**, *117*, 1378.
106. Alonso, T.; Harvey, S.; Junk, P. C.; Raston, C. L.; Skelton, B. W.; White, A. H. *Organometallics* **1987**, *6*, 2110.
107. Bogdanovic, B.; Janke, N.; Kinzelmann, H.-G.; Westeppe, U. *Chem. Ber.* **1988**, *121*, 33.
108. Bogdanovic, B.; Janke, N.; Krüger, C.; Mynott, R.; Schlichte, K.; Westeppe, U. *Angew. Chem. Int. Ed. Engl.* **1985**, *24*, 960.
109. Bogdanovic, B. *Angew. Chem. Int. Ed. Engl.* **1985**, *24*, 262.
110. Bogdanovic, B. *Acc. Chem. Res.* **1988**, *21*, 261 and references therein.

111. Bogdanovic, B.; Ritter, A.; Spliethoff, B. *Angew. Chem. Int. Ed. Engl.* **1990**, *29*, 223 and references cited therein.
112. Bogdanovic, B.; Bönnemann, H. Ger. Offen DE 3 541 633, 1987; U.S. Patent 4 713 110, 1987.
113. Bönnemann, H.; Bogdanovic, B.; Brinkman, R.; He, D. W.; Spliethoff, B. *Angew. Chem. Int. Ed. Engl.* **1983**, *22*, 728.
114. Oppolzer, W.; Schneider, P. *Tetrahedron Lett.* **1984**, *25*, 3305.
115. Oppolzer, W.; Cunningham, A. F. *Tetrahedron Lett.* **1986**, *27*, 5467.
116. Raston, C. L.; Salem, G. *J. Chem. Soc. Chem. Commun.* **1984**, 1702.
117. Rylander, P. N. *Hydrogenation Methods*; Academic Press: San Diego, 1985.
118. James, B. R. *Comprehensive Organometallic Chemistry*; Pergamon Press: Oxford, 1982; Vol. 8, Chapter 51.
119. Huffman, J. W. *Comprehensive Organic Synthesis*; Fleming, I., Ed.; Pergamon Press: Oxford, 1991; Vol. 8.
120. Brown, H. C.; Brown, C. A. *J. Am. Chem. Soc.* **1962**, *84*, 1493.
121. Brown, H. C.; Brown, C. A. *J. Am. Chem. Soc.* **1962**, *84*, 1494.
122. Brown, H. C.; Brown, C. A. *J. Am. Chem. Soc.* **1962**, *84*, 2827.
123. Brown, H. C.; Sivasankaran, K. *J. Am. Chem. Soc.* **1962**, *84*, 2828.
124. Brown, C. A. *J. Org. Chem.* **1970**, *35*, 1900.
125. Brown, C. A.; Ahuja, V. K. *J. Org. Chem.* **1973**, *38*, 2226 and references therein.
126. Brown, C. A.; Ahuja, V. K. *J. Chem. Soc. Chem. Commun.* **1973**, 553.
127. Millard, A. A.; Rathke, M. W. *J. Am. Chem. Soc.* **1977**, *99*, 4833.
128. Colon, I.; Kelsey, D. R.; *J. Org. Chem.* **1986**, *51*, 2627 and references therein.
129. Schlecht, L.; Trageser, G. Ger. 839 935, 1952; *Chem. Abstr.* **1957**, *51*, 11225i.
130. Schlecht, L.; Ackermann, K. Ger. 838 375, 1952; *Chem. Abstr.* **1957**, *51*, 11226d.
131. Schlecht, L.; Klippel, H. U.S. Patent 2 791 497, 1957; *Chem. Abstr.* **1957**, *51*, 11226f.
132. Hata, K.; Watanabe, K.; Tanaka, M. *Bull. Chem. Soc. Jpn.* **1958**, *31*, 775.
133. Krivanek, M.; Danes, V.; Nikolajenkov, V. *Collect. Czech. Chem. Commun.* **1964**, *31*, 1950.
134. Schott, H.; Wilke, G. *Angew. Chem. Int. Ed. Engl.* **1969**, *8*, 877.
135. Kalies, W.; Witt, B.; Gaube, W. *Z. Chem.* **1980**, *20*, 310.
136. Boldrini, G. P.; Savoia, D.; Tagliavini, E.; Trombini, C.; Umani-Ronchi, A. *J. Org. Chem.* **1985**, *50*, 3082.

137. a) Bönnemann, H.; Brijoux, W.; Joussen, T. *Angew. Chem. Int. Ed. Engl.* **1990**, *29*, 273. b) With a related methodology the formation of colloidal transition metals in organic media has been reported: Bönnemann, H.; Brijoux, W.; Brinkmann, R.; Dinjus, E.; Joussen, T.; Korall, B. *Angew. Chem. Int. Ed. Engl.* **1991**, *30*, 1312.
138. Gladstone, J. H. *J. Chem. Soc.* **1891**, *59*, 290.
139. Job, A.; Reich, R. *Bull. Soc. Chim. Fr.* **1923**, 1414.
140. Renshaw, R. R.; Greenlaw, C. E. *J. Am. Chem. Soc.* **1920**, *42*, 1472.
141. Kurg, R. C.; Tang, P. J. C. *J. Am. Chem. Soc.* **1954**, *76*, 2262.
142. Noller, R. C. *Org.Synth.* **1932**, *12*, 86.
143. Hota, N. K.; Willis, C. J. *J. Organomet. Chem.* **1967**, *9*, 169.
144. Miller, R. E.; Nord, F. F. *J. Org. Chem.* **1951**, *16*, 728.
145. Horii, Z.; Kugita, H.; Takeuchi, T. *J. Pharm. Soc. Jpn.* **1953**, *73*, 895; *Chem. Abstr.* **1954**, *48*, 11329g.
146. Kamenski, C. W.; Esmay, D. L. *J. Org.Chem.* **1960**, *25*, 1807.
147. Hennion, G. F.; Sheehan, J. J. *J. Am. Chem.Soc.* **1949**, *71*, 1964.
148. Shank, R. S.; Shechter, H. *J. Org. Chem.* **1959**, *24*, 1825.
149. Simmons, H. E.; Smith, R. D. *J. Am. Chem. Soc.* **1959**, *81*, 4256.
150. Smith, R. D.; Simmons, H. E. *Org.Synth.* **1961**, *41*, 72.
151. Simmons, H. E.; Cairns, T. L.; Vladuchick, S. A.; Hoiness, C. M. *Org. React.* **1973**, *20*, 1.
152. Le Goff, E. *J. Org.Chem.* **1964**, *29*, 2048.
153. Rawson, R. J.; Harrison, I. T. *J. Org.Chem.* **1970**, *35*, 2057.
154. McMurry, J. E.; Rico, J. G. *Tetrahedron Lett.* **1989**, *30*, 1169.
155. Denis, J. M.; Girard, C.; Conia, J. M. *Synthesis* **1972**, 549.
156. Clark, R. D.; Heathcock, C. H. *J. Org. Chem.* **1973**, *38*, 3658.
157. Lorimer, J. P.; Mason, T. J. *Chem. Soc. Rev.* **1987**, *16*, 239.
158. Lindley, J.; Mason, T. J. *Chem. Soc. Rev.* **1987**, *16*, 275.
159. Mason, T. J. *Ultrasonics* **1986**, *24*, 245.
160. Suslick, K. S. *Modern Synthetic Methods* **1986**, *4*, 1.
161. Suslick, K. S. *Adv. Organomet. Chem.* **1986**, *25*, 73.
162. Goldberg, Y.; Sturkovich, R.; Lukevics, E. *Heterocycles* **1989**, *29*, 597.
163. Suslick, K. S. *Science* **1990**, *247*, 1439.
164. Einhorn, C.; Einhorn, J.; Luche, J.-L. *Synthesis* **1989**, 787. A general review of sonochemistry in synthetic organic chemistry.
165. Mason, T. J.; Lorimer, J. P. *Sonochemistry, Theory, Applications, and Uses of Ultrasound in Chemistry*; Ellis Horwood: Chichester, 1988.
166. *Ultrasound, its Chemical, Physical, and Biological Effects*; Suslick, K. S., Ed.; VCH Publishers: Weinheim, 1988.
167. Ley, S. V.; Low, C. M. R. *Ultrasound in Synthesis*; Springer-Verlag: Berlin, 1989.
168. Flint, E. B.; Suslick, K. S. *Science* **1991**, *253*, 1397.

169. Maltsev, A. N. *Russ. J. Phys. Chem.* **1976**, *50*, 995.
170. Townsend, C. A.; Nguyen, L. T. *J. Am. Chem. Soc.* **1981**, *103*, 4582.
171. Cioffi, E. A.; Prestegard, J. H. *Tetrahedron Lett.* **1986**, *27*, 415.
172. Suslick, K.S.; Casadonte, D. J.; Green, M. L. H.; Thompson, M. E. *Ultrasonics* **1987**, *25*, 56.
173. Suslick, K. S.; Green, M. L. H.; Thompson, M. E.; Chatakondu, K. *J. Chem. Soc. Chem. Commun.* **1987**, 900.
174. Solov'eva, L. N. *Kolloid Zh.* **1939**, *5*, 289; *Chem. Abstr.* **1939**, *33*, 8471.
175. Walker, R.; Walker, C. T. *Nature* **1974**, *250*, 410.
176. Fry, A. J.; Herr, D. *Tetrahedron Lett.* **1978**, *19*, 1721.
177. Luche, J.-L.; Petrier, C.; Dupuy, C. *Tetrahedron Lett.* **1985**, *26*, 753.
178. Chou, T. S.; You, M. L. *Tetrahedron Lett.* **1985**, *26*, 4495.
179. Chou, T. S.; You, M. L. *J. Org. Chem.* **1987**, *52*, 2224.
180. Eliasson, B.; Edlund, U. *J. Chem. Soc. Perkin Trans. 2* **1983**, 1837.
181. Azuma, T.; Yamagida, S.; Sakurai, H.; Sasa, S.; Yoshino, K. *Synth. Commun.* **1982**, 12, 137.
182. Fujita, T.; Watanabe, S.; Suga, K.; Sugahara, K.; Tsuchimoto, K. *Chem. Ind.* **1983**, 167.
183. Suga, K.; Watanabe, S.; Fujita, T.; Tsuchimoto, K.; Hashimoto, H. *Nippon Kagaku Kaishi* **1984**, 1744; *Chem. Abstr.* **1985**, *102*, 79158.
184. Lindley, J.; Lorimer, J. P.; Mason, T. J. *Ultrasonics* **1986**, *24*, 292.
185. Lindley, J.; Mason, T. J.; Lorimer, J. P. *Ultrasonics* **1987**, *25*, 45.
186. Boudjouk, P.; Thompson, D. P.; Ohrbom, W. H.; Han, B. H. *Organometallics* **1986**, *5*, 1257.
187. Suslick, K. S.; Casadonte, D. J. *J. Am. Chem. Soc.* **1987**, *109*, 3459.
188. Petrier, C.; Luche, J.-L. *Tetrahedron Lett.* **1987**, *28*, 2347.
189. Petrier, C.; Luche, J.-L. *Tetrahedron Lett.* **1987**, *28*, 2351.
190. Suslick, K. S.; Schubert, P. F.; Goodale, J. W. *J. Am. Chem. Soc.* **1981**, *103*, 7342.
191. Suslick, K. S.; Goodale, J. W.; Schubert, P. F.; Wang, H. H. *J. Am. Chem. Soc.* **1983**, *105*, 5781.
192. Suslick, K. S.; Johnson, R. E. *J. Am. Chem. Soc.* **1984**, *106*, 6856.
193. Suslick, K. S.; *High Energy Processes in Organometallic Chemistry*; ACS Symposium Series 333: Washington, D. C., 1987; pp 191-208.
194. Suslick, K. S.; Casadonte, D. J.; Doktycz, S. J. *Solid State Ionics* **1989**, *32/33*, 444.
195. Suslick, K. S.; Doktycz, S. J. *Chem. Materials* **1989**, *1*, 6.
196. Suslick, K. S.; Doktycz, S. J. *J. Am. Chem. Soc.* **1989**, *111*, 2342.
197. Suslick, K. S.; Choe, S.-B.; Cichowlas, A. A.; Grinstaff, M. W. *Nature* **1991**, *353*, 414.
198. Renaud, P. *Bull. Soc. Chim. Fr.* **1950**, 1044.

199. Luche, J.-L.; Damanio, J. C. *J. Am. Chem. Soc.* **1980**, *102*, 7926.
200. Sprich, J. D.; Lewandos, G. S. *Inorg. Chim. Acta* **1983**, *76*, L241.
201. Sternbach, D. D.; Hughes, J. W.; Burdi, D. F.; Banks, B. A. *J. Am. Chem. Soc.* **1985**, *107*, 2149.
202. Hagiwara, H.; Uda, H. *J. Chem. Soc. Chem. Commun.* **1988**, 815.
203. Yamaguchi, R.; Hawasaki, H.; Kawanisi, M. *Synth. Commun.* **1982**, *12*, 1027.
204. Xu, L.; Tao, F.; Yu, T. *Tetrahedron Lett.* **1985**, *26*, 4231.
205. Einhorn, J.; Luche, J.-L. *J. Org. Chem.* **1987**, *52*, 4124.
206. De Nicola, A.; Einhorn, J.; Luche, J.-L. *J. Chem. Res. (S).* **1991**, 278.
207. Kuchin, A. V.; Nurushev, R. A.; Tolstikov, G. A. *Zh. Obshch. Khim.* **1983**, *53*, 2519; *Chem. Abstr.* **1984**, *100*, 103426.
208. Liou, K. F.; Yang, P. H.; Lin, Y. T. *J. Organomet. Chem* **1985**, *294*, 145.
209. Yang, P. H.; Liou, K. F.; Lin, Y. T. *J. Organomet. Chem.* **1986**, *307*, 273.
210. Knochel, P.; Normant, J. F. *Tetrahedron Lett.* **1984**, *25*, 1475.
211. Inoue, Y.; Yamashita, J.; Hashimoto, H. *Synthesis* **1984**, 244.
212. Luche, J.-L.; Petrier, C.; Lansard, J. P.; Greene, A. E. *J. Org.Chem.* **1983**, *48*, 3837.
213. Petrier, C.; de Souza-Barboza, J. C.; Dupuy, C.; Luche, J.-L. *J. Org. Chem.* **1985**, *50*, 5761.
214. Luche, J.-L.; Petrier, C.; Gemal, A. L.; Zirka, N. *J. Org. Chem.* **1982**, *47*, 3805.
215. Brown, H. C.; Racherla, U. S. *Tetrahedron Lett.* **1985**, *26*, 4311.
216. Brown, H. C.; Racherla, U. S. *Tetrahedron Lett.* **1985**, *26*, 2187.
217. Lin, Y. T. *J. Organomet. Chem.* **1986**, *317*, 277.
218. Boudjouk, P.; Han, B. H.; Anderson, K. R. *J. Am. Chem. Soc.* **1982**, *104*, 4992.
219. Masamune, S.; Murakami, S.; Tobita, H. *Organometallics* **1983**, *2*, 1464.
220. Eaborn, C.; Hitchcock, P. B.; Lickiss, P. D. *J. Organomet. Chem.* **1984**, *269*, 235.
221. Boudjouk, P.; Han, B. H. *Tetrahedron Lett.* **1981**, *22*, 3813.
222. Kim, H. K.; Matyjaszewski, K. *J. Am. Chem. Soc.* **1988**, *110*, 3321.
223. Bianconi, P. A.; Weidman, T. W. *J. Am. Chem. Soc.* **1988**, *110*, 2342.
224. Han, B. H.; Boudjouk, P. *Organometallics* **1983**, *2*, 769.
225. Han, B. H.; Boudjouk, P. *Tetrahedron Lett.* **1981**, *22*, 2757.
226. Boudjouk, P.; Han, B. H.; Jacobsen, J. R.; Hauck, B. J. *J. Chem. Soc. Chem. Commun.* **1991**, 1424.
227. Gautheron, B.; Tainturier, G.; Degrand, C. *J. Am. Chem. Soc.* **1985**, *107*, 5579.

228. Suslick, K. S.; Schubert, P. F. *J. Am. Chem. Soc.* **1983**, *105*, 6042.
229. Ley, S. V.; Low, C. M. R.; White, A. D. *J. Organomet. Chem.* **1986**, *302*, C13.
230. Itoh, K.; Nagashima, H.; Ohshima, T.; Ohshima, N.; Nishiyama, H. *J. Organomet. Chem.* **1984**, *272*, 179.
231. Rüdorff, W. *Adv. Inorg. Chem. Radiochem.* **1959**, *1*, 223.
232. Croft, R. C. *Q. Rev. Chem. Soc.* **1960**, *14*, 1.
233. Novikov, Y. N.; Vol'pin, M. E. *Russ. Chem. Rev. (Engl. Transl.)* **1971**, *40*, 733.
234. Vol'pin, M. E.; Novikov, Y. N.; Lapkina, N. D.; Kasatochkin, V. I.; Struchkov, Y. U.; Kazarkov, M. E.; Stukan, R. A.; Povitskij, V. A.; Karimov, Y. S.; Zvarikina, A. V. *J. Am. Chem. Soc.* **1975**, *97*, 3366.
235. Ebert, L. B. *Annu. Rev. Mater. Sci.* **1976**, *6*, 181.
236. Herold, A. *Intercalated Materials*; Levy, F., Ed.; Reidel: Dordrecht, 1979; pp 323-421.
237. Lalancette, J. M.; Roy, L.; Lafontaine, J. *Can. J. Chem.* **1976**, *54*, 2505.
238. Selig, H.; Ebert, L. B. *Adv. Inorg. Chem. Radiochem.* **1980**, *23*, 281.
239. West, A. R. *Solid State Chemistry and its Applications*; Wiley: New York, 1984; p 25.
240. Csuk, R.; Glänzer, B. I.; Fürstner, A. *Adv. Organomet. Chem.* **1988**, *28*, 85.
241. Boersma, M. A. M. *Catal. Rev. Sci. Eng.* **1974**, *10*, 243.
242. Kagan, H. B. *CHEMTECH* **1976**, *6*, 510.
243. Kagan, H. B. *Pure Appl. Chem.* **1976**, *46*, 177.
244. Bergbreiter, D. E.; Killough, J. M. *J. Am. Chem. Soc.* **1978**, *100*, 2126.
245. Setton, R.; Beguin, F.; Piroelle, S. *Synth. Met.* **1982**, *4*, 299.
246. Savoia, D.; Trombini, C.; Umani-Ronchi, A. *Pure Appl. Chem.* **1985**, *57*, 1887.
247. Herold, A.; Vogel, F. L. *Mater. Sci. Eng.* **1977**, *31*. Proceedings of the French-American Conference on Graphite Intercalation Compounds.
248. Estrade-Szwarckopf, H. *Helv. Phys. Acta* **1985**, *58*, 139.
249. Alazard, J. P.; Kagan, H. B.; Setton, R. *Bull. Soc. Chim. Fr.* **1977**, 499.
250. Ubbelohde, A. R. *Carbon* **1976**, *14*, 1.
251. Guerard, D.; Lagrange, P.; Herold, A. *Mater. Sci. Eng.* **1977**, *31*, 29.
252. Beguin, F.; Setton, R. *Carbon* **1972**, *10*, 539.
253. Beguin, F.; Setton, R. *J. Chem. Soc. Chem. Commun.* **1976**, 611.
254. Herinckx, C.; Perret, R.; Ruland, W. *Carbon* **1972**, *10*, 711.
255. Hannay, N. B.; Geballe, T. H.; Matthias, B. T.; Endres, K.; Schmidt, P.; MacNair, D. *Phys. Rev. Lett.* **1965**, *14*, 225.
256. Weintraub, E. U.S. Patent 922 645, 1909; *Chem. Abstr.* **1909**, *3*, 2040.

257. Fredenhagen, K.; Cadenbach, G. *Z. Anorg. Allg. Chem.* **1926**, *158*, 249.
258. Fredenhagen, K.; Suck, H. *Z. Anorg. Allg. Chem.* **1929**, *178*, 353.
259. Herold, A. *C.R. Hebd. Acad. Sci. Ser. C* **1951**, *232*, 1489.
260. Herold, A. *Bull. Soc. Chim. Fr.* **1955**, 999.
261. Hulliger, F. *Phys. Chem. Mater. Layered Struct.* **1976**, *5*, 52.
262. Podall, H.; Foster, W. E.; Giraitis, A. P. *J. Org. Chem.* **1958**, *23*, 82.
263. Lalancette, J. M.; Rollin, G.; Dumas, P. *Can. J. Chem.* **1972**, *50*, 3058.
264. Besenhard, J. O.; Witty, H.; Klein, H.-F. *Carbon* **1984**, *22*, 97.
265. Klein, H.-F.; Gross, J.; Besenhard, J. O. *Angew. Chem. Int. Ed. Engl.* **1980**, *19*, 491.
266. Guerard, D.; Herold, A. *Carbon* **1975**, *13*, 337.
267. Billaud, D.; McRae, E.; Mareche, J. F.; Herold, A. *Synth. Met.* **1981**, *3*, 21.
268. Hart, H.; Chen, B.-L.; Peng, C.-T. *Tetrahedron Lett.* **1977**, 3121.
269. Asher, R. C. *J. Inorg. Nucl. Chem.* **1958**, *10*, 238.
270. Billaud, D.; Herold, A. *Bull. Soc. Chim. Fr.* **1974**, 2715.
271. Asher, R. C.; Wilson, S. A. *Nature* **1958**, *181*, 409.
272. Billaud, D.; Herold, A. *Carbon* **1978**, *16*, 301.
273. El Makrini, M.; Furdin, G.; Lagrange, P.; Mareche, J. F.; McRae, E.; Herold, A. *Synth. Met.* **1980**, *2*, 197.
274. Guerard, D.; Herold, A. *C. R. Hebd. Acad. Sci. Ser C* **1974**, *279*, 455.
275. Guerard, D.; Chaabouni, M.; Lagrange, P.; El Makrini, M.; Herold, A. *Carbon* **1980**, *18*, 257.
276. Guerard, D.; Herold, A. *C. R. Hebd. Acad. Sci. Ser. C* **1975**, *280*, 729.
277. Guerard, D.; Herold, A. *C. R. Hebd. Acad. Sci. Ser. C* **1975**, *281*, 929.
278. El Makrini, M.; Guerard, D.; Lagrange, P.; Herold, A. *Carbon* **1980**, *18*, 203.
279. Stumpp, E.; Nietfeld, G. *Z. Anorg. Allg. Chem.* **1979**, *456*, 261.
280. Guerard, D.; Zeller, C.; Herold, A. *C. R. Hebd. Acad. Sci. Ser. C* **1976**, *283*, 437.
281. Braga, D.; Ripamonti, A.; Savoia, D.; Trombini, C.; Umani-Ronchi, A. *J. Chem. Soc. Chem. Commun.* **1978**, 927.
282. Ungurenasu, C.; Palie, M. *Synth. React. Inorg. Met.-Org. Chem.* **1977**, 7, 581.
283. Braga, D.; Ripamonti, A.; Savoia, D.; Trombini, C.; Umani-Ronchi, A. *J. Chem. Soc. Dalton Trans.* **1979**, 2026.
284. Boldrini, G. P.; Savoia, D.; Tagliavini, E.; Trombini, C.; Umani-Ronchi, A. *J. Org. Chem.* **1983**, *48*, 4108.
285. Boldrini, G. P.; Mengoli, M.; Tagliavini, E.; Trombini, C.; Umani-Ronchi, A. *Tetrahedron Lett.* **1986**, *27*, 4223.
286. Boldrini, G. P.; Savoia, D.; Tagliavini, E.; Trombini, C.; Umani-Ronchi, A. *J. Organomet. Chem.* **1985**, *280*, 307.

287. Savoia, D.; Tagliavini, E.; Trombini, C.; Umani-Ronchi, A. *J. Org. Chem.* **1982**, *47*, 876.
288. Braga, D.; Ripamonti, A.; Savoia, D.; Trombini, C.; Umani-Ronchi, A. *J. Chem. Soc. Dalton Trans.* **1981**, 329.
289. Schäfer-Stahl, H. *J. Chem. Soc. Dalton Trans.* **1981**, 328.
290. Csuk, R.; Fürstner, A.; Weidmann, H. *J. Chem. Soc. Chem. Commun.* **1986**, 775.
291. Fürstner, A.; Weidmann, H.; Hofer, F. *J. Chem. Soc. Dalton Trans.* **1988**, 2023.
292. Fürstner, A.; Hofer, F.; Weidmann, H. *J. Catal.* **1989**, *118*, 502.
293. Fürstner, A.; Hofer, F.; Weidmann, H. *Carbon* **1991**, *29*, 915.
294. Savoia, D.; Trombini, C.; Umani-Ronchi, A.; Verardo, G. *J. Chem. Soc. Chem. Commun.* **1981**, 540.
295. Savoia, D.; Trombini, C.; Umani-Ronchi, A.; Verardo, G. *J. Chem. Soc. Chem. Commun.* **1981**, 541.
296. Savoia, D.; Tagliavini, E.; Trombini, C.; Umani-Ronchi, A. *J. Org. Chem.* **1981**, *46*, 5344.
297. Savoia, D.; Tagliavini, E.; Trombini, C.; Umani-Ronchi, A. *J. Org. Chem.* **1981**, *46*, 5340.
298. Fürstner, A.; Weidmann, H. *Synthesis* **1987**, 1071.
299. Fürstner, A.; Csuk, R.; Rohrer, C.; Weidmann, H. *J. Chem. Soc. Perkin Trans. 1* **1988**, 1729.
300. Clive, D. L. J.; Murthy, K. S. K.; Zang, C.; Hayward, W. D.; Daigneault, S. *J. Chem. Soc. Chem. Commun.* **1990**, 509.
301. Fürstner, A.; Jumbam, D.; Weidmann, H. *Tetrahedron Lett.* **1991**, *32*, 6695.
302. Hennig, G. R. *Progr. Inorg. Chem.* **1959**, *1*, 125.
303. Herold, A. *Mater. Sci. Eng.* **1977**, *31*, 1.
304. Stumpp, E. *Mater. Sci. Eng.* **1977**, *31*, 53.
305. Buscarlet, E.; Touzain, P.; Bonnetain, L. *Carbon* **1976**, *14*, 75.
306. Melin, J.; Herold, A. *C. R. Hebd. Acad. Sci. Ser. C* **1975**, *280*, 641.
307. Hooley, J. G.; Sams, J. R.; Liengme, B. V. *Carbon* **1970**, *8*, 467.
308. Jadhav, V. G.; Singra, R. M.; Joshi, G. M.; Pisharody, K. P. R.; Rao, C. N. R. *Z. Phys. Chem. (Frankfurt am Main)* **1974**, *92*, 139.
309. Vangelisti, R.; Herold, A. *C. R. Hebd. Acad. Sci. Ser. C* **1975**, *280*, 571.
310. Klotz, H.; Schneider, A. *Naturwissenschaften* **1962**, *49*, 448.
311. Vangelisti, R.; Herold, A. *Mater. Sci. Eng.* **1977**, *31*, 67.
312. Novikov, Y. N.; Vol'pin, M. E. *Physica Ser. B,C (Amsterdam)* **1981**, *105*, 471. A review of syntheses and reactions of metal halide GIC.
313. Kagan, H. B.; Yamagishi, T.; Motte, J. C.; Setton, R. *Isr. J. Chem.* **1978**, *17*, 274.

314. Adams, J. M.; Thomas, J. M.; Walter, M. J. *J. Chem. Soc. Dalton Trans.* **1975**, 1459.
315. Ebert, L. B.; Huggins, R. A.; Brauman, J. I. *Carbon* **1974**, *12*, 199.
316. Hooley, J. G.; Reimer, M. *Carbon* **1975**, *13*, 401.
317. Ebert, L. B.; Selig, H. *Mater. Sci. Eng.* **1977**, *31*, 177.
318. Eichinger, G.; Besenhard, J. O. *J. Electroanal. Chem.* **1976**, *72*, 1.
319. Smith, D. J.; Fischer, R. M.; Freeman, L. A. *J. Catal.* **1981**, *72*, 51.
320. Sirokman, G.; Mastalir, A.; Molnar, A.; Bartok, M.; Sachay, Z.; Guczi, L. *J. Catal.* **1989**, *117*, 558.

PART TWO

METAL-MEDIATED REACTIONS

IV. REDUCTIONS

A. UNSATURATED SYSTEMS

1. Alkenes, Alkynes, and Aromatic Systems

Reduction is one of the more important and synthetically useful reactions, and has been extensively studied.[1-4] Among a wide number of reductive procedures, catalytic hydrogenation constitutes a common practice in chemistry,[5] and numerous, highly active metal systems have been developed. Two salient and representative examples are the heterogeneous catalysts P-1 and P-2 nickel. P-1 Nickel is a granular material having a catalytic activity comparable to that of Raney nickel. It is easily obtained by reduction of aqueous nickel(II) salts with sodium borohydride. The treatment of ethanolic nickel(II) acetate with the same reducing agent produces a nearly colloidal material denoted as P-2 nickel.[6,7] This catalyst shows a great selectivity in hydrogenation reactions, and some double bonds can be reduced in the presence of others. Reduction of alkynes gives pure *cis*-alkenes especially in the presence of 1,2-diaminoethane,[8] and conjugated dienes can be selectively semihydrogenated. Importantly, hydrogenation is usually performed without hydrogenolysis.[7]

Reactive metal powders (Chapter I) have been also obtained by reduction of metal salts with either sodium borohydride or the Willstäter procedure, reduction of metal salts by formaldehyde under alkaline conditions. These systems along with Raney nickel and Urushibara catalysts facilitate numerous hydrogenation processes.[9]

Alternatively to these heterogeneous hydrogenations,[5] reductions can be also performed by dissolving metals. Birch-type reductions using metal-ammonia solutions and related reagents have been utilized for a wide variety of organic substrates, and the scope and limitations of the method have been reviewed.[10-15]

Upon dissolution of the metal in liquid ammonia a blue homogeneous solution is usually formed, characteristic of metal cations and solvated electrons[16,17] which are indeed the ultimate reducing agent.[11,12] This solution at high concentration of metal takes on a bronze luster. In these metal-ammonia reductions, it is desirable that the ammonia be distilled. This common caution ensures that metal contaminants, particularly iron, are removed from the anhydrous ammonia. Likewise, the suitable co-solvents employed in the reduction should be freed of impurities before use. Iron catalyzes the reaction of metals (especially sodium and potassium) with ammonia and with proton donors such as alcohols. For this reason the metals employed in metal-ammonia reductions should have low transition metal content. The deliberate addition, however, of iron ions has been utilized for partial reduction of aromatic rings.[10,15] Thus, anthracene and phenanthrene are readily reduced in the 9,10 positions and the reaction is not markedly inhibited by iron, but further reduction is avoided.[18]

All alkali and alkaline-earth metals, with the exception of beryllium, are soluble to some extent in ammonia although lithium is the most widely used metal in such reductions. It is very soluble in ammonia and dissolves also in primary and secondary amines,[19,20] whereas other alkali metals are poorly soluble in such solvents. Moreover, lithium in ammonia exhibits a higher reduction potential than sodium or potassium.

Despite their low solubility in ammonia, solutions of beryllium[21] and magnesium[22,23] may be prepared by electrolysis of the corresponding metal salts. Barium and calcium[24] have been also used and can generate insoluble salts, which protect intermediates in some cases. Importantly, calcium-ammonia solutions are relatively acidic and can protonate calcium enolates.

In any event, a severe drawback of the method for some substrates is the presence of strongly basic compounds in the reaction mixture. This limitation can be overcome notably by the use of other metals. Lanthanide metals in liquid ammonia constitute a suitable alternative to alkali metals in Birch-type reductions. Lanthanide metals are also strongly electropositive, and ytterbium has reducing properties in liquid ammonia similar to those of lithium and sodium.[25]

Although europium and ytterbium were used in ammonia to reduce benzene to cyclohexa-1,3- and 1,4-diene,[11] a systematic study was not accomplished until the work by White and Larson.[25] These authors reported the reduction of various organic compounds with solutions of ytterbium in liquid ammonia. Aromatic rings, alkynes, and α,β-unsaturated ketones are readily reduced in acceptable to good yields. Reduction of benzene derivatives in a mixture of THF-*tert*-butanol as co-solvent, results in the formation of the 1,4-dihydroaromatic compounds. As in the classical Birch reduction, the rate of the reduction is largely influenced by the nature of substituents on the aromatic ring, and thus reductions are slower with alkyl and alkoxy substituents. Anthracenes are also reduced preferentially to their 9,10-dihydro-derivatives. Alkynes are reduced to *trans*-alkenes in generally good yields, using THF as co-solvent. Apparently, certain double bonds can be completely saturated with this reagent which enables selective reductions. Thus, norbornadiene was reduced to bicyclo[2.2.1]heptene, and 1-phenylpentyne gave exclusively 1-phenylpentane along with a small amount of the recovered starting material. Representative examples of these reductions are provide by equations 4.1-4.5.

In addition to this ytterbium-ammonia system, europium or ytterbium immobilized on silica are effective catalysts in the selective hydrogenation of dienes.[26] For the preparation of these catalysts, the lanthanides were suspended in liquid ammonia and then added to the silica powder. The characteristic blue color gradually disappeared by reaction of the dissolved metal with the support. Upon warming to room temperature the active Eu/SiO_2 and Yb/SiO_2 were obtained.

OCH₃-benzene $\xrightarrow[\text{THF, } t\text{-BuOH, 80\%}]{\text{Yb - NH}_3\text{ (l)}}$ 1-methoxycyclohexa-1,4-diene (4.1)

anthracene $\xrightarrow{64\%}$ 9,10-dihydroanthracene (4.2)

1-acetoxycyclohexene (OCOCH₃) $\xrightarrow{48\%}$ cyclohexyl acetate (OCOCH₃) (4.3)

norbornadiene $\xrightarrow{65\%}$ norbornene (4.4)

$$\text{Ph-C}\equiv\text{C-Ph} \xrightarrow[\text{THF, 75\%}]{\text{Yb - NH}_3\text{ (l)}} (E)\text{-PhCH=CHPh} \quad (4.5)$$

Hydrogenations on these catalysts showed relevant features: a) the catalysts discriminated between conjugated and nonconjugated double bonds. Thus 1,3-dienes are readily reduced with almost 100% selectivity, whereas the corresponding 1,4-dienes remain unaffected. Likewise, a very important property of these catalysts is that they have little or no reducing power for monoenes such as propene and 1-butene. b) Both Eu/SiO_2 and Yb/SiO_2 showed similar catalytic activity, but the latter was more effective. c) Interestingly, the lanthanide metal catalyst generated by the metal vapor condensation (Chapter II) was active for the hydrogenation of monoenes and dienes, but without selectivity between conjugated and nonconjugated dienes.[27] Ytterbium vapor catalyzes the hydrogenation of propene to propane at 323 K with 100% selectivity, and penta-1,4-diene is converted into a mixture of pentane and 1-pentene (20:80). In contrast, the supported catalysts Eu/SiO_2 or Yb/SiO_2 did not react with penta-1,4-diene but buta-1,3-diene gave smoothly a mixture of olefinic products, in which (Z)-alkenes were always favored (eq. 4.6).

$$CH_2\text{=CH-CH=}CH_2 \xrightarrow[298\text{ K}]{Eu/SiO_2,\ H_2\text{(40 torr)}} \underset{(51\%)}{CH_2\text{=CH-}CH_2CH_3} + \underset{(49\%)}{(Z)\text{-}H_3C\text{-CH=CH-}CH_3} + \underset{(trace)}{(E)\text{-}H_3C\text{-CH=CH-}CH_3} \quad (4.6)$$

Organic acids and anhydrides are also appropriate media for performing mild reductions, which are often useful alternatives to catalytic hydrogenations (see Chapter III).[28-30] Notably, the reactivity of the catalyst can be dramatically enhanced by sonication, and reductions are then carried out under milder conditions. Formic acid and palladium-on-carbon are an effective couple for the hydrogenation of a wide range of alkenes at room temperature in the presence of low intensity ultrasonic fields (cleaning bath at 50 kHz).[31] Similarly, the hydrazine-palladium-on-carbon couple is also employed for the hydrogenation of alkenes in ethanol at room temperature using an ultrasonic bath.[32]

A commercially useful example of a sonochemically enhanced catalytic reaction is the ultrasonic hydrogenation of soybean oil.[33] The procedure utilizes a three-phase non-aqueous system comprising liquid oil, hydrogen gas, and solid catalyst. The catalyst was either 1% copper chromite or 0.1% nickel catalyst (25% nickel) at 115 psi and 180 °C. Sonication at 20 kHz has considerable advantages over the currently used methods which require much longer reaction times.

Alkali metal reductions in alcoholic media are not commonly utilized for reductions although the combinations Li/EtOH, Na/EtOH, or Na/*n*-PrOH among others, are the best methods for preparing some alcohols which are difficult to obtain by hydride reductions.[14]

However, alcoholic solutions of magnesium represent a well-established procedure for the reduction of many functional groups. Magnesium and alcohols in ammonia have been used to reduce some polycyclic hydrocarbons and benzoic esters or amides.[34,35] Magnesium in methanol enables selective double bond reductions of diphenyl ethylenes, conjugated acetylenes, α,β-olefinic esters, conjugated ketones, conjugated nitriles, conjugated amides, and is useful for desulfonylation.[36-39] Addition of palladium-on-carbon to this magnesium-methanol system markedly enhances the reactivity and permits reduction of nonactivated carbon-carbon double and triple bonds.[40] One additional advantage of this modification over catalytic hydrogenation is that cyclopropyl groups, benzylic ethers, and alcohols are not affected. Yields are generally higher than 80%.

In a recent paper[39] the use of magnesium in methanol (or *d*-methanol) has been reported as a convenient method for the selective reduction of acetylenic bonds conjugated to esters (but not acids) or to two phenyl groups, to give the saturated or tetra-deuterated derivatives, respectively (eqs. 4.7-4.13). The method is also advantageous for the incorporation of deuterium in the α and β positions of long chain polyunsaturated fatty esters. Isolated acetylenes as well as alkyl substituted conjugated dienes and diynes remain unaffected. Triple and double bonds conjugated to two (but not one) phenyl groups are readily reduced as are phenyl conjugated diynes and dienes to alkenes.

$$CH_3C{\equiv}CCO_2CH_3 \xrightarrow[\text{rt, 68\%}]{\text{Mg, MeOH}} CH_3CH_2CH_2CO_2CH_3 \quad (4.7)$$

$$CH_3C{\equiv}CCO_2CH_3 \xrightarrow[\text{rt, 68\%}]{\text{Mg, MeOD}} CH_3CD_2CD_2CO_2CH_3 \quad (4.8)$$

$$CH_3(CH_2)_4C{\equiv}CCO_2H \xrightarrow[\text{or MeOD}]{\text{Mg, MeOH}} \text{no reaction} \quad (4.9)$$

$$C_6H_5C{\equiv}CC_6H_5 \xrightarrow[\text{rt, 85\%}]{\text{Mg, MeOH}} C_6H_5CH_2CH_2C_6H_5 \quad (4.10)$$

$$C_6H_5C{\equiv}CCH_3 \xrightarrow[\text{rt}]{\text{Mg, MeOH}} \text{no reaction} \quad (4.11)$$

$$C_6H_5CH{=}CHCH{=}CHC_6H_5 \xrightarrow[\text{rt, 96\%}]{\text{Mg, MeOH}} C_6H_5CH_2CH{=}CHCH_2C_6H_5 + C_6H_5CH{=}CH_2CH_2CH_2C_6H_5 \ (5.5{:}1.0) \quad (4.12)$$

$$C_6H_5CH{=}CHC_6H_5 \xrightarrow[\text{rt, 87\%}]{\text{Mg, MeOH}} C_6H_5CH_2CH_2C_6H_5 \quad (4.13)$$

Stereoselective reduction of triple bonds to *cis*-double bonds can be achieved with zinc powder in 50% aqueous *n*-propanol at reflux. However, application of the method to the complex (Z)-dienyne **1** leads to (Z,Z,*E*)-2,4,6-undecatriene (**2**), probably formed from the expected triene (**3**) by a thermal [1,7]-hydrogen shift. The desired reduction to **3** can be accomplished at room temperature without a prototropic shift with zinc activated by potassium cyanide (Scheme 4.1).[41] The method, nevertheless, is not always reproducible. This reduction could not be carried out by hydrogenation using Lindlar's catalyst or Wilkinson's catalyst or with diimide because of lack of selectivity. Similar results were obtained with this activated zinc in the presence of KCN in the synthesis of (*E*,*E*)- and (*Z*,*E*)-trienols.[42]

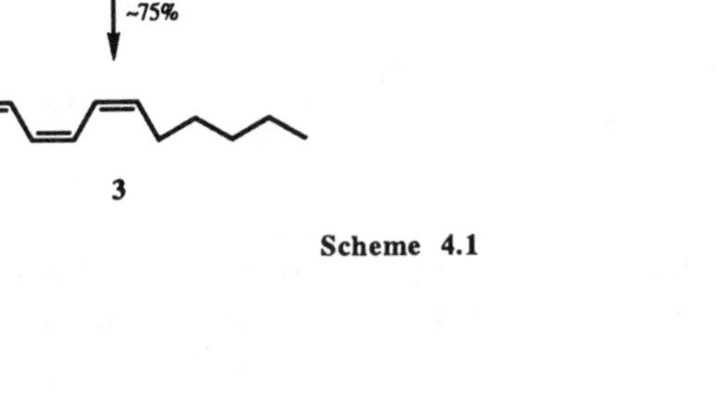

C_4H_9

Zn, n-PrOH, Δ
80%

C_4H_9

1 **2**

Zn, KCN
n-PrOH, 25 °C
~75%

3

Scheme 4.1

In addition considerable efforts have been addressed to the stereoselective reduction of alkynes in recent years. Zinc-copper couple reduces internal alkynes to (Z)-alkenes exclusively in yields generally greater than 95%.[43] Terminal alkynes are reduced to 1-alkenes by this procedure without affecting other coexisting double bonds (eq. 4.14).[44]

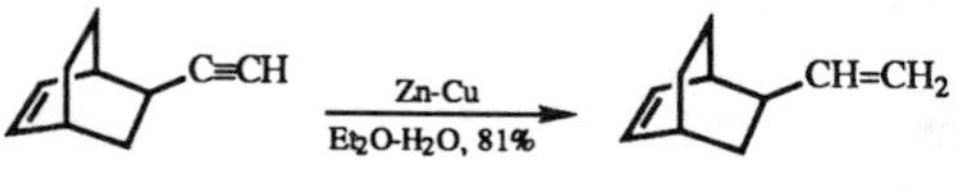

C≡CH —Zn-Cu / Et_2O-H_2O, 81%→ $CH=CH_2$ (4.14)

Alkenes substituted by one or two electronegative groups can be reduced by the zinc-copper couple of Simmons-Smith in refluxing methanol in high yields.[45] Zinc powder activated by 1,2-dibromoethane reduces conjugated diynes to (Z)-enynes in yields higher than 70%. Triple bonds conjugated with an aryl group are also reduced (80-90% yield). A more active, but less selective, reagent is obtained by activation with 1,2-dibromoethane and then with CuBr and LiBr. This activated zinc reduces monoacetylenic compounds substituted by hydroxy, *N*-alkylamino, and alkoxy groups to the (Z)-alkenes.[46]

Zinc activated by copper(II) acetate and silver nitrate effects stereoselective reduction of the dienyne **4** to the (*E*,*E*,*Z*)-triene **5** in methanol-water. Catalytic hydrogenation using Lindlar's catalyst gives **5** but only in 30% yield (eq. 4.15).[47]

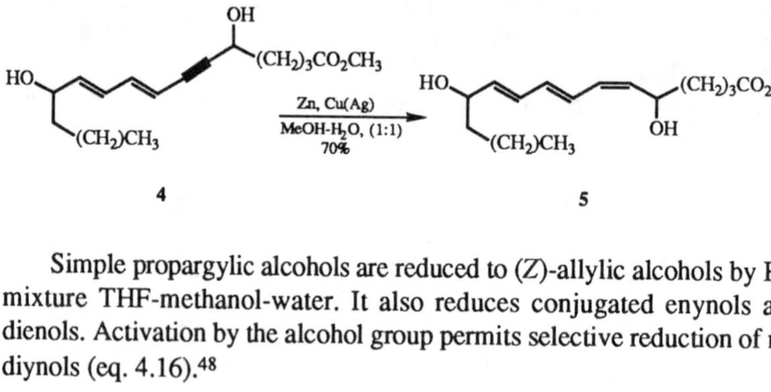

HO, OH, $(CH_2)_3CO_2CH_3$, $(CH_2)CH_3$ —Zn, Cu(Ag) / MeOH-H_2O, (1:1) 70%→ HO, $(CH_2)_3CO_2CH_3$, OH, $(CH_2)CH_3$ (4.15)

4 **5**

Simple propargylic alcohols are reduced to (Z)-allylic alcohols by Rieke-zinc in a mixture THF-methanol-water. It also reduces conjugated enynols and diynols to dienols. Activation by the alcohol group permits selective reduction of nonconjugated diynols (eq. 4.16).[48]

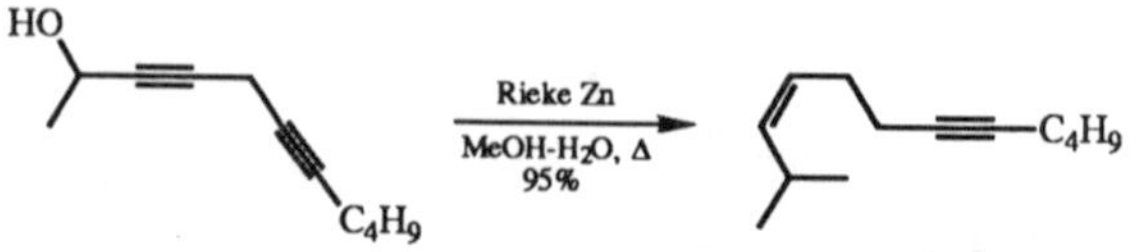

HO, C_4H_9 —Rieke Zn / MeOH-H_2O, Δ 95%→ C_4H_9 (4.16)

The incorporation of the highly reactive metal-graphites into hydrogenation procedures has resulted in fruitful and often unexpected applications. Hydrogenation of ethylene, 1-butene, butadiene, or methyl acetylene over potassium-graphite derivatives can be readily effected.[49] The intercalate $C_{36}K$ proved to be better than $C_{24}K$ which showed also more activity than C_8K.[50,51] Benzene and toluene could be hydrogenated over C_8K at 100-150 bar hydrogen pressures. Hydrogenations of activated alkenes are more efficient and thus (*E*)-stilbene was reduced to 1,2-diphenyl ethane in 90% yield.[52] Alkyl benzenes were hydrogenated to the cyclohexane derivatives, but further dealkylation also occurred, and cyclohexane and methane were isolated as by-products. Butadiene can be hydrogenated first to a mixture of (*Z*)- and (*E*)-2-butenes and then to butane.

Lamellar $C_{24}K$ is a moderately active catalyst for hydrogenation of alkenes.[53] It is a very efficient catalyst for isomerization of (*Z*)-stilbene to (*E*)-stilbene. It effects also the isomerization of alkynes and thus 2-octyne was converted to 1-octyne in 86% yield. An allene is probably the intermediate of the process because 2-decyne was transformed into a mixture of 1-decyne (20%) and 1,2-decadiene (9%). Potassium alone also effects these isomerizations, but only at higher temperatures. Benzene was converted into biphenyl slowly with $C_{24}K$ in refluxing cyclohexane. This reaction has also been reported with C_8K in DMF at 20°C in 61% yield.[54]

Palladium-graphite is a good catalyst for the hydrogenation of alkynes, alkenes, and anilines.[55] Alkenes are quantitatively converted to alkanes, and nitroarenes to aminoarenes by palladium-graphite-catalyzed hydrogenation in methanol at room temperature and atmospheric pressure. Suppression of full hydrogenation of alkynes to the corresponding alkanes was almost completely achieved (97-98%) by addition of ethylenediamine to yield predominantly (*Z*)-alkenes (*Z*:*E* = 94-98:6-2). The results obtained indicated that palladium-graphite is a suitable alternative to both palladium-on-carbon and Lindlar's catalyst (eqs. 4.17, 4.18). Palladium-graphite prepared in a different manner has been utilized for the selective reduction of alkynes or 1,3-conjugated dienes.[56]

$$HC\equiv C\text{-}C_8H_{17} \xrightarrow[\text{EDA, MeOH, 20°C}]{\text{Pd-Gr (3 mol \%), } H_2 \text{ (1 atm)}} \underset{95\%}{H_2C{=}CH\text{-}C_8H_{17}} + \underset{2\%}{CH_3CH_2C_8H_{17}} + \underset{3\%}{HC\equiv C\text{-}C_8H_{17}} \quad (4.17)$$

$$PhC\equiv C\text{-}CO_2C_2H_5 \xrightarrow[\text{EDA, MeOH, 20 °C}]{\text{Pd-Gr (3 mol \%), } H_2 \text{ (1 atm)}} \underset{92\%\ (>97\%\ Z)}{PhCH{=}CH\text{-}CO_2C_2H_5} + \underset{3\%}{PhCH_2CH_2CO_2C_2H_5} + \underset{5\%}{PhC\equiv C\text{-}CO_2C_2H_5} \quad (4.18)$$

Highly active nickel-graphite (Ni-Gr 1)[57] and partially deactivated nickel-graphite (Ni-Gr 2)[58] have also been employed as hydrogenation catalysts. Similarly to Pd-Gr, when Ni-Gr catalysts are utilized in the presence of ethylenediamine, complete hydrogenation of alkynes is inhibited and the *Z*/*E* ratio of the disubstituted alkenes is increased. Ni-Gr 1 shows a stereoselectivity comparable to that observed with Lindlar's catalyst or nickel boride.

Reduction of alkynes proceeds more slowly and with somewhat less stereoselectivity (*Z/E* ratio) with Ni-Gr 2 than with Ni-Gr 1. However, Ni-Gr 2 is particularly effective in achieving selective hydrogenations of polyfunctional compounds, simply by changing the temperature of the reaction (Scheme 4.2). Reductions are usually performed using 7-20 mol% of catalyst at 30 atm of hydrogen in methanol.

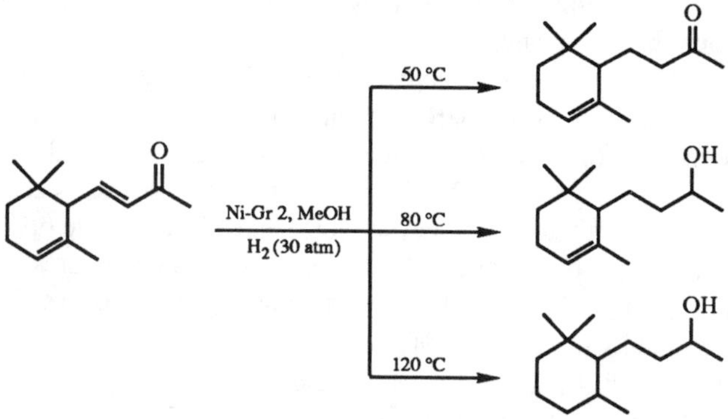

Ni-Gr 2, MeOH; H_2 (30 atm); 50 °C; 80 °C; 120 °C

Scheme 4.2

Zinc-graphite is very effective for the stereospecific reduction of alkynols in refluxing ethanol (eq. 4.19).[59]

$$CH_3(CH_2)C{\equiv}CCH_2CH_2OH \xrightarrow[\text{EtOH, }\Delta\text{, 12 h}]{\text{Zn-Gr}} CH_3(CH_2)CH{=}CHCH_2CH_2OH \quad (4.19)$$

78% (Z:E > 99:1)

The active nickel powder generated by thermal decomposition from nickel isopropoxide is a convenient catalyst for hydrogen transfer reactions (eqs 4.20-4.22).[60]

$$CH_3CH{=}CH(CH_2)_4CH_3 \xrightarrow[\text{2 h, 96\%}]{\text{Ni* (20 mol \%)}} CH_3(CH_2)_6CH_3 \quad (4.20)$$

$$CH_2{=}CH(CH_2)_8CH_3 \xrightarrow[\text{1.5 h, 93\%}]{\text{Ni* (30 mol \%)}} CH_3(CH_2)_9CH_3 \quad (4.21)$$

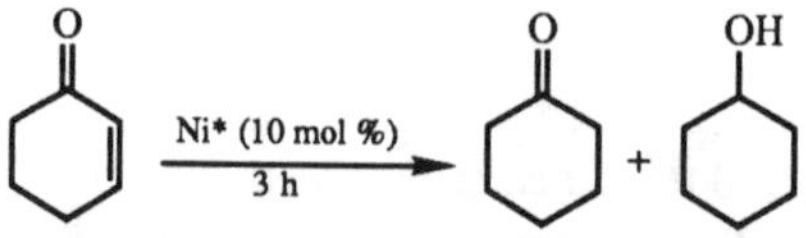

Ni* (10 mol %), 3 h; (80%) + (8%) (4.22)

While reductions with alkali and main group elements or their derivatives are well established and constitute a common resource for the synthetic chemist, transition metals or their corresponding organometallics have become increasingly important in recent years. Among them, low-valent species of groups 4, 5, and 6 metals have been utilized for the reduction of unsaturated hydrocarbons.[4,61-64] It is not clear that zero-valent metal is the active species in these reagents. Although from the stoichiometry of the reduction zero-valent states may be assumed in many cases, a mixture of the metallic element and other low oxidation states is probably closer to the truth.

Alkynes can be reduced to either alkanes (from terminal alkynes) or (Z)-alkenes (from internal alkynes) with an 1:1 equimolar mixture of $TiCl_4$-$LiAlH_4$ (eqs. 4.23-4.25).[65] Monosubstituted alkenes are readily reduced, whereas disubstituted alkenes (*e.g.* cyclooctene) are reduced only partially with a large excess of reducing agent.

Likewise, molar mixtures of lithium aluminum hydride and several transition metal chlorides also reduce alkenes, alkynes, and alkyl halides.[66] Importantly, $TiCl_3$, $CoCl_2$, and $NiCl_2$ can be employed in a catalytic fashion giving reduction products in yields higher than 94% in each case. Curiously, the reduction of 1-octene failed with a stoichiometric mixture of VCl_3 and $LiAlH_4$, but the reaction using a catalytic amount of VCl_3 gave *n*-octane in 42% yield along with a large amount of recovered alkene. Terminal and internal alkynes are reduced preferentially to alkenes with a great (Z)-selectivity.

$$CH_2{=}CH(CH_2)_5CH_3 \xrightarrow[0\ ^\circ C,\ 92\%]{TiCl_4 - LiAlH_4} CH_3(CH_2)_6CH_3 \qquad (4.23)$$

$$HC{\equiv}C(CH_2)_5CH_3 \xrightarrow[0\ ^\circ C,\ 81\%]{TiCl_4 - LiAlH_4} CH_3(CH_2)_6CH_3 \qquad (4.24)$$

$$CH_3(CH_2)_2C{\equiv}C(CH_2)_2CH_3 \xrightarrow[-40\ ^\circ C]{TiCl_4 - LiAlH_4} \underset{(73\%)}{(Z)\text{-}CH_3(CH_2)_2CH{=}CH(CH_2)_2CH_3} + \underset{(<7\%)}{(E)\text{-}CH_3(CH_2)_2CH{=}CH(CH_2)_2CH_3} + \underset{(11\%)}{CH_3(CH_2)_6CH_3} \qquad (4.25)$$

A combination of $ZrCl_4$ and $LiAlH_4$ converts terminal and strained internal olefins into alkanes.[67] However, this process presumably involves hydride species. Hydrozirconation is followed by transmetallation in the presence of excess lithium aluminum hydride. Further hydrolysis or halogenolysis gives alkanes or 1-haloalkanes in good yield.

Sato and Oshima introduced in 1982 a low-valent niobium reagent prepared by the reduction of $NbCl_5$ with $NaAlH_4$ and used it for some reductions,[68] such as the reduction of alkynes and the pinacol reductive coupling of aldehydes and ketones (see Chapter V). Reduction of internal alkynes with this low-valent niobium reagent leads to (Z)-alkenes preferentially. This result is comparable to reduction with low-valent titanium.[65]

Low-valent niobium species can be also generated by reduction of niobium(V) chloride with zinc.[69] Aluminum powder is also effective for this reduction, but ultrasonic irradiation to the reaction mixture in DME-benzene is indispensable for getting reproducible results. Both of them reduce alkynes to (Z)-alkenes with a high stereo-selectivity (usually *Z/E* ratio > 99:1). In contrast, the reaction with a reagent formed by reduction of $NbCl_5$ with magnesium affords a complex mixture containing a small amount of the desired olefin. The reductive process depends on the solvent and the bulkiness of the substituents of alkynes. A mixed solvent of THF-benzene-HMPA proved to give the best yields. Terminal alkynes polymerized in this solvent system, whereas in benzene alone cyclotrimerization takes place, and a mixture of benzene derivatives was isolated. When the reduction was conducted in DME-benzene the desired olefin was obtained in good yield (Scheme 4.3). These results contrast to the $NbCl_5$-$NaAlH_4$ reagent, which gave a complex mixture with terminal alkynes.

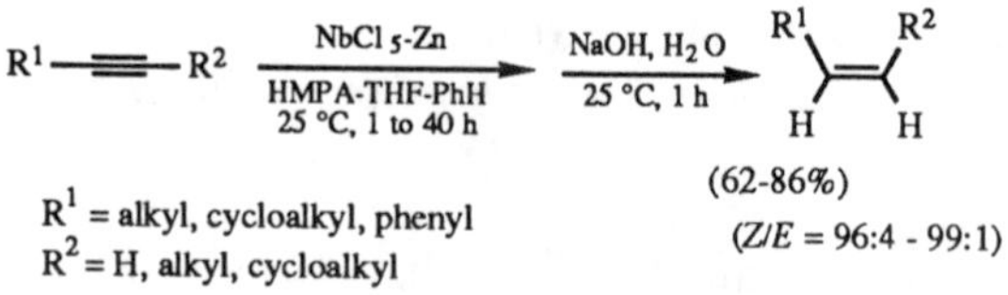

R^1—≡—R^2 —[$NbCl_5$-Zn; HMPA-THF-PhH, 25 °C, 1 to 40 h]→ —[NaOH, H_2O; 25 °C, 1 h]→ R^1(H)C=C(H)R^2 (62-86%) (*Z/E* = 96:4 - 99:1)

R^1 = alkyl, cycloalkyl, phenyl
R^2 = H, alkyl, cycloalkyl

Scheme 4.3

Similarly low-valent tantalum can be generated from tantalum(V) chloride and zinc. Complexation of alkynes with this reagent proceeds faster than those with the $NbCl_5$-Zn system. (Z)-Alkenes are also formed with excellent stereoselectivity in DME-benzene. Terminal alkynes are very reactive with the tantalum reagent, and the yield is lower than that with the niobium reagent. Remarkably, reduction of an alkyne containing a hydroxyl group was carried out in excellent yield (Scheme 4.4).

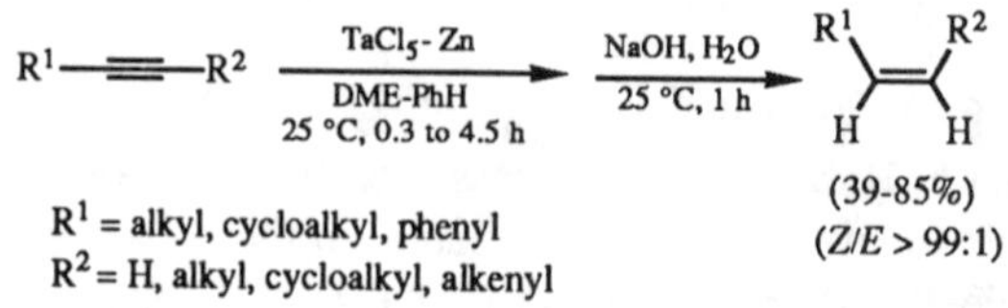

R^1—≡—R^2 —[$TaCl_5$-Zn; DME-PhH, 25 °C, 0.3 to 4.5 h]→ —[NaOH, H_2O; 25 °C, 1 h]→ R^1(H)C=C(H)R^2 (39-85%) (*Z/E* > 99:1)

R^1 = alkyl, cycloalkyl, phenyl
R^2 = H, alkyl, cycloalkyl, alkenyl

Scheme 4.4

In addition, low-valent transition metal reagents are particularly effective for partial hydrogenation of alkynes leading to (Z)-alkenes. Alternatively, partial reduction of alkynes with alkali metals (*e.g.* sodium) in liquid ammonia gives (*E*)-alkenes predominantly.

2. Carbonyl Substrates

Carbonyl compounds can be transformed to other functional groups by reductive processes using activated metals or low-valent metal species. Direct reduction of aldehydes and ketones produces alkanes and arenes.[70] These reductions are mostly performed by catalytic hydrogenation on various catalysts,[5,71] or by dissolving metals.[72] By direct elimination and related reactions alkenes can be obtained.[73] The reductive dimerization of aldehydes and ketones to afford alkenes (McMurry coupling), deserves special mention and will be dealt with in Chapter V of this book.

Alcohols are readily obtained by reduction of carbonyl compounds with metals. Most methods utilize alkali metals in protic solvents,[74-76] as they are easy alternatives to many tedious and often expensive metal hydride reductions. Reduction of ketones with metals in an alcohol is one of the earliest reductive procedures, and importantly it can proceed with stereoselectivity opposite to that obtained with metal hydrides.[76] Thus, the reduction of the steroid 3α-hydroxy-7-ketocholanic acid (**6**) to the diols **7** and **8** can be achieved by several reducing methods. The stereochemistry is strongly dependent on the nature of the reducing agent.[77] Sodium borohydride and sodium dithionite reductions afford mainly the 7α-alcohol, whereas reductions with sodium or potassium in alcohol favor the 7β-alcohol (Scheme 4.5). The same authors have described reduction of **6** to **7** and **8** in the ratio 96:4 with K, Rb, and Cs in *tert*-amyl alcohol.[78] Almost the same stereoselectivity can be obtained by addition of K, Rb, or Cs salts to reductions of sodium in *tert*-amyl alcohol.

H_3C HO H O **6** → H_3C HO H OH **7** + H_3C HO H OH **8**

Reducing System	Yield (%)	7 / 8
Na, *n*-PrOH	100	85:15
K, *t*-BuOH	100	94:6
$Na_2S_2O_4$, H_2O, $NaHCO_3$	100	7:92
$NaBH_4$, H_2O, $NaHCO_3$	100	6:94

Scheme 4.5

Transition metals and other main group elements can be conveniently utilized for carbonyl reductions. Among them, the recent low-valent antimony compounds should be mentioned. The systems $SbCl_3$-Al and $SbCl_3$-Zn in DMF-water are effective for the conversion of aldehydes to alcohols in excellent yields.[79] Deuterium-labeled alcohols can be equally obtained when the reaction is conducted in DMF-D_2O (eq. 4.26).

$$\text{RCHO} \xrightarrow[\text{DMF-H}_2\text{O or DMF-D}_2\text{O}]{\text{SbCl}_3\text{-Al or SbCl}_3\text{-Zn}} \text{RCH}_2\text{OH (or RCHDOH)} \qquad (4.26)$$

Reduction of aldehydes proceeds with $SbCl_3$-Al faster than with $SbCl_3$-Zn. With α,β-unsaturated aldehydes carbonyl reduction occurs, leaving the carbon-carbon double bonds intact. Neither unconjugated double bonds nor cyclic ketones were attacked. Ketones can be, however, reduced but the process is slower than with aldehydes. The reduction shows a great chemoselectivity and aldehydes are reduced preferentially in the presence of ketones.

Lanthanide metals represent novel and very promising reducing agents, although a systematic study should be accomplished in many cases.[80] One of the most important procedures employing lanthanide reagents is the Luche method for the selective reduction of conjugated aldehydes and ketones to allylic alcohols.[81-83] This protocol utilizes the system $CeCl_3$-$NaBH_4$ and presumably involves a low-valent cerium species. The procedure appears to be general and has been extensively utilized in selective organic chemistry, including many complex molecules and natural products. Furthermore, this combination is vastly superior to other reductants such as DIBAH, LAH, zinc borohydride, or sodium borohydride alone (eq. 4.27).

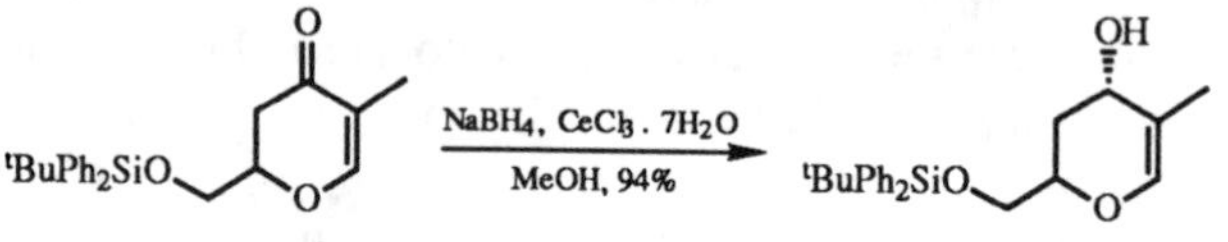

(4.27)

The excellent 1,2-selectivity is complemented with a high chemoselectivity, and many sensitive functional groups are tolerated in this reduction as demonstrated in the following example from carbohydrate chemistry (eq. 4.28).

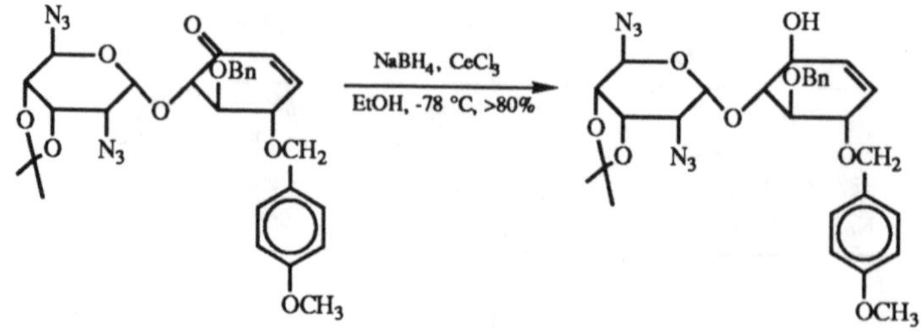

(4.28)

On the contrary, the diastereoselectivity in the reduction of chiral ketones can be variable. However, rigid bicyclic systems are reduced with high asymmetric induction resulting with hydride attack from the less hindered carbonyl face (eq. 4.29).

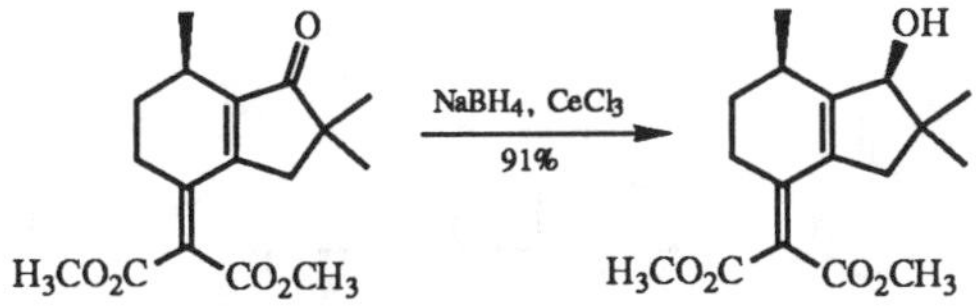

$$\xrightarrow[91\%]{NaBH_4,\ CeCl_3} \qquad (4.29)$$

Interestingly, the Luche protocol has been also applied to saturated ketones with high yields and stereoselectivities. Functional groups such as cyano, ester, or epoxy can be compatible with the reduction of ketone carbonyl groups. The procedure can even discriminate between two carbonyl groups and only one is selectively reduced. This feature is illustrated in the reduction of bifunctional molecules. Ketones can be reduced in the presence of aldehydes and conjugated or aromatic aldehydes are reduced in the presence of isolated aldehydes (eq. 4.30).

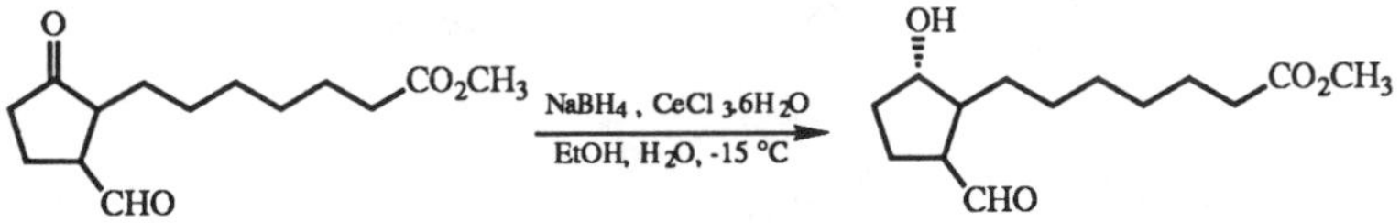

$$\xrightarrow[EtOH,\ H_2O,\ -15\ °C]{NaBH_4,\ CeCl_3 \cdot 6H_2O} \qquad (4.30)$$

In addition to cerium(III) salts, other lanthanides and hydride reducing agents have been employed for selective reduction of carbonyl groups. Thus, lanthanum(III) chloride, erbium(III) chloride, or samarium(III) iodide have proved to be useful for this purpose.[80,84-86] Again reduction of ketones can be achieved in the presence of aldehydes, and enones are preferentially reduced in the presence of isolated ketones. With regard to the hydride, $LiAlH_4$ can be utilized instead of $NaBH_4$, but lower selectivities are found for 1,2-addition to unsaturated ketones.

Saturated and conjugated ketones can be also reduced to alcohols by potassium-graphite.[87] In some cases pinacols are formed, although the alcohol is generally the predominant product. Thus, benzophenone is reduced to benzhydrol in 98% yield. α,β-Unsaturated ketones are reduced to saturated alcohols. Reduction of camphor gives predominantly the *exo*-alcohol; curiously reduction with sodium in alcohol or with potassium in the presence of graphite (but not intercalated), gives mainly the *endo*-alcohol.

A more recent reinvestigation of the reaction of benzophenone has revealed some striking points of this reduction.[88] The ratio of alcohol and pinacol products depends on the potassium-graphite/benzophenone molar ratio as well as on the solvent. In relation to the latter, polarity does not appear as the main factor, but rather the proton transfer ability. Proton-donating solvents lead to alcohols whereas formation of of pinacols occurs with non-proton-donating solvents such as benzene.

Other side reactions have been observed in reductions mediated by potassium-graphite. Thus, 3,3,5-trimethyl cyclohexanone gives carbon-carbon bond scission and carbon-carbon bond formation products (eq. 4.31).[87]

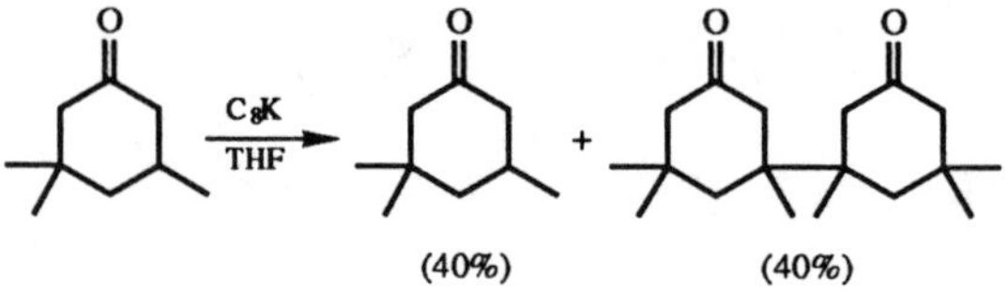

C_8K THF (4.31)

(40%) (40%)

Other metal-graphites promote also the reduction of carbonyl compounds to alcohols.[57,58] As previously mentioned, the less active Ni-Gr 2 is particularly effective for selective reductions of conjugated systems (eqs. 4.32, 4.33).[58]

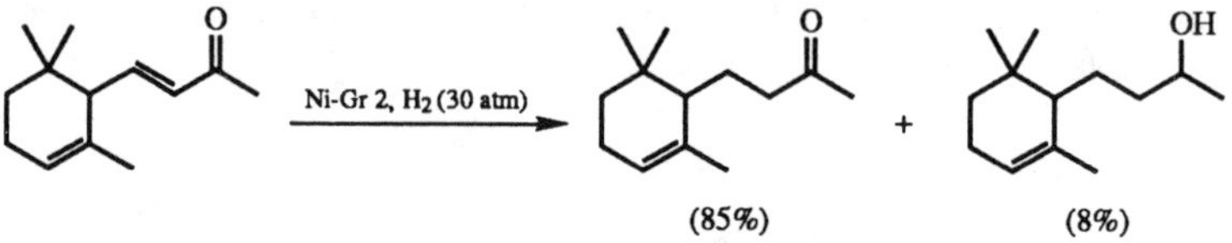

Ni-Gr 2, H_2 (30 atm) (4.32)

(85%) (8%)

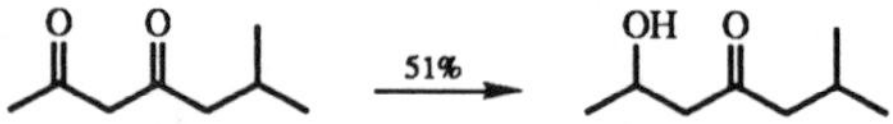

51% (4.33)

Similarly, the activated nickel by thermal decomposition of nickel isopropoxide reduces smoothly carbonyl compounds to alcohols in high yields (eq. 4.34).[60]

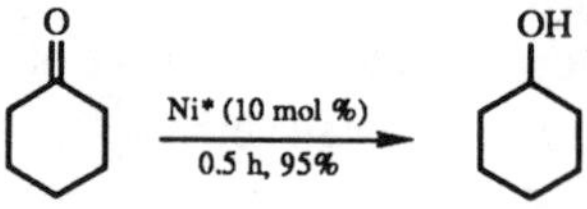

Ni* (10 mol %) 0.5 h, 95% (4.34)

3. Conjugated Reduction

Conjugated reduction of carbon-carbon double bonds without affecting the coexisting carbonyl groups can be carried out by a plethora of reductive processes.[89] These methods involve generally catalytic hydrogenation, dissolved metal reductions, or metal hydride reductions. Unfortunately, many of them are often affected by serious side reactions, especially overreductions.

The powerful activation provided by ultrasonic irradiation or by metal-graphite combinations have recently enabled mild and highly selective conjugated reductions. α,β-Unsaturated ketones, carboxylic acids, and Schiff bases can be readily reduced by potassium-graphite in THF to their corresponding saturated compounds.[90] The yields are good to excellent, employing an excess of potassium-graphite, and in the case of ketones, hexamethyldisilazane is used as co-solvent (eqs. 4.35-4.38). The process can be visualized as a heterogeneous Birch reaction with outstanding advantages. The troublesome use of ammonia or amines is avoided and organic products are easily separated from the reaction mixture by filtration. Furthermore, the alternative dissolved metal reduction often requires large amounts of HMPA, DMSO, or DMF as co-solvents.

C_8K, THF, $HN(SiMe_3)_2$; 25 °C, 10 min, 85% (4.35)

C_8K, THF, $HN(SiMe_3)_2$; 25 °C, 10 min, 77% (4.36)

HO_2C–CH=CH–CO_2H → (C_8K, THF; 55 °C, 45 min, 89%) → HO_2C–CH_2CH_2–CO_2H (4.37)

N → (C_8K, THF; 25 °C, 30 min, 92%) → N–H (4.38)

Catalytically active nickel is obtained by sonochemical reduction of nickel(II) chloride with zinc powder. This system represents a novel sonochemical hydrogenation procedure with great selectivity for the reduction of carbon-carbon double bonds in α,β-unsaturated carbonyl compounds (eq. 4.39).[91]

Zn-$NiCl_2$ (9:1), EtOH-H_2O (1:1); rt, 2.5 h,)))), 97% (4.39)

An important factor of this process is the influence of the pH, which controls the selectivity as demonstrated with the reduction of carvone (Scheme 4.6).[92]

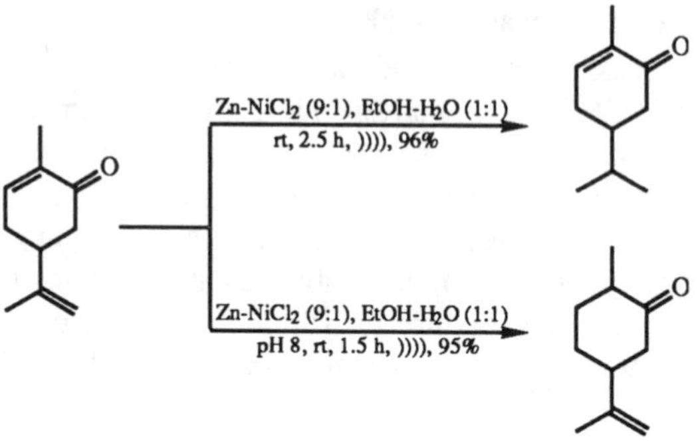

Zn-NiCl$_2$ (9:1), EtOH-H$_2$O (1:1)
rt, 2.5 h,)))), 96%

Zn-NiCl$_2$ (9:1), EtOH-H$_2$O (1:1)
pH 8, rt, 1.5 h,)))), 95%

Scheme 4.6

Also, selective reduction of α,β-unsaturated γ-dicarbonyl compounds with unactivated, powdered zinc in acetic acid at room temperature is promoted by ultrasound.[93] The method is simple, rapid, utilizes inexpensive and readily available reagents, and products are normally obtained in high yields. More importantly, isolated carbon-carbon double bonds are not affected by this reduction procedure which is a substantial advantage over alternative catalytic hydrogenations (eqs. 4.40-4.43).

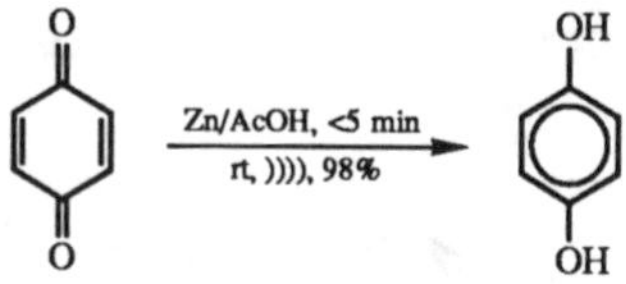

O, O — Zn/AcOH, <5 min; rt,)))), 98% — OH, OH (4.40)

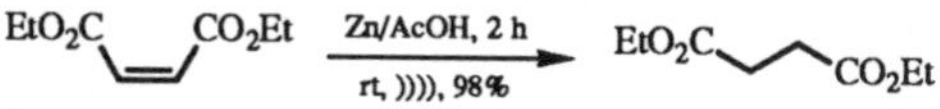

EtO$_2$C CO$_2$Et — Zn/AcOH, 2 h; rt,)))), 98% — EtO$_2$C CO$_2$Et (4.41)

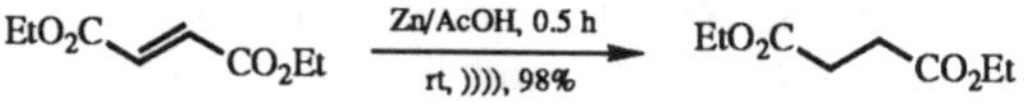

EtO$_2$C CO$_2$Et — Zn/AcOH, 0.5 h; rt,)))), 98% — EtO$_2$C CO$_2$Et (4.42)

O, O — Zn/AcOH, <5 min; rt,)))), 98% — O, O (4.43)

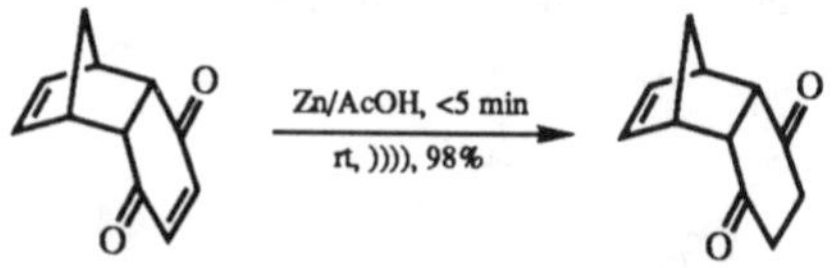

B. ORGANIC HALIDES AND RELATED SYSTEMS

Organic halides are substrates for two important reductive processes: a) reductive dehalogention to the corresponding saturated compounds,[94,95] and b) coupling reactions to give either symmetrical or unsymmetrical dimers,[96] which will be particularly treated in this Chapter. Reductive dehalogenation can be easily performed with activated metals, low-valent metals, and metal hydrides. An important related process is the reductive dehalogenation of polyhalo ketones with low-valent metals.[97]

The coupling of alkyl halides by treatment with a metal (classically with sodium) to give symmetrical products is known as the Wurtz reaction. This process is seldom used because of important serious reactions such as elimination and rearrangement. Moreover, the cross-coupling reaction affords generally a great number of products. Wurtz-type reactions can be, however, modified by changing the metal, the experimental conditions, and other factors resulting in a useful transformation.[98]

The coupling of aryl halides is called the Ullmann reaction.[96,99,100] This reaction is of enormous importance in the preparation of symmetrical and unsymmetrical biaryls. Although copper is the most widely used metal in this coupling, other activated metals and organometallics can be also employed. In fact, an Ullmann-type reaction is often a preliminary step for testing the effectiveness of any metal catalyst.

Aryl-aryl carbon bonds are formed by numerous methods requiring the presence of metal.[101] Typical examples include the reductive coupling as in the classical[100] and modified[102-104] Ullmann reactions, or the direct oxidative dehydrodimerization mediated by palladium,[105,106] vanadium,[107] copper,[108] or thallium[109] among others. More recent methods involve the use of transition metal ions as the Kharasch-type cross-coupling of aromatic Grignard reagents with aryl halides.[96]

Low-valent titanium species such as mixtures of $TiCl_4$-$LiAlH_4$,[110] $TiCl_3$-Mg,[111] or Cp_2TiCl_2-Mg,[112] are useful systems for hydrodehalogenation of alkyl halides. Low-valent titanium species generated from $TiCl_3$ and $LiAlH_4$ remove halogen atoms from *vic*-dihalides[113] and transform bromohydrins to olefins.[114] Also, benzylic halides undergo reductive dimerization with $TiCl_3$-$LiAlH_4$,[113] VCl_3-$LiAlH_4$,[115] and $CrCl_3$-$LiAlH_4$,[116] although the metal intermediates are presumably divalent species in these cases.

Semmelhack effected the direct coupling of aryl bromides and iodides by zero-valent nickel reagents with good yields.[117,118] The reaction is performed under mild conditions, but in contrast it requires stoichiometric quantities of nickel(0) reagents. Moreover, nickel(0) species are extremely air-sensitive compounds and aryl chlorides give poor yields. Several modifications were introduced,[119-121] and particularly Kumada[120] demonstrated that the procedure can be made catalytic in nickel(0) by using a stoichiometric amount of zinc, albeit aryl chlorides gave generally low yields of coupled products.

Colon and Kelsey[122] have found a general, mild procedure for the coupling of aryl chlorides by nickel and reducing metals. The nickel catalyst is generated *in situ* from nickel salts and excess reducing metal. The latter plays an essential role for success allowing the formation of biaryls from aryl chlorides in high yields. Thus, chlorobenzene is coupled in a few minutes using a catalytic amount of nickel(II) chloride, triphenylphosphine, and excess zinc in a dry, aprotic solvent at 60-80 °C under nitrogen (eq. 4.44).

$$2\ C_6H_5Cl + Zn \xrightarrow[N_2,\ 80\ ^\circ C,\ >98\%]{NiCl_2,\ PPh_3} C_6H_5-C_6H_5 \quad (4.44)$$

Nickel halides (but not the fluoride) can be used to form the nickel catalyst, with nickel(II) chloride or bromide as the most effective. Hydrated nickel salts led to an important reduction and nickel(II) oxide was completely ineffective. Various metals were attempted as reducing agents in the coupling of chlorobenzene with nickel(II) chloride, but only Zn, Mg, and Mn gave satisfactory yields of coupled products. Calcium and aluminum favored the reduction product, benzene. Nickel(II) chloride can be also reduced by sodium to give red solutions indicative of low-valent nickel species, but they favor again the reduction of chlorobenzene to benzene.

Reduction is the most serious and competitive side reaction. Acidic substrates or contamination with water or moisture produce substantial reduction to arenes. This side reaction is, however, a mild and selective method for reducing aromatic carbon-halogen bonds by the deliberate addition of water or another proton source.[123]

Activated cerium, prepared by treatment of cerium metal with iodine or mercury(II) chloride, is a very powerful reducing agent capable of reducing even alkyl fluorides.[124] Nevertheless, with this system the competitive reductive dimerization is largely favored. The alternative combination of $CeCl_3$-$LiAlH_4$ solves this problem and a variety of alkyl and aryl halides are reduced to the corresponding hydrocarbons in high yields (eqs. 4.45-4.47).[124,125] The authors suggest the participation of radical species because when $LiAlD_4$ was employed, deuterium was not incorporated in the final product.

$$\text{2-F-}C_6H_4-C_6H_5 \xrightarrow[DME,\ \Delta,\ 5\ h,\ 94\%]{LiAlH_4\text{-}CeCl_3} C_6H_5-C_6H_5 \quad (4.45)$$

$$CH_3(CH_2)_{11}F \xrightarrow[THF,\ \Delta,\ 3\ h,\ 90\%]{LiAlH_4\text{-}CeCl_3} C_{12}H_{26} \quad (4.46)$$

$$\text{Cl-}C_6H_3(\text{Cl})\text{-OH} \xrightarrow[THF,\ \Delta,\ 20\ h,\ 81\%]{LiAlH_4\text{-}CeCl_3} \text{Cl-}C_6H_4\text{-OH} \quad (4.47)$$

Many activated metals are useful for the coupling of aryl halides. Thus contrary to the classical Wurtz coupling using sodium, sodium napththalenide effects the reductive dimerization of benzyl halides in good yields (eq. 4.48).[126]

$$2\ C_6H_5\text{-}CH_2Cl \xrightarrow[80\%]{[C_{10}H_8]^-Na^+} C_6H_5\text{-}CH_2CH_2\text{-}C_6H_5 \quad (4.48)$$

Activated copper, generated simply by reduction of copper(II) sulfate with zinc dust in aqueous solution, constitutes a useful catalyst for Ullmann syntheses (eq. 4.49).[127]

$$2\ \text{2-}O_2N\text{-}C_6H_4\text{-}I \xrightarrow[240\ ^\circ C,\ 96\%]{Cu^*} \text{2,2'-}(NO_2)_2\text{-biphenyl} \quad (4.49)$$

Highly reactive metals suitable for coupling reactions can be readily obtained by reduction of metal salts by alkali metals (Rieke metals),[128] or by reduction with lithium or potassium naphthalenides (Chapter III). These reactive metals add oxidatively aryl halides to form biaryls in high yields. Thus a highly reactive copper powder, prepared by reduction of CuI with potassium naphthalenide in DME, is particularly effective for Ullmann reactions.[129] High yields of biaryls are obtained even at 85 °C, and this copper also promotes cross-coupling with allyl halides.

Similarly, reduction of a solution of $CuI.PEt_3$ with a stoichiometric amount of lithium naphthalenide in THF produces a zero-valent copper species, which adds to organic halides to give organocopper compounds. The organic halides can contain a wide variety of functional groups such as allyl, aryl, alkynyl, nitro, or cyano. The organocopper reagents can add to alkyl, aryl, and vinyl halides under very mild conditions.[130-135] Homocoupling of organic halides is very rapid with this copper. Most alkyl halides undergo self-coupling in less than 1 min after addition of the activated copper, and very little reduction product is obtained. Homocoupling can be also achieved by introducing oxygen into the reaction vessel (Scheme 4.7).

$$2\ Li + C_{10}H_8 + 2\ CuI.PEt_3 \xrightarrow[0\ ^\circ C,\ 10\ min]{THF\ or\ DME} 2\ Cu^*$$

$$2\ Cu^* + RX \xrightarrow[1\ min\ to\ 1\ h,\ THF]{0\text{-}25\ ^\circ C} CuX + RCu \begin{cases} \xrightarrow{[O_2]} R\text{-}R \\ \xrightarrow{H_2O} R\text{-}H \\ \xrightarrow{R'X} R\text{-}R' \end{cases}$$

Scheme 4.7

This activated copper also induces the coupling of aroyl chlorides in good to excellent yields, which are superior to those reported using other organometallics.[133] The reaction conditions are mild (-78 °C) and the *cis*-isomers are obtained predominantly (eq. 4.50). Aliphatic acid chlorides, however, gave a complex mixture of products. Experiments suggest that a copper(I) species is involved in these copper-mediated transformations and not copper(II) species. Interestingly, activated nickel powder formed from nickel(II) iodide, lithium, and naphthalene gave similar results in the coupling of aroyl chlorides but with the opposite stereoselectivity. The *trans*-isomer was predominantly produced.

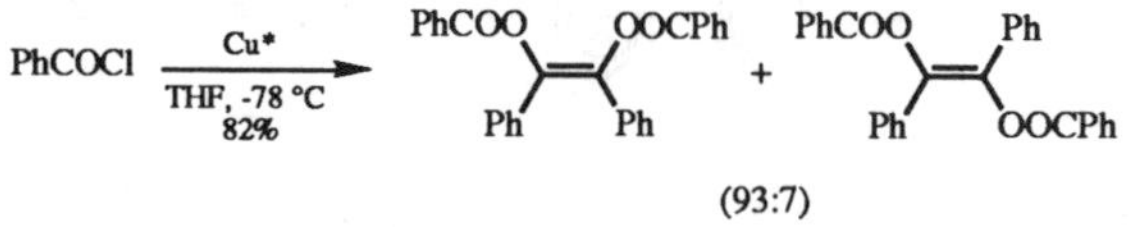

PhCOCl —(Cu*, THF, -78 °C, 82%)→ PhCOO(Ph)C=C(Ph)OOCPh + PhCOO(Ph)C=C(Ph)OOCPh (4.50)

(93:7)

Activated metallic nickel effects also dehalogenative coupling of iodobenzenes and bromobenzenes under mild conditions in moderate to good yields.[136] Metallic nickel works well for the dehalogenation of unsubstituted and 4-substituted iodobenzenes. *Ortho* substituents on the aryl groups inhibit the coupling. Thus, 2-iodomethoxybenzene gave exclusively anisole in moderate yield and no coupling product. 2-Iodonitrobenzene did not react under these conditions. These results contrast clearly with those from copper-mediated reactions, in which electron-withdrawing substituents in *ortho* positions enhance the reactivity. Bromobenzenes were less reactive than iodobenzenes, although they yielded biphenyls together with reduction products (Scheme 4.8).

NiX_2 —($Li/C_{10}H_8$, DME, Ar)→ Ni*

2 ArX —(Ni*, solvent, Δ)→ Ar-Ar + Ar-H

X = Br, I
solvent = DME, DMF, DMSO

Scheme 4.8

Similarly, this activated nickel promotes the reductive homocoupling for benzylic mono- and polyhalides (Scheme 4.9).[137]

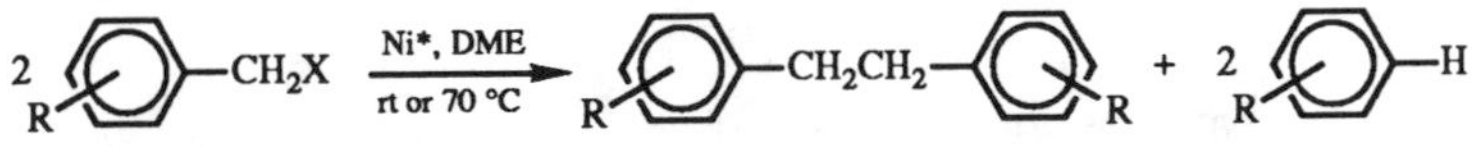

$$2\ R\text{-}C_6H_4\text{-}CH_2X \xrightarrow[\text{rt or 70 °C}]{\text{Ni*, DME}} R\text{-}C_6H_4\text{-}CH_2CH_2\text{-}C_6H_4\text{-}R + 2\ R\text{-}C_6H_4\text{-}H$$

X = Cl, Br, I
R = H, 4-CH_3, 3-OCH_3, 3-CF_3, 4-Cl, 4-Br, 4-NO_2, 4-NC, 4-$COOCH_3$

Scheme 4.9

Good yields of homocoupled products are obtained by reaction of alkyl or aryl halides with lithium wire in THF, immersed in an ultrasonic bath (50 kHz). In the absence of ultrasound little or no reaction occurs (eq. 4.51).[138,139]

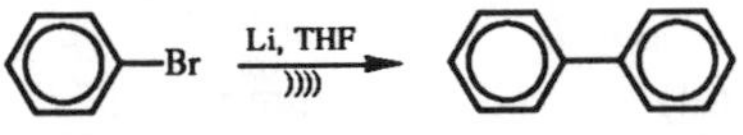

$$C_6H_5\text{-}Br \xrightarrow[\text{))))}]{\text{Li, THF}} C_6H_5\text{-}C_6H_5 \qquad (4.51)$$

Although this process appears to be a simple Wurtz-type coupling, a further investigation by Osborne on the dehalogenation of bromopyridines evidenced the presence of isomeric bipyridyls.[140] Likewise, the sonochemical coupling of bromotoluenes affords a mixture of isomeric bitolyls.[141] These results suggest the presence of radical intermediates confirmed by radical scavengers which inhibit the reaction.[141] This supports a previous suggestion of Luche on the influence of ultrasound in single electron transfer processes.[142] A more recent study has also indicated that sonication of aryl bromides produces a mixture of arenes and biaryls (eq. 4.52).[143] No dimers were detected with quinolines.

$$\text{ArX} \xrightarrow{\text{Li, THF,))))}} \underset{(0\text{-}70\%)}{\text{Ar-Ar}} + \underset{(25\text{-}100\%)}{\text{Ar-H}} \qquad (4.52)$$

Coupling of benzyl halides in the presence of copper or nickel powder, generated by lithium reduction of the corresponding halides in the presence of ultrasound, gives dibenzyls (eq. 4.53).[144] With sonication the yields are higher than those obtained in mechanically stirred reactions.

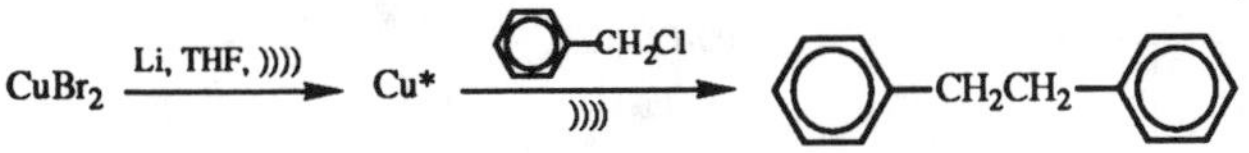

$$CuBr_2 \xrightarrow{\text{Li, THF,))))}} Cu^* \xrightarrow[\text{))))}]{C_6H_5\text{-}CH_2Cl} C_6H_5\text{-}CH_2CH_2\text{-}C_6H_5 \qquad (4.53)$$

Ullmann coupling of activated aryl halides in DMF with high intensity ultrasound (sonic horn) gave a 64-fold increase in rate over a mechanically stirred reaction (eq. 4.54).[145] Although the effect of sonication is a four-fold decrease in the particle size of the copper, this fact alone is insufficient to explain the large rate increases observed when sonication is maintained. It would appear that ultrasound assists in the breakdown of reaction intermediates and/or the desorption of products.

$$\text{2-iodonitrobenzene} \xrightarrow[\text{)))}]{\text{Cu*, DMF, 60 °C}} \text{2,2'-dinitrobiphenyl} \quad (4.54)$$

Other Ullmann-type couplings are greatly facilitated by ultrasonic irradiation. Aryl sulfonates give biaryls by treatment with *in situ* generated nickel(0) complexes (eq. 4.55).[146] The method works well for triflates (R = trifluoromethyl); for tosylates (R = 4-methylphenyl) the rate is significantly lower.

$$ArOSO_2R \xrightarrow[\text{DMF, 60 °C,))))}]{NiCl_2,\ Zn,\ PPh_3,\ NaI} Ar\text{-}Ar \quad (4.55)$$

Aryl-phosphorus bonds are readily cleaved by lithium in THF in the presence of low intensity ultrasound (cleaning bath) to give lithium dialkylphosphides, which readily couple with alkyl halides to produce phosphanes (Scheme 4.10).[147,148] Much higher rates are observed than for reactions under mechanical agitation.

$$R^1R^2PPh \xrightarrow[\text{))))}]{\text{Li, THF}} R^1R^2P^-Li^+ \begin{cases} \xrightarrow{R^3X} R^1R^2R^3P \\ \xrightarrow{Br(CH_2)_nBr} R^1R^2P(CH_2)_nPR^1R^2 \end{cases}$$

Scheme 4.10

C. DEOXYGENATION REACTIONS

Reductive elimination of epoxides to olefins can be carried out with numerous activated metals. Common methods include the less reactive forms of activated zinc such as zinc-acetic acid,[149] zinc-triphenylphosphine dibromide,[150] or zinc-copper couple (eqs. 4.56-4.58).[151]

$$\text{styrene oxide} \xrightarrow[\text{4 h}]{\text{Zn-Cu, EtOH}} \underset{(90\%)}{Ph\text{-}CH{=}CH_2} + \underset{(4\%)}{Ph\text{-}CH_2CH_3} \quad (4.56)$$

$$\textit{cis}\text{-stilbene oxide} \xrightarrow[\text{6 h}]{\text{Zn-Cu, EtOH}} \underset{(13\%)}{\textit{cis}\text{-PhCH=CHPh}} + \underset{(80\%)}{\textit{trans}\text{-PhCH=CHPh}} \quad (4.57)$$

$$\textit{trans}\text{-stilbene oxide} \xrightarrow[\text{6 h}]{\text{Zn-Cu, EtOH}} \underset{(95\%)}{\textit{trans}\text{-PhCH=CHPh}} \quad (4.58)$$

Reductive elimination of octene epoxides gives mixtures of *cis*- and *trans*-octenes, but *cis*-octenes predominate in the products from *cis*-epoxides, and *trans*-epoxides predominate from *trans*-epoxides. Some *cis* to *trans* isomerization thus appears to occur. Reduction of epoxides of octenes and of cholesterol proceeds somewhat more rapidly with zinc-copper couple than by reduction with other reductants like chromium(II)-amine complexes. The scope of this reaction was also explored with several sesquiterpene lactone epoxides and with steroidal epoxides. Sterically hindered epoxides are not reduced, and thus a steroidal 2,3-epoxide is reduced whereas an 11,12 epoxide is not reduced.

Deoxygenation of epoxides to yield alkenes is readily performed with low-valent tungsten species generated by reacting tungsten(VI) chloride with either an alkyl lithium reagent, lithium metal dispersion, or lithium iodide.[152] The reaction is a two-electron process and products can be obtained with a high degree of stereoretention. Similarly, the McMurry complex ($TiCl_3$-$LiAlH_4$) can transform epoxides into alkenes.[153] The process is non-stereospecific and the intermediacy of free radicals has been suggested (eq. 4.59).

$$CH_3(CH_2)_3\text{-epoxide-}(CH_2)_3CH_3 \xrightarrow[70\%]{TiCl_3\text{-}LiAlH_4} CH_3(CH_2)_3CH{=}CH(CH_2)_3CH_3 \quad (E{:}Z = 4{:}1) \qquad (4.59)$$

These methods, however, are vastly superior to other procedures employing zero-valent and low-valent reagents for the deoxygenation of epoxides.[63,64] Further reduction of alkenes to alkanes or rearrangement of epoxides to carbonyl compounds are common drawbacks.[154,155]

As described in Chapter II, deoxygenation of epoxides has been also carried out by metal atom co-condensation.[156] Chromium and vanadium atoms were the most effective, whereas titanium gave lower yields and nickel was considerably less reactive. Deoxygenation with chromium atoms appears to be general for epoxides. No double-bond migrations were detected, but *cis-trans* isomerization occurred and *trans*-alkenes predominated. Also, chromium atoms effected deoxygenation from heteroatom oxides, and thus dimethyl sulfoxide was reduced to dimethyl sulfide and 2,6-dimethylpyridine *N*-oxide was converted into 2,6-dimethylpyridine.[156,157] Gladysz *et al.* have also reported the deoxygenation and desulfurization of other organic compounds with transition metal atoms (Chapter II).[157,158]

The appreciable affinitiy of titanium for oxygen has been utilized in the ready deoxygenation of alcohols. 1,2-Diols (pinacols) which are indeed intermediates in the reductive coupling of carbonyls to olefins (McMurry reaction, Chapter V), are readily reduced by titanium(0) species.[159,160] 1,3-Diols undergo intramolecular reductive coupling to cyclopropanes with low-valent titanium species. The process is fairly useful because of *cis*- and *trans*-cyclopropanes are obtained along with other reduction products (eq. 4.60).[161] It has been proposed that a Ti(II) complex forms as an intermediate, which by thermal decomposition generates a diradical and then cyclizes to give cyclopropanes.[162-164]

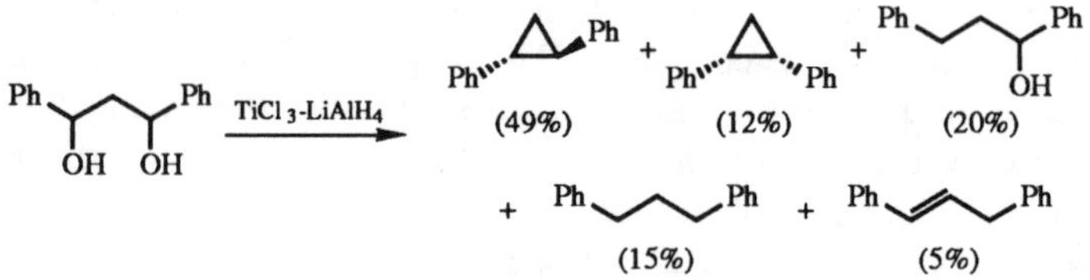

(4.60)

Benzylic and allylic alcohols can be coupled to bibenzyls and 1,5-dienes in the presence of low-valent titanium reagents (eqs. 4.61-4.64).[4,63,64,162,163,165]

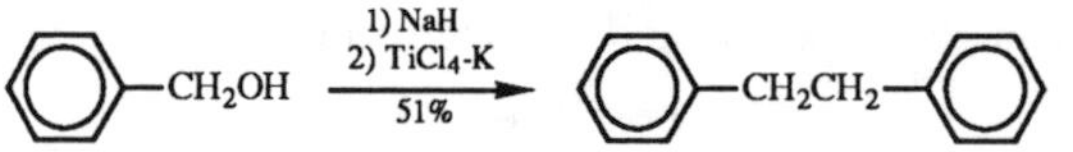

(4.61)

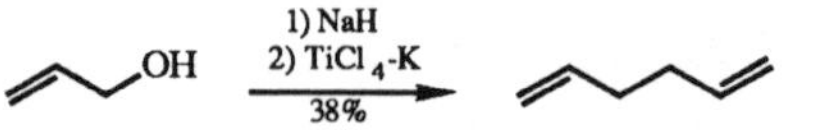

(4.62)

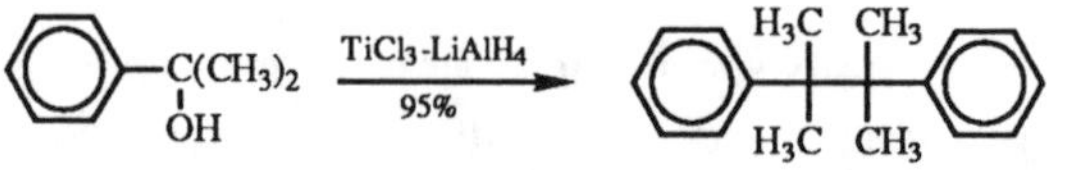

(4.63)

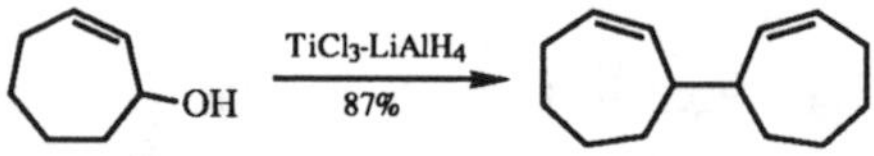

(4.64)

Allylic alcohols, ethers, or acetates can be reductively cleaved with zinc amalgam.[166] The formation of the less stable olefin is a result of protonation of an intermediate allylic zinc chloride at the more substituted end of the system (eq. 4.65).

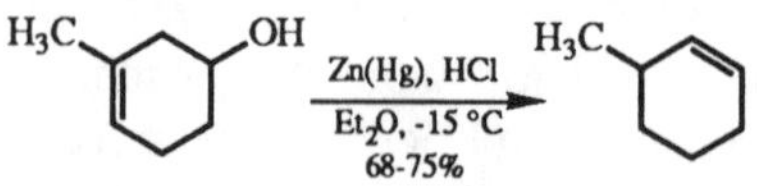

(4.65)

Metallic titanium generated from $TiCl_3$-K reduces enol phosphates to alkenes[167] (eq. 4.66). The method is preferable to that using lithium and ethylamine as reductant, and it permits the regioselective synthesis of dienes from α,β-unsaturated ketones. No overreduction is generally observed although the enol phosphate derived from isobutyrophenone is rapidly converted to isobutylbenzene.

$$\text{1-methylcyclohexen-2-yl-OPO(OC}_2\text{H}_5)_2 \xrightarrow[83\%]{TiCl_3\text{-K}} \text{1-methylcyclohexene} \quad (4.66)$$

This reduction has been extended to reduction of aryl diethyl phosphates to arenes which constitutes an indirect method for deoxygenation of phenols (eq. 4.67).[168] Yields are in the range 75-95%; Birch reduction using lithium in liquid ammonia usually proceeds in low yield.

$$\text{1-naphthol (OH)} \xrightarrow[\text{2) } ClPO(OC_2H_5)_2,\ 91\%]{\text{1) NaH}} \text{naphthyl-OPO(OC}_2\text{H}_5)_2 \xrightarrow[74\%]{Ti(0),\ THF,\ \Delta} \text{naphthalene} \quad (4.67)$$

Sulfoxides are deoxygenated to sulfides by titanium(III) chloride, as well as by the low-valent reagents $TiCl_4$-Zn[169] and $TiCl_4$-$LiAlH_4$.[170] α,β-Unsaturated sulfides can be obtained from aromatic aldehydes by reaction with sodium metal in DMSO, which leads to 2-arylethenyl methyl sulfides having the (*E*)-configuration (eq. 4.68).[171] The reagent generated *in situ* is probably dimsylsodium; the final product is then formed by reduction of the intermediate sulfoxide.

$$ArCHO \xrightarrow[THF]{DMSO,\ Na} \left[(E)\text{-Ar(H)C=C(H)SOCH}_3 \right] \xrightarrow[65\text{-}85\%]{Na} (E)\text{-Ar(H)C=C(H)SCH}_3 \quad (4.68)$$

D. REDUCTION OF NITROGEN AND SULFUR FUNCTIONAL GROUPS

A wide variety of nitrogen-based functional groups can be conveniently reduced to the corresponding amine derivatives by using dissolved metals, such as Zn, Sn, or Fe and an acid, or catalytic hydrogenation.[172-174] The use of other less usual metals or more reactive combinations has allowed selective reductions with the avoidance of side reactions and by-products.

The versatile cerium metal has been found to be highly effective for the mild reduction of nitro groups,[124] although the scope and limitations of this promising method await systematic study. Interestingly, an organic halide group is not reduced (eq. 4.69). In addition the Luche method employing a low-valent cerium reagent or other lanthanoids, is also useful for the mild conversion of α-nitroimines to nitroalkenes in good yields.[175,176]

$$\text{Cl}_4\text{C}_6\text{-NO}_2 \xrightarrow[\text{AcOH-MeOH}]{\text{Ce}} \text{Cl}_4\text{C}_6\text{-NH}_2 \qquad (4.69)$$

Reduction of α-oximino esters to the corresponding α-amino derivatives can be accomplished by a low-valent titanium reagent in a buffered solution of pH 7. L-Tartaric acid is a useful buffer, it does not induce any selectivity but affords clean reactions in good yields (eq. 4.70).[177]

$$\text{R-C(=NOH)-CO}_2\text{CH}_3 \xrightarrow[\text{MeOH-H}_2\text{O, pH 7; 64-82\%}]{\text{TiCl}_3\text{-NaBH}_4} \text{R-CH(NH}_2\text{)-CO}_2\text{CH}_3 \qquad (4.70)$$

Aromatic nitrocompounds have been reduced to amines with $TiCl_4$-Mg.[178] Also, $TiCl_3$-Mg reduces cyclohexyl isocyanide to cyclohexane and methane (eq. 4.71),[179] and this reagent also converts nitriles into alkanes.[180]

$$\text{C}_6\text{H}_{11}\text{-NC} \xrightarrow{\text{TiCl}_3\text{-Mg}} \text{C}_6\text{H}_{12} + \text{CH}_4 \qquad (4.71)$$

The highly dispersed reagent sodium naphthalenide reduces isocyanides to hydrocarbons with fewer rearranged products than with Li or Na in ammonia or THF. The only limitation is that the reduction can result in racemization.[181] Sodium naphthalenide also effects reductive decyanation, and thus *vic*-cyanohydrins can be transformed to alkenes by conversion to a methylthiomethyl ether followed by oxidation to the sulfone. Reduction of this derivative with sodium naphthalenide in HMPA results in an alkene.[182]

However, a more convenient reagent for reductive decyanation is potassium-alumina. It is prepared by melting potassium over the support with vigorous stirring. The material is sensitive to oxygen and moisture. Potassium-alumina reduces alkyl nitriles in hexane at 25 °C in 70-90% yields. The reaction is rapid for secondary and tertiary nitriles but requires about 1 hour for primary nitriles. The use of potassium alone was less efficient. The procedure was useful in a synthesis of the sex pheromone **11** (Scheme 4.11).[183]

$$\text{THPO}(CH_2)_5CH(CN)(CH_2)_2CH{=}CHC_2H_5 \ (\mathbf{9}) \xrightarrow[80\%]{K\text{-}Al_2O_3} \text{THPO}(CH_2)_8CH{=}CHC_2H_5 \ (\mathbf{10})$$

$$\xrightarrow[85\%]{AcCl,\ AcOH} H_3C\text{-}C(=O)\text{-}O(CH_2)_8CH{=}CHC_2H_5 \ (\mathbf{11})$$

Scheme 4.11

Chemical fixation of molecular nitrogen which constitutes an exciting and very promising subject can be mediated by low-valent titanium reagents. The pioneering results of van Tamelen[184] and Vol'pin[185,186] illustrate this process (eq. 4.72). Numerous carbonyl compounds can incorporate nitrogen to give mainly amines or amides by the action of titanium species.[187]

$$Mg + Cp_2TiCl_2 + N_2 \xrightarrow{THF} \xrightarrow{(C_2H_5)_2CO} (C_2H_5)_2CHNH_2 + [(C_2H_5)_2CH]_2NH \qquad (4.72)$$

α-Phenylthio ketones are desulfurized by zinc and chlorotrimethylsilane in ether with good yields (eq. 4.73).[188] The method was very useful in a synthesis of 11-deoxy-prostaglandin E and the 8,12-epimer.

$$\text{2-}(SC_6H_5)\text{-2-}((CH_2)_4CH_3)\text{cyclopentanone} \xrightarrow[84\%]{Zn,\ TMSCl} \text{2-}((CH_2)_4CH_3)\text{cyclopentanone} \qquad (4.73)$$

Carbamates and thiocarbamates can be reduced conveniently with potassium and 18-crown-6 ethers in *tert*-butylamine to the alkane and the corresponding alcohol with the former predominating (eqs. 4.74-4.75).[189-192] The method is also applicable to carbohydrate derivatives.

$$CH_3(CH_2)_{16}CH_2OC(=S)N(C_2H_5)_2 \xrightarrow[18\text{-crown-6},\ 99\%]{K,\ {}^tBuNH_2} CH_3(CH_2)_{16}CH_3 + CH_3(CH_2)_{16}CH_2OH \quad (87:13) \qquad (4.74)$$

$$(C_2H_5)_2NC(=S)O\text{-steroid} \xrightarrow{94\%} \text{steroid-H} + HO\text{-steroid} \quad (91:9) \qquad (4.75)$$

Similarly, a variety of sulfonamides is cleaved with potassium and crown ethers. The system also reductively cleaves *p*-toluenesulfonates in good yield. Sodium naphthalenide can be used for this reaction but it does not cleave mesyl amides of primary and dialkyl amines. Sulfonamides can also be cleaved by potassium in diglyme without a crown ether, but with a proton source, usually isopropanol. They are also cleaved by Na/K alloy and isopropanol in toluene but explosions can occur.[193]

α,β-Unsaturated sulfones undergo reductive cleavage of the carbon-sulfur bond with potassium-graphite in ether to afford alkenes.[194] The double bond undergoes a partial isomerization and (*Z*)- and (*E*)-2-alkenes are formed, with the latter usually predominating. These sulfones are not cleaved by Al/Hg but are reduced by lithium-ethylamine. However, a better reagent is potassium-graphite and alkenes are obtained in good yields (50-80%). Both reagents also induce partial (*E*)- to (*Z*)-isomerization, but potassium-graphite is less prone to effect this undesirable side reaction. Because the same alkene is obtained from the isomeric β,γ-sulfones, the method has been applied to the preparation of 2-alkenes by alkylation of the readily available allyl phenyl sulfone followed by cleavage with potassium-graphite (Scheme 4.12). Other alkenyl and vinylic sulfones are cleaved with potassium-graphite in 65-85% yields.[195] Sodium amalgam is not satisfactory for this carbon-sulfur bond reduction.

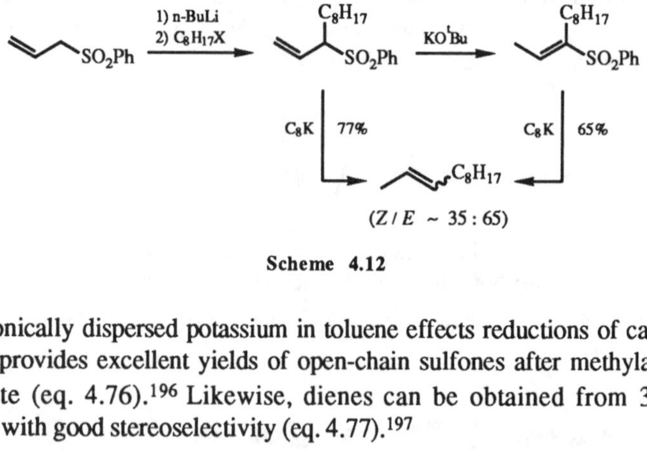

SO_2Ph —[1) n-BuLi; 2) $C_8H_{17}X$]→ C_8H_{17} / SO_2Ph —[KO^tBu]→ C_8H_{17} / SO_2Ph

C_8K 77% ; C_8K 65% → C_8H_{17}

(*Z* / *E* ~ 35 : 65)

Scheme 4.12

Ultrasonically dispersed potassium in toluene effects reductions of carbon-sulfur bonds and provides excellent yields of open-chain sulfones after methylation of the intermediate (eq. 4.76).[196] Likewise, dienes can be obtained from 3-sulfolene derivatives with good stereoselectivity (eq. 4.77).[197]

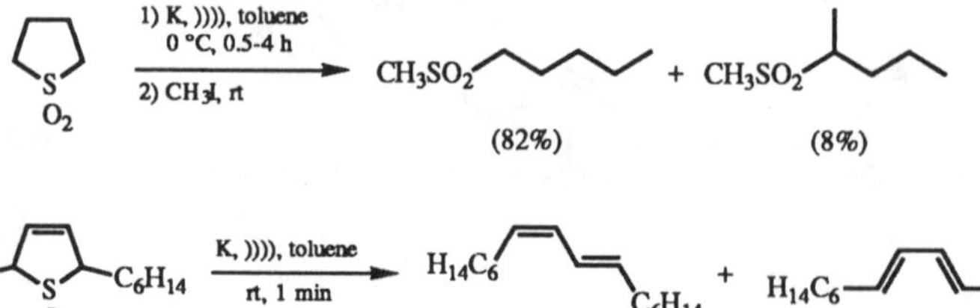

$$\text{S}(O_2) \xrightarrow[\text{2) } CH_3I\text{, rt}]{\text{1) K,)))), toluene, 0 °C, 0.5-4 h}} CH_3SO_2\text{-}(82\%) + CH_3SO_2\text{-}(8\%) \quad (4.76)$$

$$C_6H_{14}\text{-}\text{S}(O_2)\text{-}C_6H_{14} \xrightarrow[\text{rt, 1 min}]{\text{K,)))), toluene}} H_{14}C_6\text{-}\ldots\text{-}C_6H_{14} + H_{14}C_6\text{-}\ldots\text{-}C_6H_{14} \quad (4.77)$$

References

1. House, H. O. *Modern Synthetic Reactions*; W. A. Benjamin: New York, 1972; chapters 1-4.
2. Hudlicky, M. *Reductions in Organic Synthesis*; Wiley: New York, 1984.
3. March, J. *Advanced Organic Chemistry*; Wiley: New York, 1985; pp 1093-1120.
4. Pons, J.-M.; Santelli, M. *Tetrahedron* **1988**, *44*, 4295.
5. Rylander, P. N. *Hydrogenation Methods*; Academic Press: London, 1985.
6. Brown, C. A. *J. Org. Chem.* **1970**, *35*, 1900.
7. Brown, C. A.; Ahuja, V. K. *J. Org. Chem.* **1973**, *38*, 2226.
8. Brown, C. A.; Ahuja, V. K. *J. Chem. Soc. Chem. Commun.* **1973**, 553.
9. Davis, S. C.; Klabunde, K. J. *Chem. Rev.* **1982**, *82*, 153 and references cited therein.
10. Harvey, R. G. *Synthesis* **1970**, 161.
11. Birch, A. J.; Subba Rao, G. *Adv. Org. Chem.* **1972**, *8*, 1.
12. Caine, D. *Org. React.* **1976**, *23*, 1.
13. Akhrem, A. A.; Reshotova, I. G.; Titov, Yu. A. *Birch Reduction of Aromatic Compounds*; Plenum Press: New York, 1972.
14. Huffman, J. W. *Acc. Chem. Res.* **1983**, *17*, 399.
15. For a review on the partial reduction of aromatic rings by dissolving metals: Mander, L. N. *Comprehensive Organic Synthesis*; Fleming, I., Ed.; Pergamon Press: Oxford, 1991; Vol. 8.
16. Peer, W. J.; Lagowski, J. J. *J. Phys. Chem.* **1980**, *84*, 1110.
17. Thompson, J. C. *Electrons in Liquid Ammonia*; Clarendon Press: Oxford, 1976.
18. Harvey, R. G. *J. Org. Chem.* **1968**, *33*, 2570.
19. Riegel, L.; Friedel, R. A.; Wender, I. *J. Org. Chem.* **1957**, *22*, 891.
20. Benkeser, R. A.; Agnihotri, R. K.; Burrous, M. L.; Kaiser, E. M.; Mallan, J. M.; Ryan, P. W. *J. Org. Chem.* **1964**, *29*, 1313.
21. Jolly, W. L. *Prog. Inorg. Chem.* **1959**, *1*, 235.
22. Angibeaund, P.; Rivière, H.; Tchoubar, B. *Bull. Soc. Chim. Fr.* **1968**, 2937.
23. Spassky-Pasteur, A. *Bull. Soc. Chim. Fr.* **1969**, 2900.
24. Angibeaund, P.; Rivière, H. *C. R. Hebd. Acad. Sci. Ser. C* **1966**, *263*, 1076.
25. White, J. D.; Larson, G. L. *J. Org. Chem.* **1978**, *43*, 4555.
26. Imamura, H.; Konishi, T.; Sakata, Y.; Tsuchiya, S. *J. Chem. Soc. Chem. Commun.* **1991**, 1527.
27. Imamura, H.; Kitajima, K.; Tsuchiya, S. *J. Chem. Soc. Faraday Trans. 1* **1989**, *85*, 1647.

28. Owsley, D. C.; Bloomfield, J. J. *Synthesis* **1977**, 118.
29. Read, G.; Ruiz, V. M. *J. Chem. Soc. Perkin Trans. 1* **1973**, 1223.
30. Pfaltz, A.; Anwar, S. *Tetrahedron Lett.* **1984**, *25*, 2977.
31. Boudjouk, P.; Han, B. H. *J. Catal.* **1983**, *79*, 489.
32. Shin, D. H.; Han, B. H. *Bull. Korean Chem. Soc.* **1985**, *6*, 247.
33. Moulton, K. J.; Koritala, S.; Frankel, E. N. *J. Am. Oil Chem. Soc.* **1983**, *60*, 1257.
34. Markov, P.; Ivanoff, C. *Tetrahedron Lett.* **1962**, 1139.
35. Markov, P.; Ivanoff, C. *C. R. Hebd. Acad. Sci. Ser. C* **1967**, *264*, 1605.
36. Brettle, R.; Shibib, S. M. *Tetrahedron Lett.* **1980**, *21*, 2915.
37. Brettle, R.; Shibib, S. M. *J. Chem. Soc. Perkin Trans. 1* **1981**, 2912.
38. Profitt, J. A.; Ong, H. H. *J. Org. Chem.* **1979**, *44*, 3972.
39. Hutchins, R. O.; Suchismita; Zipkin, R. E.; Taffer, I. M.; Sivakumar, R.; Monaghan, A.; Elisseou, E. M. *Tetrahedron Lett.* **1989**, *30*, 55 and references cited therein.
40. Olah, G. A.; Prakash, G. K.; Arvanaghi, M.; Bruce, M. R. *Angew. Chem. Int. Ed. Engl.* **1981**, *20*, 92.
41. Näf, F.; Decorzant, R.; Thommen, W.; Wilhalm, B.; Ohloff, G. *Helv. Chim. Acta* **1975**, *58*, 1016.
42. Oppolzer, W.; Fehr, C.; Warneke, J. *Helv. Chim. Acta* **1977**, *60*, 48.
43. Sondengam, B. L.; Charles, G.; MacAkam, T. *Tetrahedron Lett.* **1980**, *21*, 1069.
44. Veliev, M. G.; Guseinov, M. M.; Mamedov, S. A. *Synthesis* **1981**, 400.
45. Sondengam, B. L.; Fomum, Z. T.; Charles, G.; MacAkam, T. *J. Chem. Soc. Perkin Trans. 1* **1983**, 1219.
46. Aerssens, M. H. P. J.; Brandsma, L. *J. Chem. Soc. Chem. Commun.* **1984**, 735.
47. Avignon-Tropis, M.; Pougny, J. R. *Tetrahedron Lett.* **1989**, *30*, 4951.
48. Chou, W.-N.; Clark, D. L.; White, J. B. *Tetrahedron Lett.* **1991**, *32*, 299.
49. Ichikawa, M.; Soma, M.; Onishi, T.; Tamaru, K. *J. Catal.* **1968**, *9*, 418.
50. Boersma, M. A. M. *Catal. Rev. Sci. Eng.* **1974**, *10*, 243.
51. Ichikawa, M.; Inoue, Y.; Tamaru, K. *J. Chem. Soc. Chem. Commun.* **1972**, 928.
52. Savoia, D.; Trombini, C.; Umani-Ronchi, A. *J. Org. Chem.* **1978**, *43*, 2907.
53. Lalancette, J.-M.; Roussel, R. *Can. J. Chem.* **1976**, *54*, 2110.
54. Béguin, F.; Setton, R. *J. Chem. Soc. Chem. Commun.* **1976**, 611.
55. Savoia, D.; Trombini, C.; Umani-Ronchi, A.; Verardo, G. *J. Chem. Soc. Chem. Commun.* **1981**, 540.
56. Lalancette, J.-M. U.S. Patent 3 804 916; *Chem. Abstr.* **1974**, *80*, 145363.

57. Savoia, D.; Tagliavini, E.; Trombini, C.; Umani-Ronchi, A. *J. Org. Chem.* **1981**, *46*, 5340.
58. Savoia, D.; Tagliavini, E.; Trombini, C.; Umani-Ronchi, A. *J. Org. Chem.* **1981**, *46*, 5344.
59. Savoia, D.; Trombini, C.; Umani-Ronchi, A. *Pure Appl. Chem.* **1985**, *57*, 1887 and references cited therein.
60. Boldrini, G. P.; Savoia, D.; Tagliavini, E.; Trombini, C.; Umani-Ronchi, A. *J. Org. Chem.* **1985**, *50*, 3082.
61. Hanson, J. R. *Synthesis* **1974**, 1.
62. McMurry, J. E. *Acc. Chem. Res.* **1974**, *7*, 281.
63. Ho, T.-L. *Synthesis* **1979**, 1.
64. Lai, Y.-H. *Org. Prep. Proced. Int.* **1980**, *12*, 363.
65. Chum, P. W.; Wilson, S. E. *Tetrahedron Lett.* **1976**, 15.
66. Ashby, E. C.; Lin, J. J. *Tetrahedron Lett.* **1977**, 4481.
67. Sato, F.; Sato, S.; Sato, M. *J. Organomet. Chem.* **1976**, *122*, C25.
68. Sato, M.; Oshima, K. *Chem. Lett.* **1982**, 157.
69. Kataoka, Y.; Takai, K.; Oshima, K.; Utimoto, K. *J. Org. Chem.* **1992**, *57*, 1615.
70. Larock, R. C. *Comprehensive Organic Transformations*; VCH: New York, 1989; pp 35-37 and references cited therein.
71. Augustine, R. L. *Catalytic Hydrogenation*; Marcel Dekker: New York, 1965.
72. Yamamura, S.; Nishiyama, S. *Comprehensive Organic Synthesis*; Fleming, I., Ed.; Pergamon Press: Oxford, 1991; Vol. 8.
73. Ref. 70, *ibidem*; pp 157-160.
74. Ref. 70, *ibidem*; pp 527-528.
75. Huffman, J. W. *Comprehensive Organic Synthesis*; Fleming, I., Ed.; Pergamon Press: Oxford, 1991; Vol. 8.
76. Pradhan, S. K. *Tetrahedron* **1986**, *42*, 6351 and references cited therein.
77. Castaldi, G.; Perdoncin, G.; Giordano, C. *Tetrahedron Lett.* **1983**, *24*, 2487.
78. Castaldi, G.; Perdoncin, G.; Giordano, C. *Angew. Chem. Int. Ed. Engl.* **1985**, *24*, 499.
79. Wang, W. B.; Shi, L. L.; Huang, Y. Z. *Tetrahedron Lett.* **1990**, *31*, 1185.
80. Molander, G. A. *Chem. Rev.* **1992**, *92*, 29 and references cited therein.
81. Luche, J.-L. *J. Am. Chem. Soc.* **1978**, *100*, 2226.
82. Luche, J.-L.; Rodriguez-Hahn, L.; Crabbé, P. *J. Chem. Soc. Chem. Commun.* **1978**, 601.
83. Gemal, A. L.; Luche, J.-L. *J. Am. Chem. Soc.* **1981**, *103*, 5454.
84. Luche, J.-L.; Gemal, A. L. *J. Am. Chem. Soc.* **1979**, *101*, 5848.
85. Gemal, A. L.; Luche, J.-L. *J. Org. Chem.* **1979**, *44*, 4187.

86. Gemal, A. L.; Luche, J.-L. *Tetrahedron Lett.* **1981**, *22*, 4077.
87. Lalancette, J.-M.; Rollin, G.; Dumas, P. *Can. J. Chem.* **1972**, *50*, 3058.
88. Tamarkin, D.; Rabinovitz, M. *Synth. Met.* **1984**, *9*, 125.
89. For a comprehensive treatment of conjugate reductions see: Ref. 70, *ibidem*; pp 8-16 and references cited therein.
90. Contento, M.; Savoia, D.; Trombini, C.; Umani-Ronchi, A. *Synthesis* **1979**, 30.
91. Petrier, C.; Luche, J.-L. *Tetrahedron Lett.* **1987**, *28*, 2347.
92. Petrier, C.; Luche, J.-L. *Tetrahedron Lett.* **1987**, *28*, 2351.
93. Marchand, A. P.; Reddy, G. M. *Synthesis* **1991**, 198.
94. Pinder, A. R. *Synthesis* **1980**, 425.
95. Ref. 70, *ibidem*; pp 18-21 and references cited therein.
96. Ref. 70, *ibidem*; pp 45-67 and references cited therein.
97. Noyori, R.; Hayakawa, Y. *Org. React.* **1983**, *29*, 163.
98. Ref. 3, *ibidem*; pp 399-400 and references therein.
99. Ref. 3, *ibidem*; pp 597-598 and references therein.
100. Fanta, P. E. *Synthesis* **1974**, 9.
101. Sainsbury, M. *Tetrahedron* **1980**, *36*, 3327.
102. Rieke, R. D.; Rhyne, L. D. *J. Org. Chem.* **1979**, *44*, 3445.
103. Cornforth, J.; Sierakowski, A. F.; Wallace, T. W. *J. Chem. Soc. Chem. Commun.* **1979**, 294.
104. Wittek, P. J.; Liao, J. K.; Cheng, C. C. *J. Org. Chem.* **1979**, *44*, 870.
105. Davidson, J. M.; Triggs, C. *Chem. Ind.* **1967**, 1361.
106. Van Helden, R.; Verberg, G. *Recl. Trav. Chim. Pays-Bas* **1965**, *84*, 1263.
107. Kupchan, S. M.; Liepa, A. J.; Rameswaran, V.; Bryan, R. F. *J. Am. Chem. Soc.* **1973**, *95*, 6861.
108. Tsuruya, S.; Kishikawa, Y.; Tanaka, R.; Kuse, T. *J. Catal.* **1977**, *49*, 254.
109. McKillop, A.; Turrell, A. G.; Young, D. W.; Taylor, E. C. *J. Am. Chem. Soc.* **1980**, *102*, 6504.
110. Mukaiyama, T.; Hayashi, M.; Narasaka, K. *Chem. Lett.* **1973**, 291.
111. Tyrlik, S.; Wolochowicz, I. *J. Chem. Soc. Chem. Commun.* **1975**, 781.
112. Nelson, T. R.; Tufariello, J. J. *J. Org. Chem.* **1975**, *40*, 3159.
113. Olah, G. A.; Prakash, G. K. S. *Synthesis* **1976**, 607.
114. McMurry, J. E.; Hoz, T. *J. Org. Chem.* **1975**, *40*, 3797.
115. Ho, T.-L.; Olah, G. A. *Synthesis* **1977**, 170.
116. Okuda, Y.; Hiyama, T.; Nozaki, H. *Tetrahedron Lett.* **1977**, 3829.
117. Semmelhack, M. F.; Ryono, L. S. *J. Am. Chem. Soc.* **1975**, *97*, 3874.
118. Semmelhack, M. F.; Helquist, P.; Jones, L. D.; Keller, L.; Mendelson, L.; Ryono, L. S.; Smith, J. G.; Stauffer, R. D. *J. Am. Chem. Soc.* **1981**, *103*, 6460.

119. Kende, A. S.; Liebeskind, L. S.; Braitsch, D. M. *Tetrahedron Lett.* **1975**, 3375.
120. Zembayashi, M.; Tamao, K.; Yoshida, J.; Kumada, M. *Tetrahedron Lett.* **1977**, 4089.
121. Takagi, K.; Hayama, N.; Sasaki, K. *Bull. Chem. Soc. Jpn.* **1984**, *57*, 1887,
122. Colon, I.; Kelsey, D. R. *J. Org. Chem.* **1986**, *51*, 2627.
123. Colon, I. *J. Org. Chem.* **1982**, *47*, 2622.
124. Imamoto, T. *Rev. Heteroat. Chem.* **1990**, *3*, 87.
125. Imamoto, T.; Takeyama, T.; Kusumoto, T. *Chem. Lett.* **1985**, 1491.
126. Güsten, H.; Horner, L. *Angew. Chem. Int. Ed. Engl.* **1962**, *1*, 455.
127. Gore, P. H.; Hughes, G. K. *J. Chem. Soc.* **1959**, 1615.
128. Rieke, R. D. *Acc. Chem. Res.* **1977**, *10*, 301.
129. Rieke, R. D.; Rhyne, L. D. *J. Org. Chem.* **1979**, *44*, 3445.
130. Ebert, G. W.; Rieke, R. D. *J. Org. Chem.* **1984**, *49*, 5280.
131. Wehmeyer, R. M.; Rieke, R. D. *J. Org. Chem.* **1987**, *52*, 5056.
132. Wu, T.-C.; Wehmeyer, R. M.; Rieke, R. D. *J. Org. Chem.* **1987**, *52*, 5057.
133. Wu, T.-C.; Rieke, R. D. *J. Org. Chem.* **1988**, *53*, 2381.
134. Ebert, G. W.; Rieke, R. D. *J. Org. Chem.* **1988**, *53*, 4482.
135. Ginah, F. O.; Donovan, Jr., T. A.; Suchan, S. D.; Pfennig, D. R.; Ebert, G. W. *J. Org. Chem.* **1990**, *55*, 584.
136. Matsumoto, H.; Inaba, S.; Rieke, R. D. *J. Org. Chem.* **1983**, *48*, 840.
137. Inaba, S.; Matsumoto, H.; Rieke, R. D. *J. Org. Chem.* **1984**, *49*, 2093.
138. Han, B. H.; Boudjouk, P. *Tetrahedron Lett.* **1981**, *22*, 2757.
139. Lash, T. D.; Berry, D. *J. Chem. Educ.* **1985**, *62*, 85.
140. Osborne, A. G.; Glass, K. J.; Staley, M. L. *Tetrahedron Lett.* **1989**, *30*, 3567.
141. Price, G. J.; Clifton, A. A. *Tetrahedron Lett.* **1991**, *32*, 7133.
142. Luche, J.-L.; Einhorn, C.; Einhorn, J.; Sinisterra, J. V. *Tetrahedron Lett.* **1990**, *31*, 4125.
143. Osborne, A. G.; Clifton, A. A. *Monatsh. Chem.* **1991**, *122*, 529.
144. Boudjouk, P.; Thompson, D. P.; Ohrbom, W. H.; Han, B. H. *Organometallics* **1986**, *5*, 1257.
145. Lindley, J.; Lorimer, J. P.; Mason, T. J. *Ultrasonics* **1986**, *24*, 292.
146. Yamashita, T.; Inoue, Y.; Kondo, T.; Hashimoto, H. *Chem. Lett.* **1986**, 407.
147. Chou, T. S.; Ying, J.-J.; Tsao, C.-H. *J. Chem. Res. (S)* **1985**, 18.
148. Chou, T. S.; Tsao, C.-H.; Hung, S. C. *J. Org. Chem.* **1985**, *50*, 4329.
149. Sharpless, K. B. *Chem. Commun.* **1970**, 1450.
150. Sonnet, P. E.; Oliver, J. E. *J. Org. Chem.* **1976**, *41*, 3279.
151. Kupchan, S. M.; Maruyama, M. *J. Org. Chem.* **1971**, *36*, 1187.

152. Sharpless, K. B.; Umbreit, M. A.; Nieh, M. T.; Flood, T. C. *J. Am. Chem. Soc.* **1972**, *94*, 6538.
153. McMurry, J. E.; Fleming, M. P. *J. Org. Chem.* **1975**, *40*, 2555.
154. Van Tamelen, E. E.; Gladysz, J. A. *J. Am. Chem. Soc.* **1974**, *96*, 5290.
155. Alper, H.; DesRoches, D.; Durst, T.; Legault, R. *J. Org. Chem.* **1976**, *41*, 3611.
156. Gladysz, J. A.; Fulcher, J. G.; Togashi, S. *J. Org. Chem.* **1976**, *41*, 3647.
157. Togashi, S.; Fulcher, J. G.; Cho, B. R.; Hasegawa, M.; Gladysz, J. A. *J. Org. Chem.* **1980**, *45*, 3044 and references cited therein.
158. Gladysz, J. A.; Fulcher, J. G.; Togashi, S. *Tetrahedron Lett.* **1977**, 521.
159. Corey, E. J.; Danheiser, R. L.; Chandrasekaran, S. *J. Org. Chem.* **1976**, *41*, 260.
160. McMurry, J. E. *Acc. Chem. Res.* **1983**, *16*, 405.
161. Baumstark, A. L.; McCloskey, C. J.; Tolson, T. J.; Syriopoulos, G. T. *Tetrahedron Lett.* **1977**, 3003.
162. Van Tamelen, E. E.; Schwartz, M. A. *J. Am. Chem. Soc.* **1965**, *87*, 3277.
163. Sharpless, K. B.; Hanzlik, R. P.; Van Tamelen, E. E. *J. Am. Chem. Soc.* **1968**, *90*, 209.
164. Hammond, G. S.; Wyatt, P.; DeBoer, C. D.; Turro, N. J. *J. Am. Chem. Soc.* **1964**, *86*, 2532.
165. McMurry, J. E.; Silvestri, M. *J. Org. Chem.* **1975**, *40*, 2687.
166. Elphimoff-Felkin, I.; Sarda, P. *Org. Synth.* **1977**, *56*, 101.
167. Welch, S. C.; Walters, M. E. *J. Org. Chem.* **1978**, *43*, 2715.
168. Welch, S. C.; Walters, M. E. *J. Org. Chem.* **1978**, *43*, 4797.
169. Drabowicz, J.; Mikolajczyk, M. *Synthesis* **1978**, 138.
170. Drabowicz, J.; Mikolajczyk, M. *Synthesis* **1976**, 527.
171. Kojima, T.; Fujisawa, T. *Chem. Lett.* **1978**, 1425.
172. Ref. 3, *ibidem*; pp 1103-1110.
173. Ref. 5, *ibidem*; chapters 7-8.
174. For reductions of nitrogen-containing organic compounds, see: Ref. 70, *ibidem*; pp 409-415, 421-425, 432-434, 437-438, and references cited therein.
175. Denmark, S. E.; Sternberg, J. A.; Lueoend, R. *J. Org. Chem.* **1988**, *53*, 1251.
176. Denmark, S. E.; Ares, J. J. *J. Am. Chem. Soc.* **1988**, *110*, 4432.
177. Hoffman, C.; Tanke, R. S.; Miller, M. J. *J. Org. Chem.* **1989**, *54*, 3750.
178. Malinowski, M. *Bull. Soc. Chim. Belg.* **1988**, *97*, 51.
179. Van Tamelen, E. E.; Rudler, H.; Bjorklund, C. *J. Am. Chem. Soc.* **1971**, *93*, 3526.

180. Van Tamelen, E. E.; Rudler, H.; Bjorklund, C. *J. Am. Chem. Soc.* **1971**, *93*, 7113.
181. Niznik, G. E.; Walborsky, H. M. *J. Org. Chem.* **1978**, *43*, 2396.
182. Marshall, J. A.; Karas, L. J. *J. Am. Chem. Soc.* **1978**, *100*, 3615.
183. Savoia, D.; Tagliavini, E.; Trombini, C.; Umani-Ronchi, A. *J. Org. Chem.* **1980**, *45*, 3227.
184. Van Tamelen, E. E.; Seeley, D.; Schneller, S. W.; Rudler, H.; Cretney, W. *J. Am. Chem. Soc.* **1970**, *92*, 5251.
185. Vol'pin, M. E.; Shur, V. B. *Organomet. React.* **1970**, *1*, 55.
186. Vol'pin, M. E. *Pure Appl. Chem.* **1972**, *30*, 607.
187. For a review of numerous organic transformations mediated by low-valent titanium reagents: Betschart, C.; Seebach, D. *Chimia* **1989**, *43*, 39. For nitrogen fixation, see pp 41-42 and references cited therein.
188. Kurozumi, S.; Toru, T.; Kobayashi, M.; Ishimoto, S. *Synth. Commun.* **1977**, *7*, 427.
189. Barrett, A. G. M.; Prokopiou, P. A.; Barton, D. H. R. *J. Chem. Soc. Chem. Commun.* **1979**, 1175.
190. Barrett, A. G. M.; Prokopiou, P. A.; Barton, D. H. R.; Boar, R. B.; McGhie, J. F. *J. Chem. Soc. Chem. Commun.* **1979**, 1173.
191. Barton, D. H. R.; Stick, R. V.; Subramanian, R. *J. Chem. Soc. Perkin Trans. 1* **1976**, 2112.
192. Barrett, A. G. M.; Prokopiou, P. A.; Barton, D. H. R. *J. Chem. Soc. Perkin Trans. 1* **1981**, 1510.
193. Ohsawa, T.; Takagaki, T.; Ikehara, F.; Takahashi, Y.; Oishi, T. *Chem. Pharm. Bull.* **1982**, *30*, 3178.
194. Savoia, D.; Trombini, C.; Umani-Ronchi, A. *J. Chem. Soc. Perkin Trans. 1* **1977**, 123.
195. Ellingssen, P. O.; Undheim, K. *Acta Chem. Scand.* **1979**, *33B*, 528.
196. Chou, T.; You, M. *Tetrahedron Lett.* **1985**, *26*, 4495.
197. Chou, T.; You, M. *J. Org. Chem.* **1987**, *52*, 2224.

V. REDUCTIVE CARBONYL COUPLING REACTIONS

A. PINACOL REACTION

1. Classical Methods

The pinacol reductive coupling of aldehydes and ketones is an old transformation in organic synthesis.[1] Griner reported as early as 1892 that unsaturated aldehydes can be coupled to give pinacols with the use of a zinc-copper couple.[2] Early procedures utilized aluminum amalgam at reflux, but yields were lower.[3,4] Also, the coupling of ketones to give pinacols can be accomplished with a variety of metal reducing agents, particularly alkali metals.[5,6]

The Gomberg-Bachmann reaction (reduction by Mg or Mg-MgI_2 mixture of saturated and aromatic ketones)[7,8] is a suitable method for producing pinacols. The reagents of this process are presumably monovalent magnesium species.[9] Similar magnesium-mediated reductions are still utilized in synthesis and have advantages over other classical procedures. Thus, the combination or magnesium-mercury(II) chloride in pyridine is very satisfactory for the reductive dimerization of indanones, tetralones, and benzosuberones (eq. 5.1).[10] In contrast, powdered magnesium reacted too violently, but lithium in ether was almost as satisfactory. Other reducing methods, such as Al/Hg or photochemical reduction were less effective. Low yields were also obtained by employing electrolytic reduction.

Mg-$HgCl_2$-py, -15 °C; HO, OH (5.1)

However, active magnesium metal prepared according to Rieke's method, has been used for the reduction of benzophenone to give benzpinacol.[11]

Recently, the Gomberg-Bachmann reaction has been applied to conjugated enones.[12] These systems can be reduced by other chemical or electrochemical methods to saturated ketones (dihydroketones) and can also dimerize to pinacols, ε-diketones, and γ-ketols (eq. 5.2).

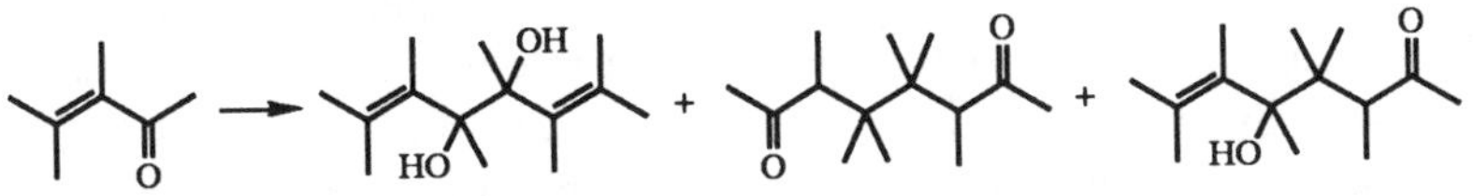

(5.2)

α,β-Enones able to have a *s-cis* conformation can be reduced by Mg or Mg-$MgBr_2$. Reductive coupling of α,β-enones by the $TiCl_4$-Mg reagent has been reported by the same authors.[13,14] Purified powdered magnesium is not efficient and it is necessary to use magnesium turnings of high purity (99.8%) or magnesium turnings in suspension in a solution of $MgBr_2$ in ether. The efficiency of purified powdered magnesium can be improved by addition of mercury(II) chloride (1% w/w) or addition of several transition metal salts (2/1000 w/w). Compounds are mainly reduced to dimers resulting from the formation of a bond between the β-carbon atoms of the α,β-enones. Dihydroketones can sometimes be obtained as by-products. The reductive coupling is very dependent on the structure (*s-cis vs. s-trans*). Thus, (+)-pulegone (a *s-cis* enone) is the more reactive enone leading to menthone and several ketol isomers. On the contrary, (-)-carvone (a *s-trans* enone) is unreactive with magnesium under various conditions. However, (-)-carvone could be reduced to pinacols or γ-ketols by using the $TiCl_4$-Mg reagent.

The mechanism appears to involve the reversible formation of a metallacycle (**2**) *via* the radical **1** (Scheme 5.1). Only *s-cis* enones can lead to the metallacycle **2** and this promotes the formation of the key intermediate **1**. This can be trapped with chlorotrimethylsilane to give the radical **3**. Both compounds **1** and **3** might be precursors of the dihydroketones or dimers, including pinacols.

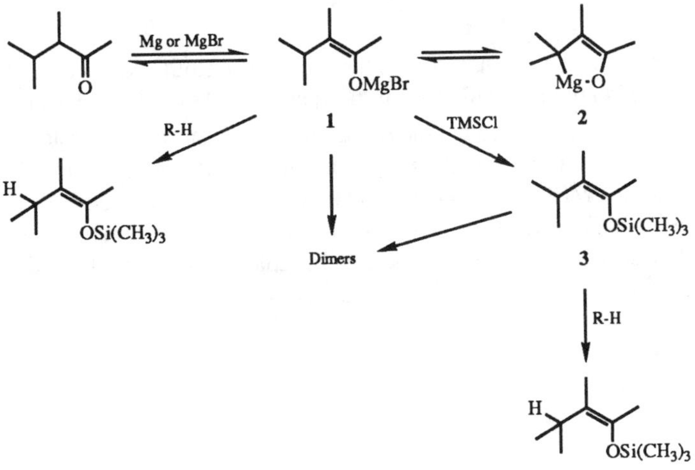

Scheme 5.1

Alkali metal reductions of saturated and unsaturated ketones in protic solvents represent also classical procedures for the preparation of pinacols among other products.[15] The solvent appears to play an important role in the stereoselectivity of the process. A study of camphor pinacolization[16] revealed that the use of Li/THF gave a single pinacol whose stereochemistry was established as *endo:endo* by X-ray analysis. With Li/NH_3 an *exo:endo* pinacol is formed as the sole product if camphor is added to Li/NH_3.

The combination of chlorotrimethylsilane and zinc is another well-known reagent for the reductive coupling of aryl and α,β-unsaturated carbonyl compounds. In a preliminary study[17] a mechanism involving a ketyl-type radical was proposed, which can lead either to pinacol products or alkenes. These must arise from an intermediate organozinc carbenoid (Scheme 5.2).

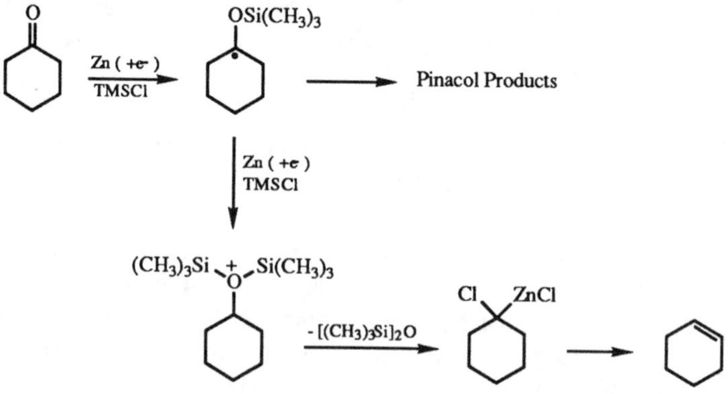

Scheme 5.2

Interestingly, these mechanistic features have been utilized by other groups[18-20] and Zn/TMSCl is now a useful reagent in organic synthesis. In a more recent study, Motherwell *et al.*[21] have extended the scope and limitations of this protocol. Reaction of benzaldehyde by the simultaneous addition method (addition of the carbonyl compound and TMSCl to a stirred suspension of zinc dust in THF) led to benzpinacol (50%) and a mixture of *cis*- and *trans*-stilbene (15%). Further experiments also demonstrated that neither benzpinacol nor its silylated derivatives were precursors of stilbene in these reactions.

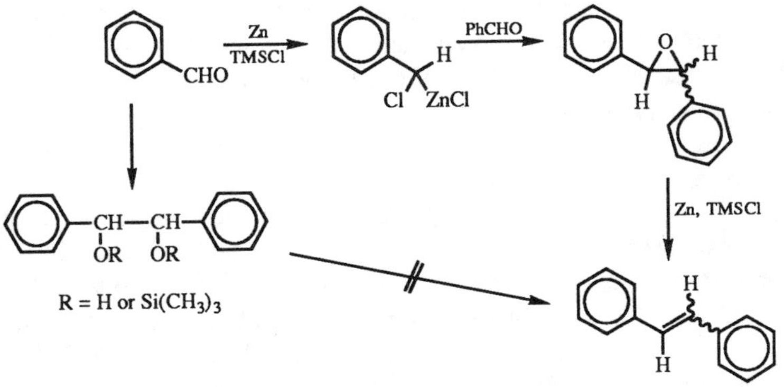

Scheme 5.3

Apparently this deoxygenative coupling reaction does not proceed *via* pinacol intermediates as in the use of low-valent metal reagents.

A possible pathway would therefore be the formation of the intermediate organozinc carbenoid, trapping by a second molecule of carbonyl compound, followed by subsequent deoxygenation of the resulting epoxide (Scheme 5.3). Since the Zn/TMSCl system is compatible with sensitive ester and halogen functions, it is therefore a mild and useful alternative to McMurry-type coupling (see subsection B) in certain cases.

Despite the intrinsic importance of pinacolization as a carbon-carbon bond-forming reaction, mechanistic studies have been very rare. Relevant discussions of this subject have been provided by McMurry[22-24] as part of his investigation of carbonyl-alkene interconversions by using low-valent titanium species. The carbonyl coupling reaction involves two steps: 1) reductive dimerization of the starting carbonyl compound to give a 1,2-diolate (pinacolate) with concomitant formation of the carbon-carbon bond, and 2) deoxygenation of the 1,2-diolate intermediate to yield an olefin (Scheme 5.4).

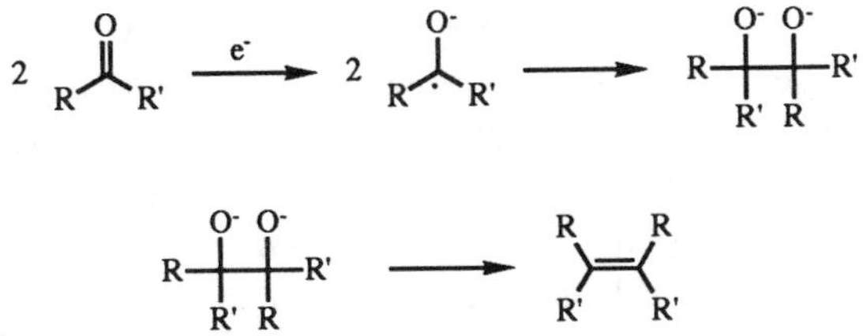

Scheme 5.4

The intermediacy of 1,2-diolates is well established and pinacols can be isolated with high yields when the coupling reaction is performed at 0 °C instead of at solvent reflux temperature. In fact, low-valent metal reagents at 0 °C (especially titanium-induced pinacol reactions) are now regarded as suitable reagents for carrying out pinacol reactions. While pinacolization can be performed with numerous reducing metals,[25,26] which are capable of adding an electron to a carbonyl group yielding an anion radical that dimerizes, the second step to afford alkenes is almost uniquely performed by low-valent titanium.

The stereochemical control, however, is not fully understood and requires a more systematic study. *Threo* and *erythro* mixtures of diols are generally obtained from intermolecular reactions. Moreover, the stereochemistry appears to be dependent on the nature of the reducing metal.[27,28] Nevertheless, McMurry and Rico[29] have found a general tendency toward intramolecular titanium-induced pinacol reactions.

Another very interesting piece of evidence is the demonstration that the two carbon-oxygen bonds of the 1,2-diolate do not break at the same time, and a mixture of *cis*- and *trans*-alkenes can be produced from a diol of known stereochemistry. Nevertheless, the reaction is very close to being concerted.[30] The two oxygen atoms of the pinacol intermediate must have a certain proximity in order to bond to a common titanium surface. But the two oxygens need not be able to bond to the same titanium atom in a five-membered ring, as previously established.[24]

2. Low-Valent Metal Reagents

Interestingly, pinacol coupling can be carried out using aqueous systems and an ammoniacal solution of titanium(III) chloride is effective for this purpose.[31] The more relevant and systematic studies of titanium-induced pinacolization in aqueous media have been reported by Clerici and Porta.[32-34] Aliphatic or aromatic aldehydes or ketones containing strongly electron-withdrawing groups are easily coupled by the action of an acidic titanium(III) chloride solution. These activated carbonyls can also give cross-coupling reactions with unactivated carbonyl compounds. This method is found to be highly effective for the preparation of unsymmetrical pinacols.[32]

Unactivated aromatic aldehydes or ketones can also be coupled with aqueous titanium(III) chloride under alkaline conditions.[33,34] The reducing power of this titanium reagent increases with increasing basicity of the medium.

However, most pinacol-forming reactions are performed by means of heterogeneous reagents in nonaqueous media. In these conditions, many metal systems can be utilized,[25] but only low-valent titanium and vanadium species are particularly effective. Low-valent tungsten was initially employed in pinacol and McMurry reactions.[35] Low-valent vanadium is usually a good reagent for pinacol reactions,[36] and can easily be formed by reduction of vanadium(III) chloride with magnesium (eq. 5.3).[37]

$$(H_3C)_2C{=}CH{-}C({=}O){-}CH_3 \xrightarrow[{}^{t}BuOH\text{-}THF]{VCl_3\text{-}Mg} (H_3C)_2C{=}CH{-}C(OH)(CH_3){-}C(OH)(CH_3){-}CH{=}C(CH_3)_2 \qquad (5.3)$$

Low-valent titanium shows an exceptional power to effect pinacol couplings under mild conditions. The reagent is generated by the reduction of titanium salts or complexes. The titanium sources include the chlorides $TiCl_3$ and $TiCl_4$, as well as the organometallics $CpTiCl_3$ and titanocene dichloride, Cp_2TiCl_2. The reducing agents are commonly alkali metals (Li, Na, or K), Zn, Mg, Mg/Hg, alkyllithiums, or metal hydrides such as $LiAlH_4$ or DIBAH. The following classical combinations should be noted: $TiCl_4$-Zn,[38] $TiCl_3$-Mg,[39] $TiCl_3$-$LiAlH_4$,[22] $TiCl_3$-K,[30,40] $TiCl_4$-Mg/Hg,[41] $TiCl_4$-Zn-pyridine,[42] and $TiCl_3$-C_8K.[43,44] In addition, divalent species from well-characterized Ti(II) organometallic complexes can be useful as well for the reductive carbonyl coupling yielding pinacols.[41,45] Importantly, when carboxylic acids, an alkyllithium reagent, and titanium(III) chloride are mixed, *vic*-diols are obtained in moderate yields.[46] These products are most likely generated by coupling of the ketone hydrates.

Some of these low-valent titanium reagents are very useful in synthesis, especially on account of their stereochemical features. Thus, the combination $TiCl_4$-Zn in THF[47] is a good reagent for the stereoselective synthesis of *erythro*- and *threo*-1,2-diols from diketo sulfides *via cis*-3,4-dihydroxy thiolanes (eq. 5.4).[48]

$$\text{Ar-CO-CH}_2\text{-S-CH}_2\text{-CO-Ar} \xrightarrow[\text{THF, 0 °C; 67-82\%}]{TiCl_4\text{-Zn}} \textit{cis}\text{-3,4-dihydroxy-3,4-diaryl thiolane} \quad (5.4)$$

The low-valent titanium species formed from $TiCl_4$-Zn/Cu converts alkyl glyoxylates into a titanium enediolate that reacts with carbonyl compounds to form α,β-dihydroxy esters (eqs. 5.5, 5.6).[49] The reaction is generally performed at low temperatures (-23 to -45 °C) in dichloromethane-DME, since DME alone promotes self-coupling of the carbonyl compound.

$$BzO_2CCHO + ArCHO \xrightarrow[80\text{-}95\%]{TiCl_4\text{-Zn/Cu},\ CH_2Cl_2\text{-DME}} BzO_2C\text{-}CH(OH)\text{-}CH(OH)\text{-}Ph \quad (5.5)$$

(*syn* / *anti* = 72-82 : 28-18)

$$BzO_2CCHO + PhCOCH_3 \xrightarrow[63\%]{TiCl_4\text{-Zn/Cu},\ CH_2Cl_2\text{-DME}} BzO_2C\text{-}CH(OH)\text{-}C(OH)(CH_3)\text{-}Ph \quad (5.6)$$

(*syn* / *anti* = 30 : 70)

However, the best titanium-based reagent for this reaction was $TiCl_3(DME)_2$-Zn/Cu,[29] which gave higher yields of pinacol products than $TiCl_4$-Mg/Hg or Ti-graphite. Several 1,2-cycloalkanediols are prepared by the titanium-induced pinacol coupling of dialdehydes. A *cis* stereochemistry is favored in six- and eight-membered rings, but in larger rings (sizes ten and above), the *trans* products predominate. Nevertheless, the coupling tends to form cycloalkenes, and in fact small-, medium-, and large-ring cycloalkenes can be obtained in good yields (eq. 5.7).[50]

CHO, CHO —Ti(0)→ [OH, OH] → (5.7)

The intermediate cyclic pinacols were isolated under mild conditions in good yields (75-89%). In addition to the stereochemical discrimination (*cis/trans* ratio) with the ring size, other features should be mentioned. Reducible functional groups such as nitro, epoxide, or sulfoxide were not compatible with low-valent titanium reagents, although nitrile, ester, and amide groups were often preserved during the reaction. Interestingly, α,β-unsaturated aldehydes were easily coupled at -50 °C and so avoided deoxygenation and the undesirable alkene formation (eqs. 5.8-5.10).

CHO, CHO —1) $TiCl_3(DME)_2$, Zn/Cu, 25 °C; 2) K_2CO_3, 85%→ OH, OH (5.8)

(cis/trans = 100:0)

CHO, CHO —1) $TiCl_3(DME)_2$, Zn/Cu, 25 °C; 2) K_2CO_3, 83%→ OH, OH (5.9)

(cis/trans = 5:95)

CHO, CHO —1) $TiCl_3(DME)_2$, Zn/Cu, -50 °C; 2) K_2CO_3, 84%→ OH, OH (5.10)

(cis/trans = 5:95)

A low-valent niobium reagent prepared by the reduction of niobium(V) chloride with $NaAlH_4$ can be utilized in the pinacol-type reductive coupling of aldehydes or ketones.[51] Similarly, aldehydes and ketones can be converted into 1,2-diols with a low-valent tantalum (or niobium) reagent, generated from $TaCl_5$ (or $NbCl_5$) and zinc.[52] Treatment of 3-phenylpropanal and cyclohexanone with $TaCl_5$-Zn in DME-benzene at 25 °C for 10 min afforded the corresponding 1,2-diols in 99% and 82%, respectively (eqs. 5.11, 5.12). Cinnamyl alcohol dimerized with loss of the hydroxyl group to give a mixture of 1,5-dienes in 73% yield by the action of the low-valent tantalum.

$$PhCH_2CH_2CHO \xrightarrow[25\ ^\circ C,\ 10\ min,\ 99\%]{TaCl_5\text{-}Zn,\ DME\text{-}benzene} PhCH_2CH_2-CH(OH)-CH(OH)-CH_2CH_2Ph \quad (5.11)$$

$$\text{cyclohexanone} \xrightarrow[25\ ^\circ C,\ 10\ min,\ 82\%]{TaCl_5\text{-}Zn,\ DME\text{-}benzene} \text{1,1'-bicyclohexyl-1,1'-diol (HO, OH)} \quad (5.12)$$

3. Other Metal-mediated Pinacolic Reactions

Among the wide number of reagents capable of forming pinacol products, some novel and promising reagents deserve to be considered. Among these are magnesium-based reagents such as Mg-TMSCl,[53] and particularly magnesium-graphite[54] which reduces a variety of carbonyl compounds, both aldehydes and ketones to *vic*-diols. The method is rapid, proceeds at room temperature, and is superior to other pinacolic reductions (eq. 5.13). At reflux no change in yields was detected. Aluminum, a powerful one-electron donor[4] was applied in the form of mercury-free aluminum-graphite, but this system was found to be unsuitable for pinacolic reductions.

$$\text{cyclohexanone} \longrightarrow \text{1,1'-bicyclohexyl-1,1'-diol (HO, OH)} \quad (5.13)$$

Method	Temperature	Time (h)	Yield (%)
Al/Hg	reflux	4	55
Mg/Hg	reflux	—	35
$TiCl_4$-Zn	ambient	—	24
Cl-I_2	ambient	14	95
$TiCl_4$-Mg/Hg	0 °C	0.5	93
Mg-C_8K	ambient	4	84

Traditionally, many carbonyl coupling reactions have been carried out with Mg or Al systems, although the metals used most commonly are the early electron-rich transition metals.[26] Lanthanides and actinides, however, have been scarcely utilized despite their very promising electronic characteristics. They exhibit a great oxophilicity, forming strong metal-oxygen bonds. The availability of 4f and 5f orbitals would be of great benefit in some organic transformations, which fail even with most transition metals. Additionally, lanthanides and actinides have large coordination numbers and high kinetic lability. Thus far, only cerium, samarium, and ytterbium species have been employed to form pinacol products. Remarkably, these activated or low-valent metals can be prepared oxidatively from lanthanide metal, whereas most transition metal systems are reductively generated from metal halides.

Low-valent cerium species are effective in synthesizing pinacols from ketones or aldehydes.[55] The method is applicable not only to aromatic but also to aliphatic carbonyl compounds. Ketones and aromatic aldehydes provide excellent yields of coupled products. In contrast, benzophenone was unreactive under the conditions utilized, and cyclododecanone was not converted into the corresponding pinacol but to cyclododecanol in 70% yield. Sensitive functional groups such as esters, nitriles, and alkenyl halides are compatible with this cerium-mediated pinacolization. The authors utilized the Ce-I_2 combination preferentially (eq. 5.14), although activated cerium can be obtained equally by other procedures.

$$\mathrm{RR'C{=}O} \xrightarrow[\text{THF, Ar, 0 °C to rt}]{\text{Ce (2 mmol)},\ I_2\text{ (mmol)}} \mathrm{R'C(R)(OH)-C(R)(OH)R'} \qquad (5.14)$$

Both iodine and 1,2-diiodoethane should serve as entrainers of the metal surface, and presumably form cerium diiodide, more reactive than cerium metal. Similarly, samarium and ytterbium diiodide are easily prepared by the reaction of samarium or ytterbium metal with 1,2-diiodoethane in THF.[56-58] Also, low-valent cerium can be generated by the reaction of cerium with iodobenzene, the reduction of cerium(III) iodide with potassium, or the reaction of cerium with $TiCl_4$. The latter is similar to the reductive coupling with $TiCl_4$-Zn.[38] Good yields are generally obtained with these systems. By contrast, cerium metal or cerium(III) iodide alone gave very poor yields. It has been suggested that low-valent cerium is probably a divalent species,[55] but this point is not absolutely clear in all cases.

Diaryl ketones react with activated ytterbium to give reactive organometallics. Addition of epoxides affords diols (eq. 5.15),[59,60] although side products are diaryl carbinols and the regioisomeric 1,3-diols. Moreover, the protocol is restricted to diaryl ketones alone, so the method has limited synthetic value.

$$\text{Ph}_2\text{C=O} \xrightarrow[\text{rt. 0.5 h}]{\text{Yb, THF-HMPA}} [\text{Ph}_2\text{C(O)Yb}] \xrightarrow{\text{cyclohexene oxide, rt, 2 h, 67\%}} \text{2-(HO)Ph}_2\text{C-cyclohexanol} \quad (5.15)$$

Importantly, the organoytterbium reagent is effective for pinacolic cross-coupling reactions[59,61] which are extremely difficult to achieve. Ytterbium diaryl ketone dianions, formed from ytterbium metal and diaryl ketones, are coupled with aldehydes and ketones to provide *vic*-diols in high yields (eq. 5.16).

$$\text{Ph}_2\text{C=O} \xrightarrow[\text{THF}]{\text{Yb}} \xrightarrow[\text{rt, 15 min, 82\%}]{CH_3CHO} \text{Ph}_2\text{C(OH)-CH(OH)CH}_3 \quad (5.16)$$

In addition to the great number of zinc-based pinacolizations,[25] a new and extremely mild method for coupling aromatic aldehydes and ketones using Zn-$ZnCl_2$ has been recently reported.[62] This system was found to be a useful reagent for the reduction of activated olefins[63] and for the reduction of ketones.[64] The reagent produces *vic*-diols, both in solution and in the solid state.[62] It can be used at room temperature and is not sensitive to oxygen (eqs. 5.17, 5.18).

$$\text{PhCHO} \xrightarrow[\text{50\% aq. THF, rt, 3 h}]{\text{Zn-ZnCl}_2} \underset{39\%}{\text{PhCH}_2\text{OH}} + \underset{11\%}{\text{Ph-CH(OH)-CH(OH)-Ph}} \quad (5.17)$$

$$\text{PhCHO} \xrightarrow[\text{no solvent, 3 h}]{\text{Zn-ZnCl}_2} \underset{\text{trace}}{\text{PhCH}_2\text{OH}} + \underset{46\%}{\text{Ph-CH(OH)-CH(OH)-Ph}} \quad (5.18)$$

Reduction to the corresponding alcohol is a serious limitation, especially with aldehydes in solution. As the water content decreases, the alcohol/pinacol ratio decreases. The pinacol product is favored always in the solid state. This can be attributed to a high-concentration effect, so that the intermolecular reaction would occur more easily to produce mainly the coupling product. The coupling reaction of aromatic ketones is more selective and only *vic*-diols are produced (eq. 5.19). In this case, the reaction in aqueous THF is more effective than in the solid state. Heating for a long time is even necessary in the solid state in many cases.

$$\text{NC-C}_6\text{H}_4\text{-COCH}_3 \xrightarrow[\text{aq. THF, 3 h, rt, 100\%}]{\text{Zn-ZnCl}_2} \text{NC-C}_6\text{H}_4\text{-C(OH)(CH}_3\text{)-C(OH)(CH}_3\text{)-C}_6\text{H}_4\text{-CN} \quad (5.19)$$

B. McMURRY-TYPE REACTIONS

1. Titanium Reagents

The reductive coupling of carbonyl or dicarbonyl compounds to alkenes or cycloalkenes by low- or zero-valent titanium is generally referred to as the McMurry reaction. Since the discovery of this synthetically useful procedure by three independent groups,[22,38,39] several extended reviews[23,24,26,36,65-70] and numerous papers have appeared in the literature. More importantly, this methodology has been found to be particularly useful in synthesizing many natural products.

As previously mentioned, low-valent titanium shows an exceptional ability to effect carbonyl couplings that yield either pinacols or olefins,[23,24,67,68] although other metals can be utilized with greater or less success.[26] Pinacolates are the true intermediates in these titanium-induced transformations, and in fact olefins are invariably obtained from the corresponding diols by heating or after prolonged reaction times.

Notably, the versatility of titanium-mediated carbonyl coupling reactions is not limited to the reductive dimerization of aldehydes and ketones, as other important organic processes can be effected. These include reductions,[36,68] nitrogen fixation,[68] coupling of imine derivatives to ethylenediamines,[68] keto-ester coupling,[71] methylenation, and other olefination reactions.[67,68]

A reagent derived from $TiCl_3$-$LiAlH_4$ was initially prepared by McMurry and Fleming,[22] and it was capable of coupling both aromatic and aliphatic substrates. The Mukaiyama[38] and Tyrlik[39] reagents utilized $TiCl_4$-Zn and $TiCl_3$-Mg, respectively. These low-valent titanium species, however, appeared limited to aromatic cases. The $TiCl_3$-$LiAlH_4$ system is useful in dehalogenations and coupling of organic halides,[36] and Walborsky and Wüst[72] utilized this reagent in the preparation of 1,3-dienes by a 1,4-reductive elimination (eqs. 5.20-5.22).

$TiCl_3$-$LiAlH_4$, THF, 70% (5.20)

$TiCl_3$-$LiAlH_4$, THF, 83% (5.21)

$TiCl_3$-$LiAlH_4$, THF, 62% or 78% (5.22)

In any event, the reductive coupling of carbonyls to olefins with $TiCl_3$-$LiAlH_4$ tends to give erratic yields. McMurry and his associates introduced other low-valent titanium reagents such as $TiCl_3$-K[40] or $TiCl_3$-Li.[73] The resulting active titanium metal was prepared by Rieke's method, that is, by reaction of $TiCl_3$ slurried in THF or DME with potassium or lithium. In general, this Ti(0) gives reproducible results. Yields are somewhat higher with metal prepared with potassium, but lithium is easier to handle. 1,2-Diols are also reduced to olefins (eqs. 5.23-5.25).

2 (cyclohexanone) $\xrightarrow[85\%]{TiCl_3\text{-K, DME}}$ (cyclohexylidenecyclohexane) (5.23)

(1,1'-dihydroxy-bicycloheptyl diol) $\xrightarrow[85\%]{TiCl_3\text{-K, DME}}$ (olefin) (5.24)

2 $^iPr_2C{=}O$ $\xrightarrow[40\%]{TiCl_3\text{-K, DME}}$ $^iPr_2C{=}C^iPr_2$ (5.25)

The method can be employed for the preparation of unsymmetrical olefins, if one component is used in excess. Highest yields are obtained for diaryl ketones (eq. 5.26). In some cases mixed coupling can occur with an equimolar mixture of carbonyls (eq. 5.27).[73]

$Ph_2C{=}O$ + $(H_3C)_2C{=}O$ (4 : 1) $\xrightarrow[94\%]{TiCl_3\text{-K, DME}}$ $Ph_2C{=}C(CH_3)_2$ (5.26)

$Ph_2C{=}O$ + (cyclohexanone) (1 : 1) $\xrightarrow{TiCl_3\text{-K, DME}}$ $Ph_2C{=}$(cyclohexylidene) (78%) + $Ph_2C{=}CPh_2$ (19%) + (cyclohexylidenecyclohexane) 6% (5.27)

Interestingly, the coupling of 1-methyl-2-adamantanone with Ti(0), prepared from $TiCl_3$-K, results in the formation of the highly hindered *trans*-1-methyl-2-

adamantylidene-1-methyladamantane (4) in 52% yield.[74] The *cis*-isomer is not formed because it has much higher energy. This reductive coupling also fails with highly strained olefins.[67]

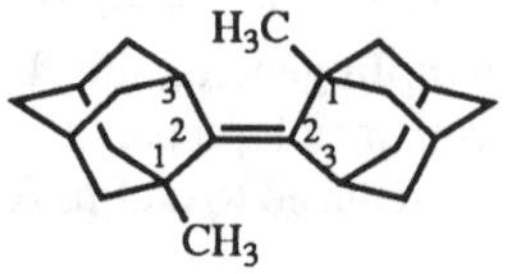

4

Further studies of this reagent[30] demonstrated that the optimum ratio of $TiCl_3$/K/ketone is 4:14:1. Intramolecular dicarbonyl coupling to cyclic alkenes is also possible; in this case the combination of $TiCl_3$ with a Zn/Cu couple in DME is the most efficient reagent.[50] The importance of this cyclization is that yields are satisfactory, even excellent, for all ring sizes unlike the Thorpe-Ziegler or acyloin cyclizations, in which yields are low in the formation of seven- to eleven-membered rings (eqs. 5.28-5.30).

However, early studies from various research groups reported low yields and unreproducible results with these low-valent titanium reagents in some cases. It was clear that the reactivity of the reagent depended on its method of preparation. Although the reagent prepared from $TiCl_3$-Zn/Cu is highly reactive and gives reproducible results, a further modification using the reduction of the crystalline blue solvate $TiCl_3(DME)_{1.5}$ with zinc-copper couple affords an extraordinary active Ti(0) reagent, which allows most couplings in very high yields with excellent reproducibility.[29,75] With this reagent, the reaction takes place equally well on many substrates, be they saturated or unsaturated, aliphatic or aromatic, ketone or aldehyde. Furthermore, the reaction can be applied in an inter- or intramolecular sense to afford acyclic or cyclic alkenes, respectively. Importantly, cycloalkenes are obtained with a preferential (*E*)-configuration.

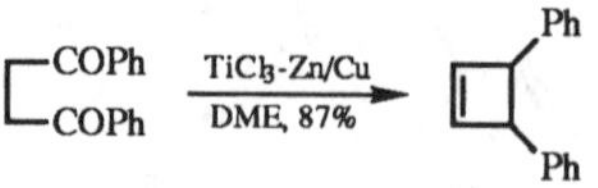

COPh / COPh —($TiCl_3$-Zn/Cu, DME, 87%)→ Ph / Ph (5.28)

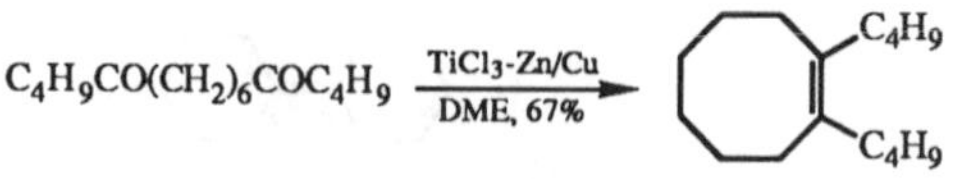

$C_4H_9CO(CH_2)_6COC_4H_9$ —($TiCl_3$-Zn/Cu, DME, 67%)→ C_4H_9 / C_4H_9 (5.29)

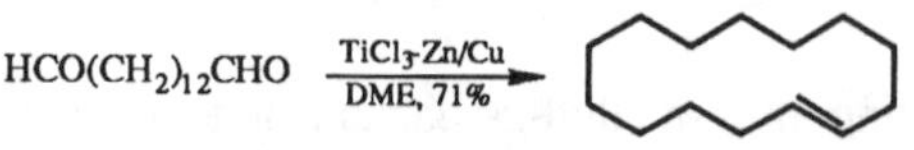

$HCO(CH_2)_{12}CHO$ —($TiCl_3$-Zn/Cu, DME, 71%)→ (5.30)

Other modifications such as $TiCl_4$-Zn-pyridine[42] and the reactive titanium-graphite,[43,44] generated by reduction of $TiCl_3$ with potassium-graphite (eq. 5.31), have also been useful in some cases.

$$\text{Graphite} + \text{K} \xrightarrow[\text{Ar, 15 min}]{150\ ^\circ\text{C}} \text{C}_8\text{K} \xrightarrow[\text{Ar, reflux, 2 h}]{\text{TiCl}_3\text{, THF}} \text{Ti-C}_8\text{K} \qquad (5.31)$$

Thus, titanium-graphite promotes the reductive coupling of carbonyl compounds to alkenes in refluxing THF with good yields (eq. 5.32).[43]

$$\text{PhC(O)CH}_3 \xrightarrow[\text{THF, reflux, Ar, 12 h; 86\%}]{\text{Ti-C}_8\text{K (4 equiv)}} \text{Ph(H}_3\text{C)C=C(CH}_3\text{)Ph} \quad (E/Z = 2{:}1) \qquad (5.32)$$

Again, it should be emphasized that the reagents are a form of low-valent titanium. From the stoichiometry of the reduction, it is possible to deduce formal oxidation states for the reagent. Studies by Geise *et al.*[76-78] confirm that the active species obtained by reduction of $TiCl_3$ with excess of Li, K, or Mg is Ti(0), but suggest that the active species generated by reduction of $TiCl_3$ with $LiAlH_4$ (2:1) is Ti(I). This ratio was found to be optimum for the coupling, and a slight excess of titanium(III) is required because of impurities in $TiCl_3$. Anyway, Ti(I) species would provide a mixture of Ti(0) and Ti(II) by disproportionation. The latter is probably a more real situation in the reaction mixture. Geise and co-workers[76-78] have shown that the active species is almost Ti(0), and if the $TiCl_3$ is completely reduced prior to coupling, the method used is not important. In addition, Betschart and Seebach[68] have reviewed from 1971 to 1989 the reagents obtained by reduction of a number of Ti(III) and Ti(IV) salts with a variety of reductants. All are probably forms of metallic, soluble Ti(0), and have similar potential for organic synthesis. The fact that some reagents appear to be ineffective for the McMurry coupling indicates complex redox processes in solution, where one would expect the participation of Ti(0) and Ti(II) species among others.

2. Scope and Limitations

In recent years, the exhaustive studies of carbonyl coupling reactions by McMurry[23,24] and others[26,36,67,68] have made possible to gain an insight into the mechanism of the McMurry reaction,[24] and to ascertain the scope, limitations, and possible developments of this outstanding procedure.

A crucial point has been the exact knowledge of functional-group compatibility. Thus, acetal, alcohol (with the exception of allylic alcohol), alkene, alkyl silane, amine, ether (with the exception of epoxides), halide (but not 1,2-dihalide), sulfide, and vinyl silane groups are compatible with the titanium-induced carbonyl coupling, and generally they remain unaffected. Depending on experimental conditions and particularly on reaction times, other important functional groups such as alkyne, amide, ester, or ketone appear to be semicompatible.[24,67,68]

In contrast, several functional groups are completely incompatible and undergo rapid reduction with low-valent titanium or vanadium (see Chapter IV).[36,68,79] This feature can be exploited for the reduction of these groups in preparative syntheses. Allylic and benzylic alcohols are rapidly coupled to afford 1,5-hexadienes and bibenzyls.[80-82] 1,2-Diols (pinacols) are easily reduced to alkenes as previously evidenced,[30] and epoxides are also converted into alkenes.[83] 1,3-Diols undergo reductive coupling to cyclopropanes,[84-86] enediones and quinones are reduced to saturated diketones and hydroquinones, respectively,[87] and halohydrins are converted to alkenes.[88] Aryl *gem*-dihalides are reductively coupled to olefins,[81] and α-halo ketones are dehalogenated to yield ketones.[89] Aromatic nitro compounds are reduced to anilines,[90] and aliphatic nitro compounds[91] and oximes[92] are transformed into imines. Sulfoxides can be reduced to yield sulfides.[93]

Ti (0)

n = 1,2,3

(5.33)

As mentioned, intramolecular coupling of keto esters[71] is a useful variation of this titanium-mediated reaction, and gives cycloalkanones (eq. 5.33). It only works well with five-, six-, and seven-membered rings. The keto-ester coupling has been employed in syntheses of natural products such as capnellene[94] and isocaryophyllene.[95] Likewise, an enone-aldehyde coupling has been recently reported[96] that describes the synthesis of the ring system of mevinolin (eq. 5.34), a fungal metabolite used for treatment of hypercholesterolemia.

$TiCl_3$ - C_8K, DME
86%

(5.34)

The reagent was titanium-graphite and the McMurry coupling was highly sensitive to the $C_8K/TiCl_3$ ratio and to the relative amount of substrates and titanium reagent, but consistently high yields can be obtained by use of a large excess of the low-valent titanium generated from $TiCl_3$ and C_8K in the ratio 2:1.

Aromatic acyloxycarbonyl and acylamidocarbonyl compounds can be readily cyclized upon treatment with titanium-graphite (4 equiv.) to give (benzo)furans and indoles (eqs. 5.35, 5.36).[97]

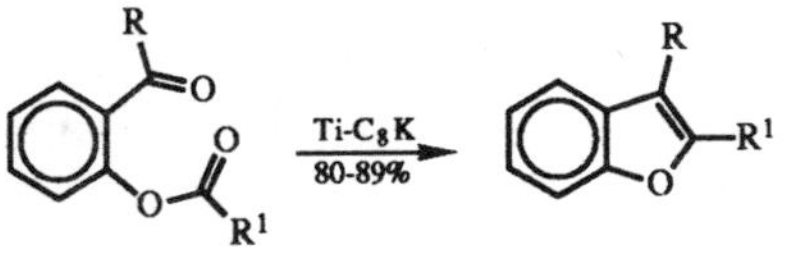

Ti-C_8K, 80-89% (5.35)

R = H, CH_3, Ph
R^1 = CH_3, Ph

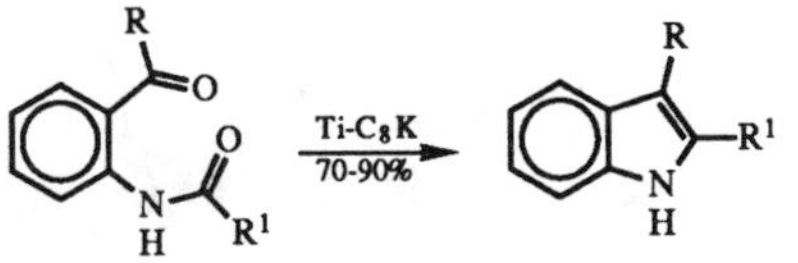

Ti-C_8K, 70-90% (5.36)

R = R^1 = CH_3, Ph

The yield increases with the number of aromatic rings in the substrates. Surprisingly, all attempts to extend this strategy to the synthesis of non-aromatic heterocycles resulted in acyl migration on hydrolysis. The method constitutes another important extension of the original keto-ester cyclization.[71]

The reductive elimination of allylic diols is a convenient and stereoselective method for the synthesis of optically active trienols from optically active α-hydroxy esters.[98-100] The protocol is very attractive since many natural products contain a chiral allylic trienol unit, wherein the conjugated triene has either an all-*trans* or a *trans-trans-cis* configuration (eqs. 5.37, 5.38).

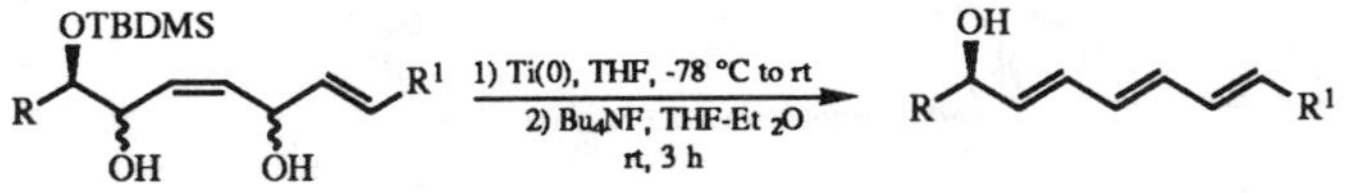

1) Ti(0), THF, -78 °C to rt; 2) Bu_4NF, THF-Et_2O, rt, 3 h (5.37)

1) Ti(0), THF, -78 °C to rt; 2) Bu_4NF, THF-Et_2O, rt, 3 h (5.38)

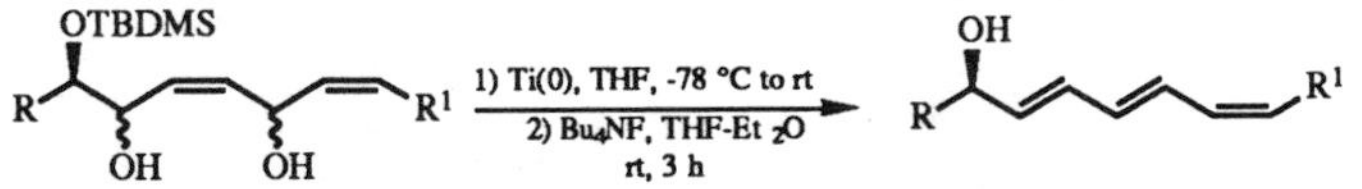

The low-valent titanium reagent was prepared by reduction of $TiCl_3$ (4 equiv.) and $LiAlH_4$ (2 equiv.) in THF.

In connection with the synthesis of natural products, low-valent titanium coupling has been successfully used for the preparation of numerous compounds.[24,67] Also, the pinacolization and direct coupling to olefins have been utilized for the synthesis of novel, fascinating unusual molecules such as the betweenanenes, circulenes, cyclic tetrapyrroles, or ferrocene cyclophanes.[24,67,68] Remarkably, the keto-aldehyde carbonyl coupling reaction has enabled the preparation of *in*-bicyclo[4.4.4]tetradecane,[75] a precursor of *in,out*-bicyclo[4.4.4]tetradecane and of the *in*-bicyclo[4.4.4]-1-tetradecyl cation. This substance is the first stable organic compound containing a three-center, two-electron C-H-C bond.[101]

N-Alkyl imines can be coupled with $TiCl_4$-Mg-$HgCl_2$ to give substituted ethylenediamines (eq. 5.39).[102,103]

2 [N-alkyl imine] —Ti (0)→ [HN NH diamine] (5.39)

N-Silyl imines derived from aromatic aldehydes are efficiently coupled with $TiCl_4$-Mg to yield after hydrolysis, primary ethylendiamines (eq. 5.40).[104]

2 Ar-C=N-Si$(CH_3)_3$ —$TiCl_4$- Mg, 50-79%→ Ar-CH(NH_2)-CH(NH_2)-Ar (5.40)

meso / *d,l* > 1

Similarly, the low-valent reagent generated from $TiCl_4$-Mg system is also useful to obtain ethylenediamines from either tertiary amines or iminium salts derived from aromatic aldehydes (eq. 5.41).[68,104,105]

Interestingly, enantiomerically pure, cyclic diaryl amines can be obtained with a related methodology starting from α-amino acids (eq. 5.42).[68,104]

(5.41)

X = H, CH_3, OCH_3, Cl, Br, CN

> 90%
meso / *d,l* ~1:1

(5.42)

1) n-BuLi
2) ArCHO
3) $TiCl_4$
4) $TiCl_4$ - Mg
5) H^+, workup

Other important dimerization processes have been carried out with low-valent titanium species.[68] The combination Cp_2TiCl_2-Mg has been utilized with several carbonyl functions.[106] The behavior of α,β-unsaturated aldehydes should be noted. At low temperatures (-78 °C) pinacol reaction takes place, whereas at room temperature the reductive dimerization leads to bis(allyl)ethers (Scheme 5.5).

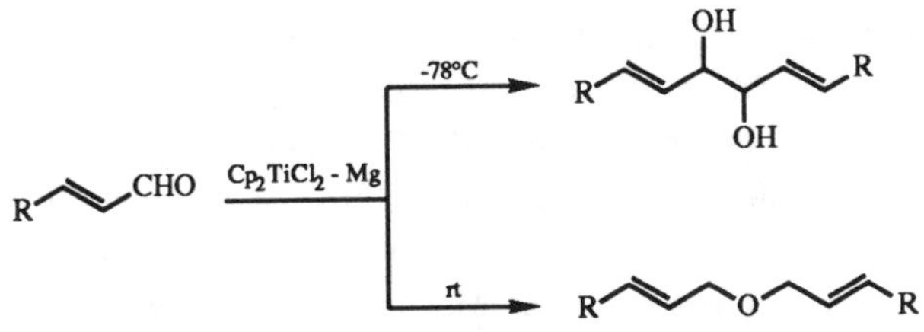

Scheme 5.5

Phenyl isocyanate and phenyl isothiocyanate react with Grignard reagents in the presence of Cp_2TiCl_2 at room temperature to afford reduction-coupling products.[107] The corresponding *N*-methyl-*N*,*N'*-diphenylurea or thiourea were isolated with no or little addition product (eq. 5.43).

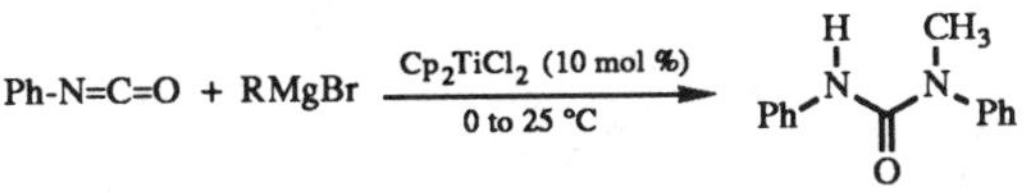

$$\text{Ph-N=C=O} + \text{RMgBr} \xrightarrow[\text{0 to 25 °C}]{Cp_2TiCl_2\ \text{(10 mol \%)}} \text{PhNH-C(=O)-N(CH}_3\text{)Ph} \qquad (5.43)$$

3. McMurry Coupling with Metals Other than Titanium

Low-valent titanium effects carbonyl couplings very smoothly and it is with this reagent that most reactions have been performed. Curiously, the first report on a McMurry-type coupling was a paper by Schreibmann[4] in 1970. This author isolated alkenes from the reaction of benzophenone, acetophenone, and benzaldehyde with Al/Hg, though yields were not given.

Tungsten is, along with titanium, the metal utilized in reductive ketone-coupling reactions, but low-valent tungsten is considerably less effective than titanium reagents. Sharpless *et al.*[35] employed a combination of tungsten(VI) chloride and butyl lithium for this reaction. A low-valent tungsten generated from WCl_6-$LiAlH_4$ promotes the coupling as well. However, experimental evidences indicated that the reductive coupling could proceed *via* a carbene intermediate. Indeed, a Fischer carbene complex effected the coupling of benzaldehyde to stilbene in 42% yield.[108] Remarkably, a true low-valent tungsten may be formed by electroreduction of tungsten(VI) chloride. The metal species effected reductive coupling of benzaldehyde to stilbene quantitatively.[109] The coupling is quite solvent dependent and works well in THF, DMSO, or DMF. Instead, dichloromethane gave styrene and probably this solvent produces carbenes under these electrochemical conditions.[110] The method is limited to aromatic systems and thus aliphatic aldehydes gave low yields of products. Nevertheless, the yields can be enhanced when the aldehyde is present during the electroreduction of WCl_6.

A comparative study of the effectiveness of low-valent metals on the coupling of benzophenone to tetraphenylethylene was reported by Geise *et al.*[111] A number of metal halides were reduced with $LiAlH_4$ (with the exception of $MgCl_2$ and $ZnCl_2$ which were reduced with potassium), and then utilized as reagents in the ketone-olefin coupling. Reagents derived either from $TiCl_3$ or $TiCl_4$ gave a high yield (> 90%) of tetraphenyl ethylene, whereas V, Nb, Zr, Mo, and W species gave very modest yields, in the range 8-30%. The rest of the reduced metals were ineffective and no product could be isolated.

Mention should be made of a tantalum-assisted process, in which alkynes are converted into allylic alcohols.[112] A low-valent tantalum reagent prepared by reduction of $TaCl_5$ with zinc in DME/benzene adds to alkynes to form a complex that reacts with aldehydes to form (*E*)-allylic alcohols (eq. 5.44). The regioselectivity is determined by the bulkiness of the groups on the alkyne and of the substituents of the aldehyde.

$$n\text{-}C_5H_{11}C{\equiv}C\text{-}C_5H_{11}\text{-}n \xrightarrow[\text{DME-benzene}]{TaCl_5\text{-}Zn} \left[n\text{-}C_5H_{11}\text{-}C{=}C\text{-}C_5H_{11}\text{-}n \; (TaLn) \right] \xrightarrow[\substack{2)\ NaOH,\ H_2O \\ 96\%}]{1)\ Ph(CH_2)_2CHO} (n\text{-}C_5H_{11})HC{=}C(C_5H_{11}\text{-}n)CH(OH)(CH_2)_2Ph \quad (5.44)$$

Reduction of uranium(IV) chloride in DME by sodium-potassium alloy and a catalytic amount of naphthalene (5-10% based on Na/K alloy) yields a black slurry of activated uranium.[113] Reaction of this active uranium with allyl iodide at room temperature produces 1,5-hexadiene quantitatively, and benzophenone in refluxing DME yields up to 50% tetraphenyl ethylene (eqs. 5.45, 5.46).

$$CH_2{=}CHCH_2I \xrightarrow[\sim 100\%]{U^*,\ DME,\ rt} CH_2{=}CHCH_2CH_2CH{=}CH_2 \quad (5.45)$$

$$2\ PhC(O)Ph \xrightarrow[>50\ °C]{U^*,\ DME,\ reflux} Ph_2C{=}CPh_2 \quad (5.46)$$

A more convenient procedure uses lithium naphthalenide-TMEDA as reducing agent in hydrocarbon solvents.[114] This method is now preferred since active uranium reacts with oxygenated solvents such as THF or DME. Activated uranium (and thorium) couples aryl ketones at 70 °C to afford tetraaryl ethylenes. At higher temperatures (140 °C) the formation of metal hydride species occurs and tetraaryl ethanes can be also isolated.

It has already been stated that vaporized metals can produce pinacols and olefins by co-condensation reactions at low temperatures (Chapter II). The reaction is, from a practical viewpoint, of small interest and products are usually isolated in low yields.[26,115,116] Magnesium atoms effect the coupling of cycloalkyl ketones to yield pinacols. Carbenoid species are presumably involved in this process. Other metal vapors, Ti, Cr, Co, Ni, Nd, and U were co-condensed with cyclohexanone. Products of ketone-olefin coupling were hardly detected, and only titanium vapors gave pinacol in good yield (71%). The rest of elements favored aldol products, particularly Co, Cr, Nd, and U. Moreover, Co and Ni yielded no appreciable amounts of pinacol.

References

1. Fittig, R. *Liebigs Ann. Chem.* **1859**, *110*, 23.
2. Griner, *Ann. Chim. Phys.* **1892**, *26*, 369.
3. Böeseken, J.; Cohen, W. D. *Z. Chem.* **1915**, *I*, 1375.
4. Schreibmann, A. A. P. *Tetrahedron Lett.* **1970**, 4271.
5. For classical procedures of the pinacol reaction: Kropf, H.; Thiem, J. *Houben Weyl-Methoden der organischen Chemie*; Georg Thieme Verlag: Stuttgart, 1973; Vol. 6, chapter IB, 2.
6. Adams, R.; Adams, E. W. *Org. Synth. Coll. Vol. I* **1941**, 459.
7. Gomberg, M.; Bachmann, W. E. *J. Am. Chem. Soc.* **1927**, *49*, 236.
8. For a concise and good treatment on the Gomberg-Bachmann reaction: March, J. *Advanced Organic Chemistry*; Wiley: New York, 1985; pp 1110-1112.
9. For a review on reductions involving unipositive magnesium: Rausch, M. D.; McEwen, W. E.; Kleinberg, J. *Chem. Rev.* **1957**, *57*, 417.
10. Majerus, G.; Yax, E.; Ourisson, G. *Bull. Soc. Chim. Fr.* **1967**, 4143.
11. Rieke, R. D. *Top. Curr. Chem.* **1975**, *59*, 1.
12. Pons, J.-M.; Santelli, M. *Tetrahedron Lett.* **1988**, *29*, 3679.
13. Pons, J.-M.; Santelli, M. *Tetrahedron Lett.* **1986**, *27*, 4153.
14. Pons, J.-M.; Santelli, M. *J. Org. Chem.* **1989**, *54*, 877 and references cited therein.
15. Pradham, S. K. *Tetrahedron* **1986**, *42*, 6351.
16. Pradham, S. K.; Thakker, K. R.; McPhail, A. T. *Tetrahedron Lett.* **1987**, *28*, 1813.
17. Motherwell, W. B. *J. Chem. Soc. Chem. Commun.* **1973**, 935.
18. Hodge, P.; Khan, M. N. *J. Chem. Soc. Perkin Trans. 1* **1975**, 809.
19. Smith, C. L.; Arnett, J.; Ezike, J. *Tetrahedron Lett.* **1980**, *21*, 653.
20. Corey, E. J.; Pyne, S. G. *Tetrahedron Lett.* **1983**, *24*, 2821.

21. Banerjee, A. K.; Sulbaran de Carrasco, M. C.; Frydrych-Houge, C. S. V.; Motherwell, W. B. *J. Chem. Soc. Chem. Commun.* **1986**, 1803.
22. McMurry, J. E.; Fleming, M. P. *J. Am. Chem. Soc.* **1974**, *96*, 4708.
23. McMurry, J. E. *Acc. Chem. Res.* **1983**, *16*, 405.
24. McMurry, J. E. *Chem. Rev.* **1989**, *89*, 1513.
25. For an excellent account on pinacolization reagents: Larock, R. C. *Comprehensive Organic Transformations*; VCH: New York, 1989; pp 547-548 and references cited therein.
26. Kahn, B. E.; Rieke, R. D. *Chem. Rev.* **1988**, *88*, 733.
27. Mundy, B. P.; Srinivasa, R.; Kim, Y.; Dolph, T.; Warnet, R. J. *J. Org. Chem.* **1982**, *47*, 1657.
28. Mundy, B. P.; Bruss, D. R.; Kim, Y.; Larsen, R. D.; Warnet, R. J. *Tetrahedron Lett.* **1985**, *26*, 3927.
29. McMurry, J. E.; Rico, J. G. *Tetrahedron Lett.* **1989**, *30*, 1169; *ibidem* 1173.
30. McMurry, J. E.; Fleming, M. P.; Kees, K. L.; Krepski, L. R. *J. Org. Chem.* **1978**, *43*, 3255.
31. Karrer, P.; Yen, Y.; Reichstein, I. *Helv. Chim. Acta* **1930**, *13*, 1308.
32. Clerici, A.; Porta, O. *J. Org. Chem.* **1989**, *54*, 3872 and references cited therein.
33. Clerici, A.; Porta, O. *Tetrahedron Lett.* **1982**, *23*, 3517.
34. Clerici, A.; Porta, O. *J. Org. Chem.* **1985**, *50*, 76.
35. Sharpless, K. B.; Umbreit, M. A.; Nieh, M. T.; Flood, T. C. *J. Am. Chem. Soc.* **1972**, *94*, 6538.
36. Pons, J.-M.; Santelli, M. *Tetrahedron* **1988**, *44*, 4295 and references cited therein.
37. Pons, J.-M.; Zahra, J. P.; Santelli, M. *Tetrahedron Lett.* **1981**, *22*, 3965. Recently, Pedersen and Park have introduced a low-valent vanadium-mediated pinacol cross-coupling reaction to afford 1,2-diols, but authors suggest the intermediacy of vanadium(II) ions: Park, J.; Pedersen, S. F. *Tetrahedron* **1992**, *48*, 2069.
38. Mukaiyama, T.; Sato, T.; Hanna, J. *Chem. Lett.* **1973**, 1041.
39. Tyrlik, S.; Wolochowicz, M. I. *Bull. Soc. Chim. Fr.* **1973**, 2147.
40. McMurry, J. E.; Fleming, M. P. *J. Org. Chem.* **1976**, *41*, 896.
41. Corey, E. J.; Danheiser, R. L.; Chandrasekaran, S. *J. Org. Chem.* **1976**, *41*, 260.
42. Lenoir, D. *Synthesis* **1977**, 553.
43. Boldrini, G. P.; Savoia, D.; Tagliavini, E.; Trombini, C.; Umani-Ronchi, A. *J. Organomet. Chem.* **1985**, *280*, 307.
44. Fürstner, A.; Weidmann, H. *Synthesis* **1987**, 1071.
45. Chen, T. L.; Chan, T. H.; Shaver, A. *J. Organomet. Chem.* **1984**, *268*, C1.

46. Axelrod, E. H. *J. Chem. Soc. Chem. Commun.* **1970**, 451.
47. Nakayama, J.; Machida, H.; Saito, R.; Hoshino, M. *Tetrahedron Lett.* **1985**, *26*, 1983.
48. Nakayama, J.; Yamaoka, S.; Hoshino, M. *Tetrahedron Lett.* **1987**, *28*, 1799.
49. Mukaiyama, T.; Sugimura, H.; Ohno, T.; Kobayashi, S. *Chem. Lett.* **1989**, 1401.
50. McMurry, J. E.; Kees, K. L. *J. Org. Chem.* **1977**, *42*, 2655.
51. Sato, M.; Oshima, K. *Chem. Lett.* **1982**, 157.
52. Kataoka, Y.; Takai, K.; Oshima, K.; Utimoto, K. *J. Org. Chem.* **1992**, *57*, 1615.
53. Chan, T. H.; Vinokur, E. *Tetrahedron Lett.* **1972**, 75.
54. Csuk, R.; Fürstner, A.; Weidmann, H. *J. Chem. Soc. Chem. Commun.* **1986**, 1802.
55. Imamoto, T.; Kusumoto, T.; Hatanaka, Y.; Yokoyama, M. *Tetrahedron Lett.* **1982**, *23*, 1353.
56. Girard, P.; Namy, J. L.; Kagan, H. B. *J. Am. Chem. Soc.* **1980**, *102*, 2693.
57. Kagan, H. B.; Namy, J. L.; Girard, P. *Tetrahedron* **1981**, *37*, 175.
58. Girard, P.; Couffignal, R.; Kagan, H. B. *Tetrahedron Lett.* **1981**, *22*, 3959.
59. Hou, Z.; Takamine, K.; Aoki, O.; Shiraishi, H.; Fujiwara, Y.; Taniguchi, H. *J. Chem. Soc. Chem. Commun.* **1988**, 668.
60. Takaki, K.; Tanaka, S.; Beppu, F.; Tsubaki, Y.; Fujiwara, Y. *Chem. Lett.* **1990**, 1427.
61. Hou, Z.; Takamine, K.; Aoki, O.; Shiraishi, H.; Fujiwara, Y.; Taniguchi, H. *J. Org. Chem.* **1988**, *53*, 6077.
62. Tanaka, K.; Kishigami, S.; Toda, F. *J. Org. Chem.* **1990**, *55*, 2981.
63. Toda, F.; Iida, K. *Chem. Lett.* **1978**, 695.
64. Toda, F.; Tanaka, K.; Tange, H. *J. Chem. Soc. Perkin Trans. 1* **1989**, 1555.
65. Welzel, P. *Nachr. Chem. Tech. Lab.* **1983**, *31*, 814.
66. Dang, Y.; Geise, H. J. *Janssen Chim. Acta* **1989**, *6*, 3.
67. Lenoir, D. *Synthesis* **1989**, 883.
68. Betschart, C.; Seebach, D. *Chimia* **1989**, *43*, 39.
69. Ref. 25, *ibidem*; pp 160-162 and references cited therein.
70. Ho, T.-L. *Polarity Control for Synthesis*; Wiley: New York, 1991; p 56.
71. McMurry, J. E.; Miller, D. D. *J. Am. Chem. Soc.* **1983**, *105*, 1660.
72. Walborsky, H. M.; Wüst, H. H. *J. Am Chem. Soc.* **1982**, *104*, 5807.
73. McMurry, J. E.; Krepski, L. R. *J. Org. Chem.* **1976**, *41*, 3929.
74. Lenoir, D.; Frank, R. *Tetrahedron Lett.* **1978**, 53.
75. McMurry, J. E.; Lectka, T.; Rico, J. G. *J. Org. Chem.* **1989**, *54*, 3748.

76. Dams, R.; Malinowski, M.; Geise, H. J. *Bull. Soc. Chim. Belg.* **1981**, *90*, 1141.
77. Dams, R.; Malinowski, M.; Westdorp, I.; Geise, H. J. *J. Org. Chem.* **1982**, *47*, 248.
78. Dams, R.; Malinowski, M.; Geise, H. J. *Transition Met. Chem.* **1982**, *7*, 37.
79. McMurry, J. E. *Acc. Chem. Res.* **1974**, *7*, 281.
80. McMurry, J. E.; Silvestri, M. *J. Org. Chem.* **1975**, *40*, 2687.
81. Olah, G. A.; Prakash, G. K. S. *Synthesis* **1976**, 607.
82. Ho, T.-L.; Olah, G. A. *Synthesis* **1977**, 170.
83. McMurry, J. E.; Fleming, M. P. *J. Org. Chem.* **1975**, *40*, 2555.
84. Rodewald, L. B.; DePuy, C. H. *Tetrahedron Lett.* **1964**, 2951.
85. Van Tamelen, E. E.; Schwartz, M. A. *J. Am. Chem. Soc.* **1965**, *87*, 3277.
86. Sharpless, K. B.; Hanzlik, R. P.; Van Tamelen, E. E. *J. Am. Chem. Soc.* **1968**, *90*, 209.
87. Blaszczak, L. C.; McMurry, J. E. *J. Org. Chem.* **1974**, *39*, 258.
88. McMurry, J. E.; Hoz, T. *J. Org. Chem.* **1975**, *40*, 3797.
89. Ho, T.-L.; Wong, C. M. *Synth. Commun.* **1973**, *3*, 237.
90. Knecht, E.; Hilbert, E. *Ber. dtsch. chem. Ges.* **1903**, *36*, 166.
91. McMurry, J. E.; Melton, J. *J. Org. Chem.* **1973**, *38*, 4367.
92. Timms, G. H.; Wildsmith, E. *Tetrahedron Lett.* **1971**, 195.
93. Ho, T.-L.; Wong, C. M. *Synth. Commun.* **1973**, *3*, 37.
94. Iyoda, M.; Kushida, T.; Kitami, S.; Oda, M. *J. Chem. Soc. Chem. Commun.* **1987**, 1607.
95. McMurry, J. E.; Miller, D. D. *Tetrahedron Lett.* **1983**, *24*, 1885.
96. Clive, D. L. J.; Murthy, K. S. K.; Wee, A. G. H.; Prasad, J. S.; DaSilva, G. V. J.; Majewski, M.; Anderson, P. C.; Evans, C. F.; Haugen, R. D.; Heerze, L. D.; Barrie, J. R. *J. Am. Chem. Soc.* **1990**, *112*, 3018.
97. Fürstner, A.; Jumbam, D. N.; Weidmann, H. *Tetrahedron Lett.* **1991**, *32*, 6695.
98. Solladié, G.; Hutt, J. *J. Org. Chem.* **1987**, *52*, 3560.
99. Solladié, G.; Girardin, A. *Tetrahedron Lett.* **1988**, *29*, 213.
100. Solladié, G.; Hamdouchi, C. *Synlett* **1989**, 66.
101. For a review on this topic: McMurry, J. E.; Lectka, T. *Acc. Chem. Res.* **1992**, *25*, 47 and references cited therein.
102. Mangeney, P.; Grosjean, F.; Alexakis, A.; Normant, J. F. *Tetrahedron Lett.* **1988**, *29*, 2675.
103. Mangeney, P.; Tejero, T.; Alexakis, A.; Grosjean, F.; Normant, J. F. *Synthesis* **1988**, 255.
104. Betschart, C.; Schmidt, B.; Seebach, D. *Helv. Chim. Acta* **1988**, *71*, 1999.

105. Betschart, C.; Seebach, D. *Helv. Chim. Acta* **1987**, *70*, 2215.
106. Schobert, R. *Angew. Chem. Int. Ed. Engl.* **1988**, *27*, 855.
107. Zhang, Y.; Jiang, J.; Zhang, Z. *Tetrahedron Lett.* **1988**, *29*, 651.
108. Fujiwara, Y.; Ishikawa, R.; Akiyama, F.; Teranishi, S. *J. Org. Chem.* **1978**, *43*, 2477.
109. Petit, M.; Mortreux, A.; Petit, F. *J. Chem. Soc. Chem. Commun.* **1984**, 341.
110. Gilet, M.; Mortreux, A.; Folest, J. C.; Petit, F. *J. Am. Chem. Soc.* **1983**, *105*, 3876.
111. Dams, R.; Malinowski, M.; Geise, H. J. *Bull. Soc. Chim. Belg.* **1982**, *91*, 149.
112. Takai, K.; Kataoka, Y.; Utimoto, K. *J. Org. Chem.* **1990**, *55*, 1707.
113. Rieke, R. D.; Rhyne, L. D. *J. Org. Chem.* **1979**, *44*, 3445.
114. Kahn, B. E.; Rieke, R. D. *Organometallics* **1988**, *7*, 463.
115. Wescott, Jr., L. D.; Williford, C.; Parks, F.; Dowling, M.; Sublett, S.; Klabunde, K. J. *J. Am. Chem. Soc.* **1976**, *98*, 7853.
116. Togashi, S.; Fulcher, J. G.; Cho, B. R.; Hasegawa, M.; Gladysz, J. A. *J. Org. Chem.* **1980**, *45*, 3044.

VI. ADDITION REACTIONS TO CARBONYL COMPOUNDS

A. BARBIER-TYPE REACTIONS

1. Introduction

The Barbier reaction is an old, very attractive one-step procedure for the preparation of alcohols from halides and carbonyl compounds in the presence of magnesium (eq. 6.1).[1] This method is a variation of an even older procedure developed by Saytzeff and Wagner, who employed zinc metal.[2]

$$\text{R-X} + \text{Mg} + R^1R^2C{=}O \xrightarrow[\text{2) } H_2O]{\text{1) Solvent}} R-C(R^1)(R^2)-OH \qquad (6.1)$$

It was Barbier who had Grignard repeat some experiments on the preparation of a tertiary alcohol from a mixture of methyl iodide, magnesium, and methyl heptyl ketone. Grignard treated the halide with magnesium first and carried out the reaction in ether. The process was a complete success.[3] Grignard's doctoral dissertation (1901) reported the synthesis of alcohols, acids, and hydrocarbons using such organo-magnesium reagents. The method, a two-step procedure, was rapidly and successfully exploited throughout the world, and at the time of Grignard's death in 1935, more than 6,000 papers reporting applications of the Grignard reaction had been published. It must be realized that the Grignard (or Barbier-Grignard) reaction will never be obsolete, a privilege shared by a very small number of organic reactions.

Although Grignard's reaction is a two-stage synthetic procedure, it offers some advantages over the Barbier reaction. Thus, a suitable start of the reaction is accomplished when only magnesium metal and the organic halide are present, and the number of by-products decreases considerably. However, the Barbier synthesis is preferred in many cases, and has found a wide application with allyl halides.[4] Also, the Barbier reaction is particularly useful for the preparation of other organometallic reagents in a one-step transmetallation (eqs. 6.2-6.4).

$$Sn + Mg + EtBr \longrightarrow Et_4Sn \quad (6.2)$$

$$RX + Mg + B_2H_6 \longrightarrow R_3B \quad (6.3)$$

$$CH_3I + Mg/Ga \longrightarrow (CH_3)_3Ga \quad (6.4)$$

A discussion of the mechanism was reported by Molle and Bauer,[5] although this point has given rise to much controversy. Two basic mechanisms have been proposed for this apparently simple reaction. It is currently assumed that an intermediate organometallic reagent is generated *in situ* and rapidly trapped by the carbonyl compound (eq. 6.5).

$$R^1R^2C{=}O + R^3X + M \longrightarrow [R^3M] \longrightarrow R^1R^2C(OH)R^3 \quad (6.5)$$

A possible alternative should involve radical intermediates, such as a radical anion (**1**), a free radical (**2**), or a radical pair (**3**) following different single electron transfer processes (eq. 6.6). Each of these species can react with the carbonyl compound or the ketyl radical anion derived from it.

$$\text{R-X} + \text{M} \xrightarrow{\text{SET 1}} \underset{\mathbf{1}}{[\text{R-X}]^{\dot{-}}\,\text{M}^{+}} \xrightarrow{\text{-MX}} \underset{\mathbf{2}}{[\text{R}\cdot]} \longrightarrow \underset{\mathbf{3}}{[\text{R}\cdot\text{M}\cdot]} \xrightarrow{\text{SET 2}} \text{R-M} \quad (6.6)$$

This radical pathway has been examined recently by Luche *et al.*,[6] who performed a theoretical study of the Barbier reaction. The authors studied the model of formaldehyde and methyl iodide undergoing SET processes. Calculations indicated that the first SET takes place on the halide to give a molecular complex, in which the formaldehyde approaches methyl iodide from the side opposite to the C-I bond. A clear consequence would be the configuration inversion at the halide carbon center. The authors carried out an experiment with an optically active alkyl halide, and the resulting formation of an alcohol with complete inversion of configuration at the chiral center was observed. This suggests that the reaction of the radical anion **1** with the carbonyl compound is favored.

A second SET leads to the separation of products. Barbier reactions of unsaturated carbonyl compounds must involve in the first place the formation of the carbonyl radical anion, and thus proceed through path B (Scheme 6.1). With saturated carbonyl compounds SET 1 should be the formation of the halide radical anion (path A). In addition, the formation observed of Wurtz or pinacol by-products is in agreement with these hypotheses.

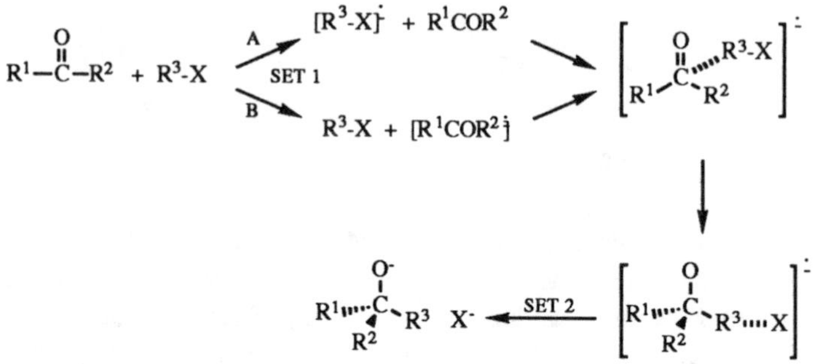

$R^1{-}C(=O){-}R^2 + R^3\text{-}X \xrightarrow{\text{A (SET 1)}} [R^3\text{-}X]^{\cdot -} + R^1COR^2$

$R^1{-}C(=O){-}R^2 + R^3\text{-}X \xrightarrow{\text{B (SET 1)}} R^3\text{-}X + [R^1COR^2]^{\cdot -}$

$\longrightarrow [R^1C(=O)R^2 \cdots R^3\text{-}X]^{\cdot -} \longrightarrow [R^1R^2C(O)\cdots R^3 \cdots X]^{\cdot -} \xrightarrow{\text{SET 2}} R^1R^2C(O^-)R^3 \; X^-$

Scheme 6.1

2. Activated Metals

Numerous metals are capable of effecting Barbier reactions.[4,7] In general, metal activation greatly benefits the reaction times and promotes the mildness of the process. Zinc metal was used for the synthesis of alcohols in the early history of this reaction (Wagner-Saytzeff reaction),[2] but it was completely replaced by magnesium in the Barbier and Grignard reactions. However, it is remarkable that a zinc-assisted Barbier-type reaction —the Reformatsky reaction (see subsection B)— constitutes an important one-step procedure in organic synthesis.

Lithium-mediated Barbier reaction is the best alternative to magnesium, and the yields of many different products are higher in most cases than those obtained with Grignard reagents using a two-step procedure. Moreover almost no side reactions take place, and aldehydes, ketones, or esters can be equally used as substrates. Organolithium compounds are true intermediates in this one-step reaction.[8,9]

Lithium- and zinc-promoted allylation of carbonyl compounds has been improved by several methods, particularly by means of ultrasonic irradiation, with which Luche and his associates have carried out important contributions.

Luche and Damanio reported[10] significant improvements both in yields and simplification of experimental techniques over conventional methodology in the Barbier reaction, when the reactions were performed in a simple ultrasonic cleaning bath at 50 kHz (eq. 6.7). Reactions are usually complete within 15-40 min and alcohols are obtained in 76-100% yield. The relevant advantages of this method are that the reactions can be conducted in commercial undried THF and side products derived from reduction or enolization, which are common in conventional methodology, are largely suppressed. Even benzyl halides give yields greater than 95% and very little Wurtz coupling.

$$R^1\text{-X} + R^2\text{-C(=O)-}R^3 \xrightarrow[2)\ H_2O]{1)\ Li,\ THF,\ ))))} R^1R^2C(OH)R^3 \quad (6.7)$$

R^1 = alkyl, aryl, vinyl

Similarly, organocopper reagents were generated by sonication at 50 kHz of copper(I) compounds, organic halide, and lithium sand in THF-ether, and then were allowed to react with α-enones (eq. 6.8).[11]

$$\text{cyclohexenone} + CH_3(CH_2)_2CH_2Br \xrightarrow[2)\ H_3O^+,\ 89\%]{1)\ Li,\ CuI,\ Et_2O\text{-THF},\ )))),\ 0\ ^\circ C} \text{3-}^n\text{Bu-cyclohexanone} \quad (6.8)$$

Also, α-enones and α-enals give 1,4-addition products by reaction with ultrasonically generated arylzinc compounds from lithium, aryl halide, and zinc bromide (eq. 6.9).[12] The reaction requires the presence of catalytic amounts of a nickel complex. The method can be extended to alkyl halides, but a more intense sonication and a better medium for cavitation (toluene-THF) are needed.[13,14]

$$\text{Ar-Br} \xrightarrow[))))]{Li,\ ZnBr_2,\ THF} Ar_2Zn \xrightarrow[-40\ ^\circ C]{Ni(acac)\ (cat.),\ \text{enone},\ ))))} \text{Ar-CH}_2CH_2\text{-C(=O)-} \quad (6.9)$$

Alkali metals and magnesium react with aryl halides and alkyl isocyanates under the sonochemical Barbier conditions to give secondary aryl amides (eq. 6.10).[15] Good yields are obtained with magnesium and lower with lithium. The reaction with sodium proceeds slowly but can be greatly accelerated by the addition of HMPA.

$$\text{Ar-X} + \text{R-N=C=O} \xrightarrow[2)\ H_2O]{1)\ M,\ THF,\ ))))} \text{Ar-C(=O)-NHR} \quad (6.10)$$

Interestingly, the intermediates can also be *ortho*-lithiated when the metal is sodium, but complications take place with magnesium owing to Mg-Li exchange. The *ortho*-lithiated intermediates react with a variety or electrophiles to give *ortho*-substituted aryl amides (eq. 6.11).

$$\text{PhBr} + \text{R-N=C=O} \xrightarrow[))))]{Na,\ THF} \left[\text{Ph-C(=NR)-O}^-\ Na^+\right] \xrightarrow[2)\ E^+]{1)\ RLi} \text{2-E-C}_6H_4\text{-C(=O)NHR} \quad (6.11)$$

E = RNCO, RCHO

For perfluoroalkyl halides, the use of sonochemically generated perfluoroalkylzinc reagents is preferred to Li or Mg, which are normally used in the Barbier reaction. Perfluoroalkyl derivatives of these metals undergo rapid β-elimination to give alkenes. It is necessary to use a higher boiling solvent such as DMF that allows the more powerful cavitation needed to activate the zinc (eq. 6.12).[16,17]

$$R_F\text{-}X + R^1R^2C{=}O \xrightarrow[2)\ H_3O^+]{1)\ Zn,\ DMF,\ ))))} R_F R^1 R^2 C\text{-}OH \qquad (6.12)$$

An alternative route to perfluoroalkyl alcohols involves the Barbier reaction of perfluoroaldehydes with *in situ* sonically generated alkyl or allyl Grignard reagents (eq. 6.13).[18]

$$CF_3CHO + RX \xrightarrow[2)\ H_2O]{1)\ Mg,\ ))))} CF_3RCHOH \qquad (6.13)$$

R = alkyl, allyl

A zinc-mediated Barbier reaction between perfluoroalkyl iodides and chiral arene metal carbonyls has made it possible to obtain chiral perfluoroalcohols with 30-66% optical purity (Scheme 6.2).[19]

$$(o\text{-}CH_3C_6H_4CHO)Cr(CO)_3 + R_FI \xrightarrow[2)\ H_3O^+]{1)\ Zn,\ DMF,\ ))))} (o\text{-}CH_3C_6H_4C^*HR_FOH)Cr(CO)_3 \xrightarrow{h\nu} o\text{-}CH_3C_6H_4C^*HR_FOH$$

Scheme 6.2

Organometallics derived from zinc metal have been extensively utilized in Barbier-Grignard-type carbonyl additions of allylic halides.[20] Numerous variations in substrates can be introduced to obtain products other than alcohols. Allyl halides react with nitriles in the presence of zinc to give allyl ketones, but yields are generally low. The use of a zinc-silver couple has been found to improve yields to 70-80% in reactions with allyl bromide. The reaction proceeds better in benzene or THF than in ether. The process involves an attack at the more substituted position (eq. 6.14).[21]

$$CH_3CH_2CN + (CH_3)_2C{=}CHCH_2Br \xrightarrow[71\%]{Zn\text{-}Ag,\ C_6H_6} CH_3CH_2COC(CH_3)_2CH{=}CH_2 \qquad (6.14)$$

In a related protocol, the so-called three-component condensation, β-hydroxy nitriles can be obtained in excellent yield from a mixture of an alkyl halide, acrylonitrile, and a carbonyl compound in the presence of zinc powder (eq. 6.15).[22] The reaction can be performed with other variations (eqs. 6.16, 6.17).

$$(CH_3)_2CHI + CH_2{=}CHCN + (CH_3)_2CO \xrightarrow[98\%]{Zn,\ \Delta} (CH_3)_2CHCH_2CH(CN)C(OH)(CH_3)_2 \qquad (6.15)$$

$$(CH_3)_2CHI + CH_2{=}CHCN + (CH_3CO)_2O \xrightarrow[\Delta,\ 81\%]{Zn,\ CH_3CN} (CH_3)_2CHCH_2C(CN){=}C(OH)CH_3 \qquad (6.16)$$

$$CH_2{=}CHCN + CH_3CO(CH_2)_2CH_2I \xrightarrow[42\%]{Zn,\ \Delta} \text{1-methyl-2-cyanocyclohexan-1-ol } (CH_3,\ OH,\ CN) \qquad (6.17)$$

The organozinc reagent, prepared from a β-iodoalanine derivative by sonication with zinc-copper couple, reacts with acid chlorides in the presence of palladium(II) catalysts to afford enantiomerically pure protected γ-keto α-amino acids (eq. 6.18).[23]

$$ICH_2CH(NHBoc)CO_2Bz \xrightarrow[))))]{Zn\text{-}Cu,\ C_6H_6} \xrightarrow[\text{2) } RCOCl,\ ))))\ \ 75\text{-}90\%]{\text{1) } Pd^{2+}\ (cat.)} RCOCH_2CH(NHBoc)CO_2Bz \qquad (6.18)$$

In recent years, Knochel's organocopper-zinc reagents have proved to be useful and versatile tools in modern organic synthesis.[24-35] These reagents are compatible with a wide variety of functional groups, in contrast to organocopper compounds which are difficult to prepare containing reactive substituents. Also, organocopper-zinc reagents can be prepared in the presence of relatively acidic hydrogens.[29]

$$PhthN(CH_2)_3Cu(CN)ZnI \xrightarrow[65\%]{PhCH{=}CHCHO,\ BF_3.OEt_2} PhthN(CH_2)_3CH(OH)CH{=}CHPh \qquad (6.19)$$

Organozinc iodides (or bromides) are readily formed by the reaction of alkyl or aryl iodides with zinc. This metal is simply activated by treatment with 1,2-dibromoethane and chlorotrimethylsilane.[24,25] The organozinc halides, RZnI, undergo transmetallation to RCu(CN)ZnI on reaction with CuCN.2LiCl, a soluble salt generated by reaction of CuCN with anhydrous LiCl in THF. The resulting dimetallic reagents are comparable to organocopper reagents in reactions with electrophiles and in additions to unsaturated substrates (eq. 6.19),[33] (Scheme 6.3).[25]

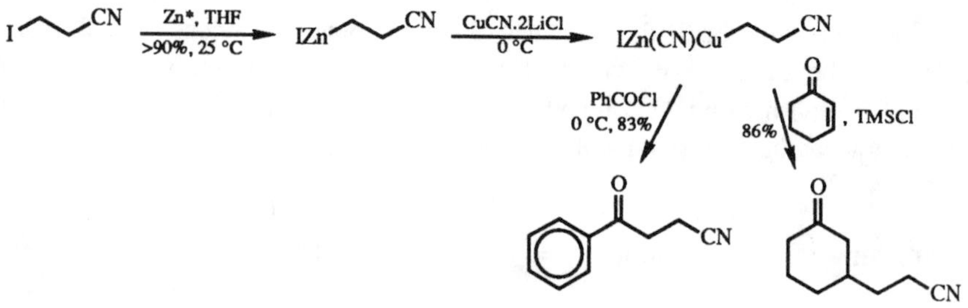

$$\mathrm{I(CH_2)_2CN} \xrightarrow[>90\%,\ 25\ ^\circ C]{Zn^*,\ THF} \mathrm{IZn(CH_2)_2CN} \xrightarrow[0\ ^\circ C]{CuCN.2LiCl} \mathrm{IZn(CN)Cu(CH_2)_2CN}$$

PhCOCl, 0 °C, 83% → $\mathrm{PhCO(CH_2)_3CN}$

cyclohexenone, TMSCl, 86% → 3-(2-cyanoethyl)cyclohexanone

Scheme 6.3

Allylcopper-zinc reagents can be prepared directly by reaction of vinyl copper reagents with (iodomethyl)zinc iodide, the Simmons-Smith reagent. These allylcopper-zinc reagents do not couple with an alkyl iodide or benzyl bromide, but react rapidly with electrophiles such as aldehydes, ketones, or imines (eq. 6.20).[34]

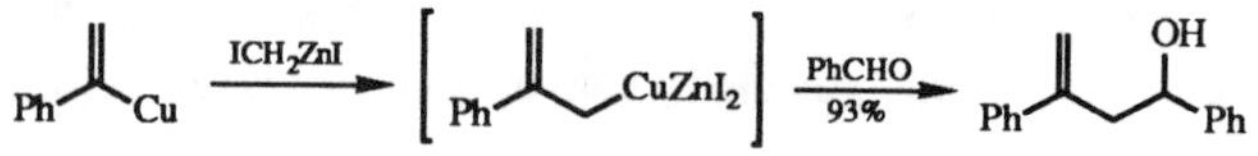

$$\mathrm{Ph(C{=}CH_2)Cu} \xrightarrow{ICH_2ZnI} \left[\mathrm{Ph(C{=}CH_2)CH_2CuZnI_2}\right] \xrightarrow[93\%]{PhCHO} \mathrm{Ph(C{=}CH_2)CH_2CH(OH)Ph} \qquad (6.20)$$

Rieke *et al.*[36] have employed the versatility of these bimetallic reagents, but activated zinc was obtained previously by reduction of a zinc halide with lithium naphthalenide in THF or DME. The resulting species, RCu(CN)ZnBr, couple directly with acid chlorides or undergo conjugate addition to enones. Reaction of the copper-zinc reagents with allylic halides results mainly in S_N2' substitution (eq. 6.21).

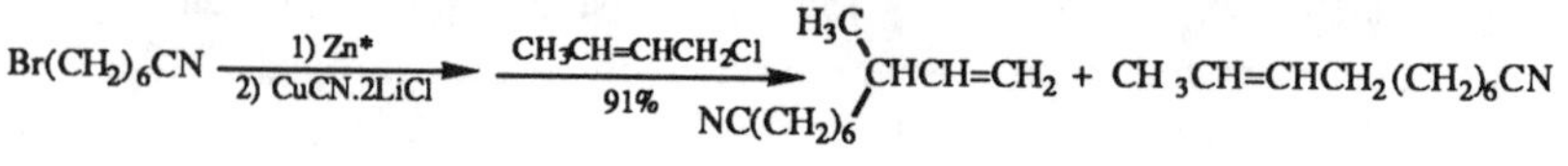

$$\mathrm{Br(CH_2)_6CN} \xrightarrow[2)\ CuCN.2LiCl]{1)\ Zn^*} \xrightarrow[91\%]{CH_3CH{=}CHCH_2Cl} \mathrm{NC(CH_2)_6CH(CH_3)CH{=}CH_2} + \mathrm{CH_3CH{=}CHCH_2(CH_2)_6CN} \qquad (6.21)$$

(97 : 3)

Although zinc-mediated allylations are usually performed in ethereal solvents, a systematic study of the scope and limitations in other media has not been achieved. Some organic transformations using zinc metal have been carried out in an aqueous solvent system with remarkable success. These processes will be treated later (see subsection VIA. 3).

Japanese chemists have examined the zinc-promoted allylation of carbonyl compounds using DMF as the solvent.[37] The reaction rate and yield are greatly increased under mild and simple reaction conditions. More importantly, this method does not require any anhydrous solvent or an inert atmosphere, and zinc (usually as plates) was used without any preactivation (eq. 6.22). Aldehydes and ketones work equally well, although an experiment of competitive chemoselectivity was not reported. α,β-Unsaturated carbonyl compounds gave mainly 1,2-addition. Mention should be made of a study of the nature of the solvent in this Zn-promoted allylation. The reaction between cyclohexanone and allyl bromide gave good yields (86-99%) when the solvents were amides, lactams, and oxazolines, but not THF.

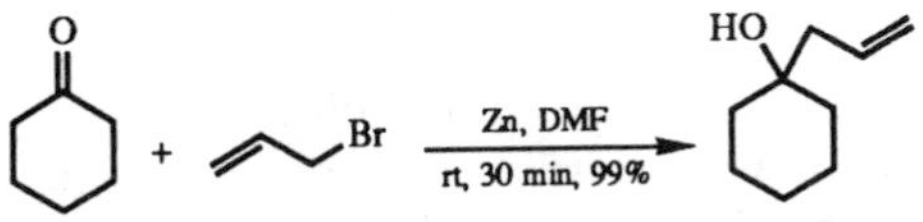

(6.22)

Other metals have been used to perform Barbier-type reactions and the list goes on and on. In the presence of metallic tin, both allyl iodide and bromide react with carbonyl compounds at room temperature to form homoallylic alcohols in 75-90% yield.[38] Presumably a diallyltin dihalide is formed initially. Other tin-based metal systems have also been reported for allylations of carbonyl compounds.[39,40] Organometallics derived from manganese and the low-valent species (probably as divalent metals) $MnCl_2$-$LiAlH_4$ or $CrCl_3$-$LiAlH_4$ were introduced by Hiyama and Nozaki.[41-44]

Butsugan *et al.*[45] reported the first example of a Barbier-Grignard-type allylation of aldehydes using metallic antimony, which gives high yields of homoallylic alcohols in a regiospecific fashion (eq. 6.23). The best solvent system was a THF-HMPA mixture, and the same reactions in THF or DMF alone gave lower yields.

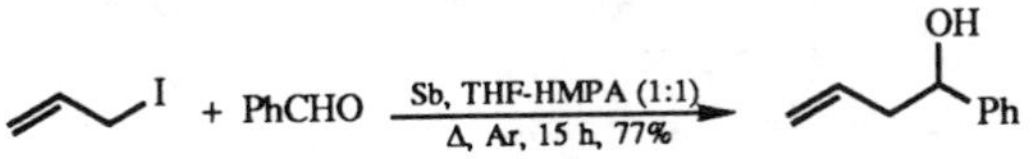

(6.23)

The process can be applied to aromatic and aliphatic aldehydes. The reaction is chemoselective and ketones react very sluggishly. Furthermore, ester and cyano groups remain unchanged. Allylic bromides and phosphates are less reactive than allylic iodides, and thus poor yields of coupled products were obtained. This difficulty can be overcome by the addition of one equivalent of lithium iodide (eq. 6.24).

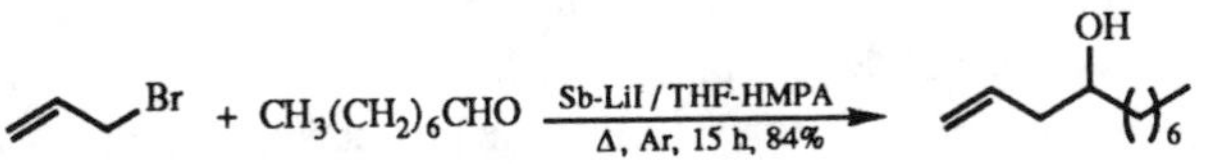

(6.24)

Active zero-valent antimony generated from $SbCl_3$-Fe or $SbCl_3$-Al induces allylation of aldehydes with allylic halides at room temperature to give high yields of homoallylic alcohols with high regio- and chemoselectivity.[46,47] The less reactive allyl bromides react with the same aldehydes at a higher temperature. Again, addition of sodium iodide allows the reaction at room temperature. With iron as the reducing agent, DMF was found to be the suitable solvent, whereas with metallic aluminum the process was also effective in DMF-water (eq. 6.25).

$$\text{RCHO} + \text{CH}_2\text{=CHCH}_2\text{Br} + \text{NaI} \xrightarrow[\text{SbCl}_3\text{-Al, DMF-H}_2\text{O; rt, 80-90\%}]{\text{SbCl}_3\text{-Fe, DMF or}} \text{RCH(OH)CH}_2\text{CH=CH}_2 \quad (6.25)$$

The authors suggest the intermediacy of an allylantimony reagent formed through the oxidative addition of an allyl halide to zero-valent antimony. Also, the reaction proceeds with a preferential *erythro* selectivity, which is consistent with a noncyclic transition state.[48]

Allylic halides react with aldehydes in the presence of bismuth metal and bismuth salt using the reductive combinations $BiCl_3$-Zn, $BiCl_3$-Fe, or $BiCl_3$-Al to give homoallylic alcohols in high yields.[49] The reaction occurs under mild conditions with high chemo-, regio-, and stereoselectivity. Although a commercially available massive bismuth can be utilized, bismuth powder is more convenient and effective. Bismuth-mediated allylation can be carried out following several experimental procedures.

A) Treatment of an allylic halide with bismuth metal in DMF and further addition of an aldehyde.

B) A Barbier-type reaction: a mixture of an allylic halide and an aldehyde is treated with Bi(0) in DMF. This was found the best solvent for this addition reaction. When THF was used, the yields of homoallylic alcohols decreased and were not reproducible. Allyl bromide and 3-phenyl propionaldehide react, however, in aqueous THF using a Bi(0)-Al(0) system in the presence of a catalytic amount of hydrobromic acid (eq. 6.26). Interestingly, in the absence of water the yield decreases to 47%. The reaction does not proceed at all without Bi(0) and is sluggish when Al(0) is not used.

$$\text{CH}_2\text{=CHCH}_2\text{Br} + \text{PhCH}_2\text{CH}_2\text{CHO} \xrightarrow[\text{THF-H}_2\text{O, rt, 90\%}]{\text{Bi-Al / HBr (cat.)}} \text{PhCH}_2\text{CH}_2\text{CH(OH)CH}_2\text{CH=CH}_2 \quad (6.26)$$

C) Reduction of $BiCl_3$ with Zn in THF and the resulting black suspension reacts with a mixture of an allylic halide and an aldehyde. In contrast to methods A and B which give very poor yields with ketones, in this method acetophenone affords 56% yield of product.

D) The combination $BiCl_3$-Fe in THF also mediates allylation of aldehydes. This system generates gradually a black powder, whereas reduction with Zn metal produces a black powder immediately.

E) The reduction of $BiCl_3$ with Al in THF gives no homoallylic alcohol. However, a Barbier-type allylation of aldehydes is easily effected in THF-water. The use of $BiCl_3$ is essential and poor results are obtained with the Bi(0)-Al(0) system.

Moreover, $BiCl_3$ does not promote the allylation in the absence of Al(0), and this metal alone gives none of the desired products. It is noteworthy that only a catalytic amount of $BiCl_3$ is needed to perform this reaction (eq. 6.27).

Br + PhCHO —[$BiCl_3$ (10 mol %) - Al; THF-H_2O, rt, 82%]→ Ph–CH(OH)–CH₂CH=CH₂ (6.27)

With regard to the stereoselectivity, the reaction of benzaldehyde with crotyl bromide (*cis/trans* mixture 30:70) was investigated. Method D gave the best *erythro* selectivity and yield, but the selectivity is dependent on the solvent (eq. 6.28). The *erythro* selectivity can be enhanced in THF by the addition of two equivalents of $BF_3.OEt_2$.

PhCHO + crotyl Br —[$BiCl_3$-Fe; solvent]→ Ph (OH) + Ph (OH) (6.28)

Solvent	Yield (%)	Erythro / Threo
DMF	94	90 : 10
MeOH	81	85 : 15
EtOH	100	89 : 11
THF-$BF_3.OEt_2$	92	92 : 8

Methods B(DMF), C(THF), and E(THF-water) also provided good *erythro* selectivity (77:23-85:15). Tin-mediated allylation, however, favors the *threo*-isomer but with poor selectivity. An exceptional *threo* selectivity is found with the Nozaki-Hiyama reagent ($CrCl_3$-$LiAlH_4$).[44]

A low-valent titanium reagent generated by reduction of $TiCl_4$ with aluminum in THF effects allylation of imines with allyl bromide, even when used in catalytic amounts (0.05 equiv).[50] This $TiCl_4$-Al combination presumably forms Al(III) and Ti(0), which reacts with the allyl halide to afford an allyltitanium intermediate as the reactive species (eq. 6.29).

PhCH=N–Bz + Br —[Al, $TiCl_4$; THF, 83%]→ Ph–CH(NHBz)–CH₂CH=CH₂ (6.29)

Allylic indium halides can be easily prepared from indium metal and the corresponding allylic bromide. They react with satisfactory chemo-, regio-, and stereoselectivity.[51-54] Very recently, allylic indium reagents have been utilized for the carbometallation (carboindation) of alkynols, and the procedure was applied to the one-step synthesis of the naturally-occurring monoterpene alcohol, yomogi alcohol (eq. 6.30).[55]

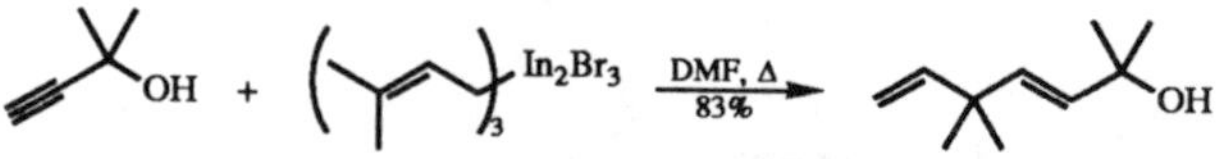

OH + In_2Br_3 DMF, Δ 83% OH (6.30)

Besides magnesium, other alkaline-earth metals have been also employed in Barbier reactions. For example, tertiary aliphatic alcohols are better obtained using calcium metal in a one-step procedure.[56] Rieke and his associates have prepared a form of highly reactive calcium by reduction of calcium halides in THF with lithium biphenylide.[57] This activated calcium reacts with organic halides, including fluorides, to form organocalcium reagents in high yields. These undergo Grignard-type additions, and thus reactions with cyclohexanone provide access to tertiary alcohols. Transmetallation with Knochel's copper(I) complex[27] CuCN.2LiBr forms calcium cuprate reagents, which undergo several cross-coupling reactions (eq. 6.31).

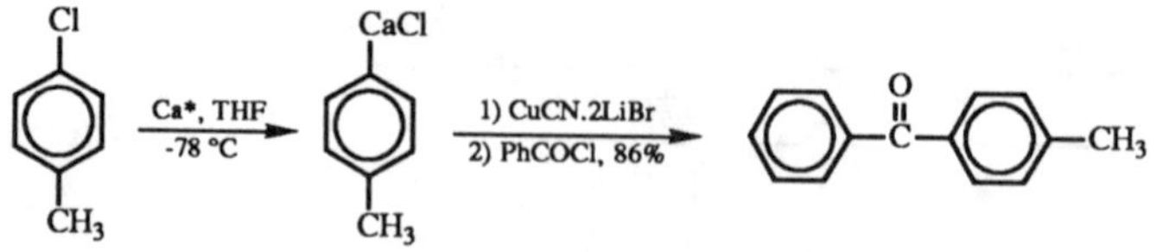

Cl, CH_3 — Ca*, THF, -78 °C → CaCl, CH_3 — 1) CuCN.2LiBr 2) PhCOCl, 86% → O, C, CH_3 (6.31)

Similarly, highly reactive barium has been prepared by reduction of barium iodide with lithium biphenylide,[58] and has then been exposed to allylic chlorides. The resulting organobarium reagents react with carbonyl compounds in a few minutes to afford homoallylic alcohols with high α-selectivity and with retention of stereochemistry of the starting halide. Not only aldehydes, but also ketones were found to be reactive toward this organometallic system (Scheme 6.4).

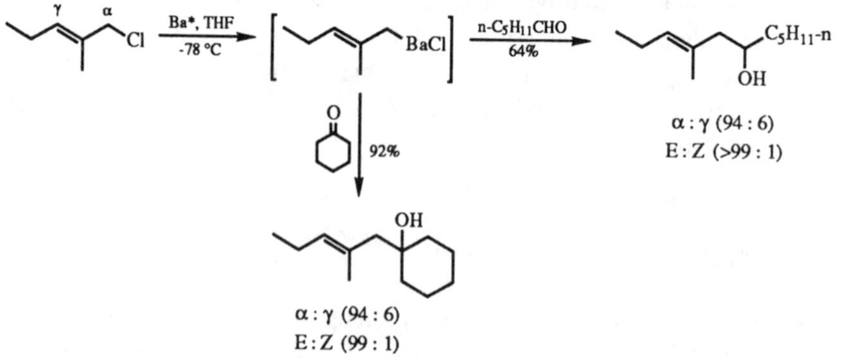

γ α Cl — Ba*, THF, -78 °C → [BaCl] — n-C_5H_{11}CHO, 64% → C_5H_{11}-n, OH

α : γ (94 : 6)
E : Z (>99 : 1)

O, 92% → OH

α : γ (94 : 6)
E : Z (99 : 1)

Scheme 6.4

Also, metallic nickel was easily prepared in DME by reduction of nickel(II) iodide with lithium and naphthalene.[59] This finely divided metallic nickel induces Grignard-type addition of benzyl halides to 1,2-diketones. A wide range of functional groups such as halogen, cyano, or methoxycarbonyl groups can be employed in the starting halide. The addition reaction seems to involve oxidative addition of benzyl halides to nickel metal to afford benzylnickel halide intermediates. Further insertion of 1,2-diketone into the carbon-nickel bond gives hydroxy ketones after hydrolysis (eq. 6.32).

$$4\text{-}CH_3O\text{-}C_6H_4\text{-}CH_2Br + Ph\text{-}CO\text{-}CO\text{-}Ph \xrightarrow[2)\ H_2O]{1)\ Ni^*,\ DME,\ 85\ ^\circ C} CH_3O\text{-}C_6H_4\text{-}CH_2\text{-}C(Ph)(OH)\text{-}CO\text{-}Ph \quad (6.32)$$

Although lanthanide-promoted Barbier-type reactions are usually performed with samarium(II) iodide[60] and to a less extent with other lanthanide salts[61] in homogeneous reactions, metals have been also employed. Samarium metal in combination with diiodomethane effects an interesting iodomethylation process (eqs. 6.33, 6.34).[62-64] Aldehydes and ketones are easily alkylated at room temperature or below under these conditions. Conjugated aldehydes and ketones give only 1,2-addition products. As an additional advantage, the method exhibits a high diastereoselectivity.

$$\text{cyclopropyl methyl ketone} + CH_2I_2 \xrightarrow[0\ ^\circ C,\ 62\%]{Sm,\ THF} \text{cyclopropyl-C(OH)(CH_3)-CH_2I} \quad (6.33)$$

$$\text{2-cyclohexen-1-one} + CH_2I_2 \xrightarrow[0\ ^\circ C,\ 55\%]{Sm,\ THF} \text{1-(iodomethyl)cyclohex-2-en-1-ol} \quad (6.34)$$

Acid halides couple with aldehydes and ketones in the presence of samarium(II) iodide to give α-hydroxy ketones.[65] This protocol is usually applied in THF, but low yields of products are frequently obtained owing to the THF attack on the acid halide. This limitation can be overcome by the *in situ* preparation of samarium(II) iodide from samarium metal in acetonitrile (eq. 6.35).

$$PhCOCl + PhCOCH_3 \xrightarrow[2)\ H_3O^+,\ 78\%]{1)\ Sm,\ ICH_2CH_2I,\ CH_3CN} Ph\text{-}CO\text{-}C(OH)(CH_3)Ph \quad (6.35)$$

Metal-graphites have been utilized in Barbier-type reactions and other related alkylation processes. Hart *et al.*[66] have described the preparation of dispersions of sodium or potassium on charcoal, as well as of sodium on graphite and alumina. The materials are highly reactive, pyrophoric, and presumably dispersed forms of the metals, without intercalation in the case of graphite supported reagents. Hart and co-workers studied these supported reagents for the alkylation of ketones in hexane. However, mono- and dialkylation products along with the corresponding alcohol and pinacol are usually obtained.

Tin-graphite undergoes oxidative addition with allylic bromides in THF at room temperature to give diallyltin dibromides, which add to aldehydes.[67] The reaction proceeds with allylic rearrangement and provides homoallylic alcohols in good yields. With α,β-unsaturated aldehydes, selective 1,2-addition is observed in agreement with other Barbier-Grignard-type reactions (eqs. 6.36, 6.37). The strategy was applied to a simple synthesis of racemic Artemisia alcohol (eq. 6.38).

Br + ($)_6$CHO —Sn-Gr, THF / 25 °C, 93%→ OH (6.36)

Br + CHO —Sn-Gr, THF / 25 °C, 86%→ OH (6.37)

Br + CHO —Sn-Gr, THF / 25 °C, 89%→ OH (6.38)

These Italian researchers have introduced an enantioselective synthesis of homoallylic alcohols by forming *in situ* a chiral tin(IV) complex.[68] The diallyltin dibromide, generated by treatment with tin-graphite, reacted with benzaldehyde in the presence of the sodium salt of L-(+)-diethyl tartrate to afford (*S*)-1-phenyl-3-buten-1-ol in 71% ee (eq. 6.39).

$(\text{allyl})_2SnBr_2$ + PhCHO —L-(+)-DET / NaH, THF, -40 °C, 65%→ OH (6.39)

3. Reactions in Aqueous Media

Barbier-Grignard-type reactions have been successfully carried out in aqueous media, enhancing the importance of these classical processes. Most reactions utilize a zinc metal system, and the application of ultrasonic irradiation makes it possible to perform reactions under extremely mild conditions, with high yields, and with unusual substrates. Zinc has been known to be non-polluting and readily available, although organozinc reagents are less reactive than other organometallics toward the carbonyl group. Apparently, these reactions do not involve the intermediacy of organozinc reagents owing to their inherent reactivity with water, but rather a radical pathway on the zinc surface.

In the presence of zinc (or tin), allyl halides undergo 1,2-addition to carbonyl compounds.[69-71] The reaction is facilitated by sonication, and aldehydes react more rapid than ketones. In fact in molecules containing both functionalities, reaction takes place almost exclusively at the aldehyde group without affecting a preexisting ketone group. Allylation with zinc powder gives the highest yields in a mixture of saturated aqueous ammonium chloride-THF (5:1).[69] Similar results are obtained with tin powder in THF-water (1:5).[70]

Luche *et al.*[72] found that a Zn-Cu couple prepared by sonication of Zn and CuI in ethanol-water (9:1) allows conjugate addition of alkyl halides to enones and enals (eqs. 6.40, 6.41). The order of reactivity is R-I > R-Br, and *tert* > *sec* >> primary. As solvent, THF-water or pyridine-water mixtures, or even pure water can be used.

$$CH_2{=}CHCHO + I(CH_2)_2OH \xrightarrow[\text{py-H}_2\text{O, 70\%}]{\text{Zn-Cu,))))}} \text{(2-hydroxytetrahydropyran)} \quad (6.40)$$

$$\text{(bicyclic enone)} + ICH(CH_3)CH_2CH{=}CH_2 \xrightarrow[\text{py-H}_2\text{O, 82\%}]{\text{Zn-Cu,))))}} \text{(bicyclic ketone)}{-}CH_2CH(CH_3)CH_2CH{=}CH_2 \quad (6.41)$$

In further studies, a highly active Zn-Cu couple was prepared by sonication of a mixture of Zn and CuI (3:1 ratio) in an alcohol-water mixture (~65:35). A black heavy suspension is formed which should be used at once.[73] This couple and sonication effect addition of alkyl iodides to enones.[74] Highest yields are obtained with isopropanol-water or *n*-propanol-water mixtures as solvent for primary iodides. Ethanol-water is the solvent of choice for *tert*- or *sec*-iodides (eq. 6.42). Yields are decreased with α-methyl substituted enones and are particularly depressed by a β-methyl substituent.

$$\text{(cyclohexenone)} + (CH_3)_2CHI \xrightarrow[\text{EtOH-H}_2\text{O, 95\%}]{\text{Zn-Cu,))))}} \text{(3-substituted cyclohexanone)}{-}CH(CH_3)_2 \quad (6.42)$$

This sonochemical addition of alkyl halides can be extended to a variety of α,β-unsaturated carbonyl compounds and nitriles (eq. 6.43).[75]

$$RX + CH_2{=}CH{-}Y \xrightarrow[R^1OH\text{-}H_2O]{\text{Zn-Cu,))))}} R{-}CH_2CH_2{-}Y \quad (6.43)$$

X = Br, I

Y = CHO, COR, CO_2R, $CONR_2$, CN

In addition, the method has also been applied to epoxyalkyl halides.[76,77] The reactions with 3,4-epoxyalkyl halides constitute a rapid and satisfactory access to cyclopropylmethanols, which are found in many natural products (eq. 6.44).[77]

$$CH_3CH_2CH_2CH(Br)CH_2C(CH_3)(O)CH_2 \xrightarrow[\text{EtOH-H}_2\text{O, 60\%}]{\text{Zn-Cu,))))}} CH_3CH_2CH_2\text{-cyclopropyl-}C(CH_3)CH_2OH \quad (6.44)$$

One unusual feature of Luche allylation[69-71] of aldehydes is the use of THF-water as solvent. Recently, Wilson and Guazzaroni[78] have effected this allylation with allyl halides and zinc dust in an aqueous ammonium chloride solution with a solid organic support in place of THF. The support can be reverse-phase C-18 silica gel. One advantage of this new system is that a hydroxyl group does not require protection (eqs. 6.45, 6.46).

$$n\text{-}C_6H_{13}CHO + CH_3CH{=}CHCH_2Cl \xrightarrow[\text{C-18 SiO}_2\text{, 88\%}]{\text{NH}_4\text{Cl, Zn, H}_2\text{O}} n\text{-}C_6H_{13}CH(OH)CH(CH_3)CH{=}CH_2 \quad (6.45)$$

$$\text{2-hydroxycyclohexanone} + CH_2{=}CHCH_2Br \xrightarrow[\text{C-18 SiO}_2\text{, 90\%}]{\text{NH}_4\text{Cl, Zn, H}_2\text{O}} \text{1-}(CH_2CH{=}CH_2)\text{-cyclohexane-1,2-diol (OH, OH)} \quad (6.46)$$

Chan and co-workers have reported directed crossed aldol-type reactions in aqueous media.[79] The reactions involve a α-halocarbonyl compound and an aldehyde using metallic zinc or tin to promote the reaction in water (eqs. 6.47, 6.48).

$$PhC(O)CH(Br)CH_3 + PhCHO \xrightarrow[\text{rt, 82\%}]{\text{Zn, H}_2\text{O}} PhCH(OH)CH(CH_3)C(O)Ph \quad (6.47)$$

(Erythro / threo 71 : 29)

$$PhC(O)CH(Br)CH_3 + PhCHO \xrightarrow[80\ ^\circ\text{C, 82\%}]{\text{Sn, H}_2\text{O}} PhCH(OH)CH(CH_3)C(O)Ph \quad (6.48)$$

(Erythro / threo 47 : 53)

Experimental results discount the formation of classical metal enolates as intermediates. Side products derived from self-condensation of the aliphatic aldehydes, which would be expected from metal enolates in aqueous media, were not obtained. Also, free metal enolates should be extremely unstable under aqueous conditions. The authors have proposed the intermediacy of a radical anion species, generated by a SET process from the α-halocarbonyl compound. The carbon-carbon bond formation must occur prior to the formation of the free metal enolate, and presumably at the zinc (or tin) metal surface (Scheme 6.5).

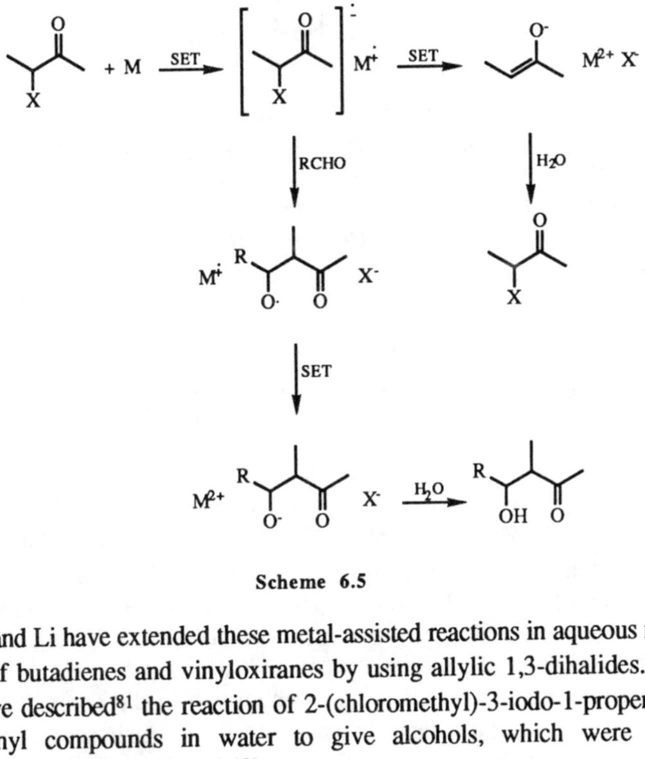

+ M —SET→ [ ]$^{\cdot-}$ M^{+} —SET→ M^{2+} X^-

RCHO ↓ H_2O ↓

M^{+} X^-

SET ↓

M^{2+} X^- —H_2O→ OH O

Scheme 6.5

Chan and Li have extended these metal-assisted reactions in aqueous media to the synthesis of butadienes and vinyloxiranes by using allylic 1,3-dihalides.[80] The same authors have described[81] the reaction of 2-(chloromethyl)-3-iodo-1-propene with zinc and carbonyl compounds in water to give alcohols, which were cyclized to methylenetetrahydrofurans (eq. 6.49).

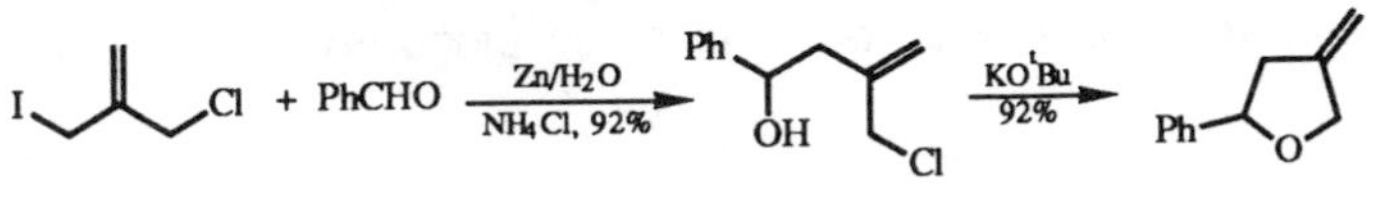

I–CH₂C(=CH₂)CH₂–Cl + PhCHO —Zn/H_2O, NH_4Cl, 92%→ Ph–CH(OH)–CH₂C(=CH₂)CH₂Cl —KO^{t}Bu, 92%→ Ph-(methylenetetrahydrofuran) (6.49)

The presence of water is critical for the success of this reaction. In THF or diethyl ether there was little or no formation of the alcohol. Furthermore, addition of water to the organic solvent resulted in the formation of the alcohol accompanied by the reduction product of the aldehyde, benzyl alcohol. The procedure works equally well with aldehydes and ketones, although the process is highly chemoselective and aldehydes react in presence of ketones. Conjugated carbonyl compounds give mainly 1,2-addition products. Again, a single electron transfer process on the metallic zinc surface has been invoked as the possible reaction mechanism.

One relevant aspect of this process is that the combination of this 1,3-dihalide with zinc can be considered as a functional equivalent of trimethylenemethane (TMM) (eq. 6.50).

(6.50)

These 1,3-dipolar TMM species introduced by Trost[82] have proved to be useful tools in organic synthesis and can react with a carbonyl function in a [3+2] fashion to afford cyclopentannulation products. There is a considerable interest in novel alternatives for the preparation of TMM derivatives. A very recent procedure utilizes a Barbier-type reaction between 3-chloro-2-chloromethylpropene and a carbonyl compound with lithium powder and a catalytic amount of naphthalene (eq. 6.51).[83]

1) Li, $C_{10}H_8$ (cat.), THF, -78 °C
2) H_2O, 67%

(6.51)

Likewise, a zinc-promoted 2-chloroallylation of aldehydes and ketones with 2,3-dichloropropene in a two-phase system of water and toluene containing a small amount of acetic acid has been described.[84] The method offers an efficient and mild synthesis of 3-chlorohomoallylic alcohols. The reaction proceeds well without adding acetic acid, but it requires longer reaction times. The process is very simple, gives high yields of products, and can be applied on an industrial scale.

An aqueous medium appears to be essential and in the absence of water no reaction occurs. The reaction is highly chemoselective and aldehydes are more reactive than ketones, except cyclohexanone. Efficient syntheses from ketones, however, are performed simply by increasing the amounts of 2,3-dichloropropene and zinc. α,β-Unsaturated carbonyl compounds produce the 1,2-addition products exclusively (eqs. 6.52-6.53).

PhCHO + Zn, H_2O-toluene, 45 °C, 1 h, 95% (6.52)

Zn, H_2O-toluene, 45 °C, 1 h, 91% (6.53)

Although metallic zinc (with or without activation) has proved to be the most effective promoter of Barbier-type reactions in an aqueous solvent system, other metals have been used in allylations and alkylations of carbonyl compounds. Thus, tin metal was successfully employed in the ultrasonically assisted Luche allylation.[69-71] Chinese chemists have utilized tin metal in aqueous THF in the reaction of allyl bromides with α,β-unsaturated aldehydes, ketones, and quinones (eq. 6.54).[85]

Br + Ph CHO —Sn, THF-H_2O, 71%→ Ph ... OH (6.54)

Also, Schmid and Whitesides have reported[86] a carbon-carbon bond formation in aqueous ethanol using allyl bromide and tin metal with the aid of ultrasound. The process was applied to the diastereoselective transformation of unprotected carbohydrates to higher carbon sugars (eq. 6.55).

CHO, HO, OH, OH, CH_2OH + Br —Sn,)))), EtOH-H_2O (9:1)→ —five steps→ AcO, AcO, AcO, O, OCH_3, OAc (43%) (6.55)

Bimetallic combinations such as Sn-Al,[39,87] $SnCl_2$-Al,[88,89] and $BiCl_3$-Al[49] have been utilized in aqueous media. In addition to indium-mediated allylations in organic solvents,[51-54] Chan and Li have described organometallic reactions with indium metal in water.[90] Compared with zinc and tin, allylation reactions of aldehydes and ketones with indium proceed smoothly without the need of any promoter (eq. 6.56). Allyl iodides or bromides can be used equally well, as can even the less reactive allyl chlorides but the reaction needs longer reaction times. Aldehydes react rapidly whereas allylation of ketones requires prolonged reaction times. Hydroxyl groups do not require protection and acid sensitive groups such as acetal remain unaffected during allylation of the carbonyl group. Furthermore, side products such as alcohols or pinacols, which are frequently observed with zinc or tin, are completely absent with indium.

PhCHO + Br —In, H_2O, rt, 97%→ Ph (OH) (6.56)

Again, the authors propose that organometallic reactions must occur by SET processes on the indium surface. The reaction was extended to the synthesis of 2-

methylene-γ-lactones by reaction of carbonyl compounds with 2-bromomethyl acrylate.[90] The same authors have reported a concise and elegant synthesis of the sialic acid KDN in a few steps from D-mannose.[91] The key step was an indium-mediated coupling reaction in water.

Waldmann has investigated asymmetric Barbier reactions in aqueous media using α-keto amides of proline benzyl ester as chiral auxiliaries.[92] These react with allyl halides in the presence of zinc dust and pyridinium toluenesulfonate in a THF-water mixture (Scheme 6.6). The resulting α-hydroxy amides are obtained in high yields and with good diastereoselectivity. The chiral auxiliary can be easily removed from the amides by treatment with methyl lithium in THF yielding α-hydroxy ketones. It should be pointed out that valine benzyl ester as chiral auxiliary instead of proline benzyl ester, gave a very poor stereoselectivity.

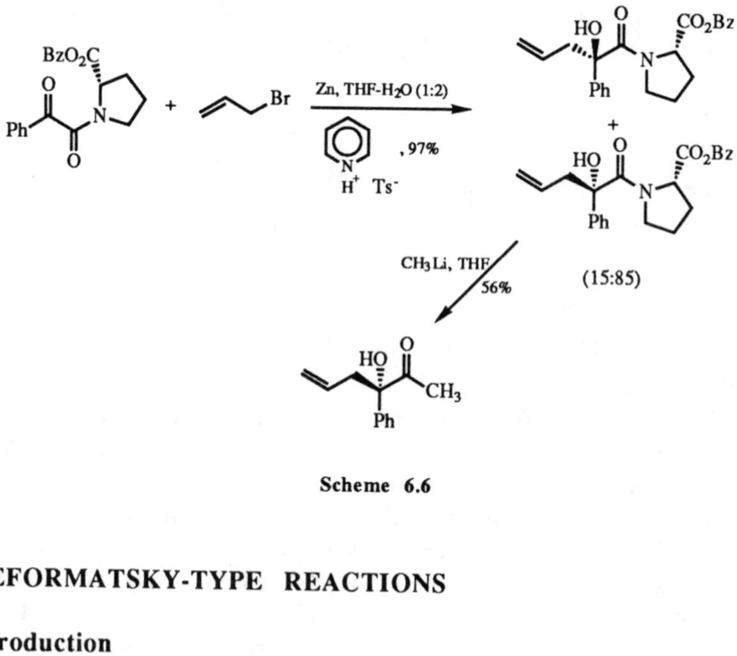

BzO_2C Ph O O N + Br Zn, $THF-H_2O$ (1:2) N H^+ Ts^- , 97% HO O CO_2Bz N Ph + HO O CO_2Bz N Ph (15:85) CH_3Li, THF 56% HO O CH_3 Ph

Scheme 6.6

B. REFORMATSKY-TYPE REACTIONS

1. Introduction

The Reformatsky reaction is a zinc-mediated process for the preparation of β-hydroxy esters from haloacetates and aldehydes or ketones (eq. 6.57). Since its discovery[93] chemists realized the importance of the reaction and the classical conditions were soon expanded to other alkyl 2-haloalkanoates as precursors of Reformatsky donor reagents, and electrophiles other than aldehydes or ketones were also employed. The successive reviews that have appeared in the literature evidence the increasing interest in this old transformation.[94-100]

$$R^1C(O)R^2 + XCH_2CO_2Et \xrightarrow[2)\ H_3O^+]{1)\ Zn} R^1R^2C(OH)CH_2CO_2Et \qquad (6.57)$$

A recent review[99] published in 1989 gives a concise and modern interpretation of this reaction, emphasizing the importance of metal reagents as well as other unusual conditions. As Fürstner has conspicuously mentioned,[99] the great versatility and applicability of this process have caused different reactions now to be considered as Reformatsky or Reformatsky-type reactions. In general, metal insertions into carbon-halogen bonds activated by carbonyl or carbonyl-derived groups in vicinal or vinylogeous positions are called Reformatsky-type reactions.

A similar transformation, called the Blaise reaction,[101] is carried out on nitriles. β-Keto esters are obtained in good yield by slow addition of α-bromoesters to nitriles in the presence of activated zinc (eq. 6.58).[100,102,103] An excess of the ester is necessary because some undergoes self-condensation rather than addition. Also, the reaction of α-bromoacetates gives generally low yields.

$$R\text{-}CN + R^1\text{-}CH(Br)\text{-}COOR^2 \xrightarrow[2)\ 70\text{-}95\%]{1)\ Zn,\ THF} R(NH_2)C{=}C(R^1)COOR^2 \xrightarrow[80\text{-}85\%]{H_3O^+} R\text{-}C(=O)\text{-}CH(R^1)\text{-}COOR^2 \qquad (6.58)$$

2. The Reformatsky Reagent

Early mechanistic considerations of the nature of intermetallic species in Reformatsky reactions centered on two possible monomeric forms of the reagent (**4** and **5**).[95,96]

$$H_2C{=}O \rightarrow Zn(Br)\text{-}CH_2\text{-}C(=O)OEt \quad \textbf{4}$$

$$H_2C{=}O \rightarrow Zn(Br)\text{-}O\text{-}C(OEt){=}CH_2 \quad \textbf{5}$$

In view of these structures, the chemical evidence favored the enolate form **5**. Further studies, however, by X-ray analysis evidenced dimeric structures in THF solution and in the solid state.[104,105] Thus, the zinc enolate of *tert*-butyl acetate was isolated as a crystalline compound having a cyclic dimeric structure where the zinc atoms are tetracoordinated with the bromide ion, the anionic carbon atom of the ester, the carbonyl oxygen of the second ester molecule, and THF (**6**).

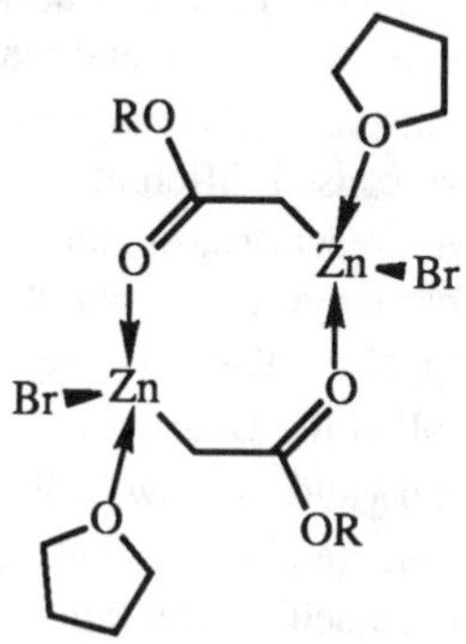

6

This arrangement is favored in the most polar solvents. The coordination with THF molecules could be replaced by the carbonyl oxygen of the electrophile, indicating again that the usual coordination number for zinc in solution is four. NMR data also support these structures and confirm an exclusive *C*-metallation.[105-107] As expected, sterically hindered esters destabilize the dimeric structure. In contrast, the more reactive zinc enolates of ketones present a different molecular arrangement. The majority of analytical methods also support the strong evidence for dimeric structures of the Reformatsky reagent; however ^{1}H- and ^{13}C-NMR data suggest the existence of *O*-metallated species in a closer analogy to alkali metal enolates.[108,109]

A theoretical study of the Reformatsky reagent by Dewar and Merz is in agreement with the aforementioned evidence, and also provides information on the reaction mechanism.[110] Although the MNDO method utilized predicts a coordination number around zinc to be three, further calculations with tetracoordinated zinc do not differ appreciably. The results agree with the experimental evidence that the reagent is a dimer in solution or in the crystal. Moreover, dimerization appears as an exothermic process by about 5 kcal/mol.

Previous mechanistic studies suggested two possible pathways for Reformatsky reactions.[104] In the first, the methylene carbon is attached to carbonyl carbon *via* a 1,3-sigmatropic reaction. In the second, the carbonyl carbon is attached to the transannular methylene. Dewar and Merz[110] found a transition state for the sigmatropic process but not for the latter, presumably due to a greater energetic cost. The authors also concluded that the Reformatsky reaction occurred *via* the monomeric species resulting from the dissociation of the dimeric complex by the action of a carbonyl compound (eq. 6.59). This implies the conversion of *C*-metallated into *O*-metallated species *via* a six-electron cyclic transition state.

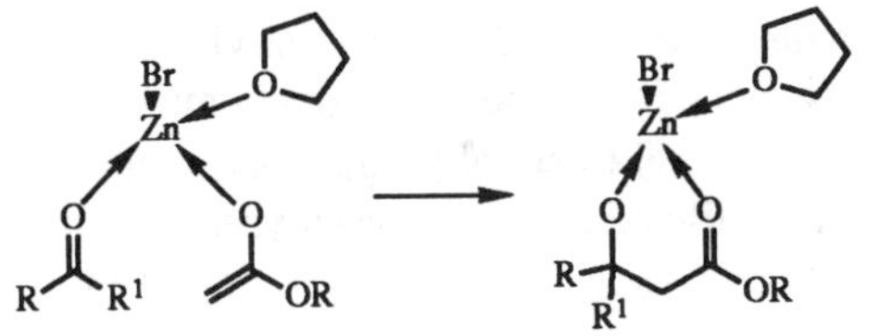

(6.59)

The intrinsic carbon-carbon bond formation resembles a [3,3]-sigmatropic process. It should be pointed out that some authors described how the Reformatsky reaction could proceed by a pericyclic mechanism.[111,112] Since the transition state of Reformatsky reactions adopts a chair conformation in a clear analogy with the Cope or Claisen rearrangements, the Reformatsky synthesis can be considered as a metallo-Claisen-type reaction.[99] This consideration along with the existence of a six-electron cyclic transition state with the carbonyl compound are consistent with the stereochemical features of Reformatsky reactions. More importantly, the effect of coordination on zinc involves the saturation of its empty orbitals by π-electron donation. This offers an explanation of the difficulties found in performing Reformatsky reactions with electrophiles other than carbonyl compounds, which have a less electrodonating effect.

3. The Role of Zinc Activation

Zinc for Reformatsky reactions can be activated by a plethora of procedures.[99,113] These include the simplest depassivating methods (see Chapter III) employing washing of the zinc, addition of iodine, 1,2-dibromoethane, metal halides, chlorotrimethylsilane, or the formation of zinc-copper or zinc-silver couples. With alkyl dihaloacetates, amalgams of very low zinc, magnesium, or calcium concentration were needed in the presence of a carbonyl compound.[114,115] Actually these substrates are better utilized with zinc/silver-graphite.[116] However, the most powerful activation methods for zinc, particularly Rieke-zinc, sonication, and zinc-graphites are preferred in Reformatsky reactions and result in high to excellent yields, good reproducibility, and faster reactions. Finely divided zinc allows Reformatsky reactions to proceed under milder conditions with uncommon donors and/or electrophiles, and finds applications in more complex Reformatsky reactions, especially intramolecular versions and stereoselective processes.

In many cases, the preparation of a zinc intermetallic couple is good enough for these reactions.[117-119] Both zinc-copper[120-124] and zinc-silver[125] couples are very effective for the metallation of iodo esters, ketones, and nitriles having the halogen atom in various positions. These processes require the presence of solvent mixtures containing aprotic dipolar solvents such as DMA, DMF, DMSO, or HMPA. Further addition of a carbonyl compound allows the so-called “remote Reformatsky reactions” to take place. Reactions with acid chlorides, aryl, allyl, and vinyl halides or sulfonates need catalysis by copper(I) salts or palladium(0) complexes. Thus, organozinc reagents derived from ethyl esters of β- and γ-iodo acids were prepared by using a Zn-Cu couple, and then utilized for the Pd-catalyzed coupling reaction with an acid chloride giving γ- and δ-ketoesters, respectively (eq. 6.60).[120]

$$I(CH_2)_nCO_2Et \xrightarrow[C_6H_6\text{-DMA}]{Zn\text{-}Cu} [\, IZn(CH_2)_nCO_2Et \,] \xrightarrow[Pd(0),\ C_6H_6,\ 72\text{-}100\%]{RCOCl} RCO(CH_2)_nCO_2Et \qquad n = 2, 3 \tag{6.60}$$

In contrast, aldehydes and ketones can be readily added in the presence of tris(isopropoxy)titanium chloride[122] or chlorotrimethylsilane[123] (eq. 6.61). Again, the organozinc intermediates were generated by reaction with Zn-Cu couple in toluene-DMA. DMA may be replaced as the co-solvent by N-methylpyrrolidone (NMP), but HMPA retards this reaction. In the presence of 1 equivalent of chlorotrimethylsilane the organozinc reagents add to aldehydes to form alcohols. This reaction can be utilized to obtain γ-, δ-, and ε-hydroxy esters from β-, γ-, and δ-zinc esters.

$$IZn(CH_2)_nCO_2Et + ArCHO \xrightarrow[\text{Toluene-DMA},\ 80\text{-}95\%]{TMSCl\ (1\ equiv)} ArCH(OH)(CH_2)_nCO_2Et \qquad n = 2, 3, 4 \tag{6.61}$$

Rieke-zinc[126,127] promotes reactions of ethyl bromoacetate with several aldehydes and ketones at room temperature. Yields of β-hydroxy esters are usually greater than 95% (eq. 6.62).

$$BrCH_2CO_2Et + \text{cyclohexanone} \xrightarrow[97\%]{\text{Rieke-Zn},\ Et_2O} \text{1-hydroxycyclohexyl-}CH_2CO_2Et \tag{6.62}$$

Similarly, reduction of zinc halides by sodium or lithium naphthalenide affords a finely dispersed and very reactive zinc metal, which is highly desirable for Reformatsky reactions.[128] The main disadvantage of these reductive procedures is that the reduction of zinc halides takes several hours. By contrast, zinc-graphites or zinc activated by sonication are generated rapidly and Reformatsky reactions are usually complete within a few minutes.

Reduction of anhydrous zinc halides by potassium-graphite generates a highly dispersed zinc on the graphite surface with an exceptional reactivity, and Reformatsky reactions proceed generally at room temperature.[129] The reactivity of this metal can be considerably enhanced by doping with a silver salt to afford zinc/silver-graphite.[130,131] With this reagent Reformatsky reactions proceed at low temperatures, even at -78 °C, and the less reactive α-haloalkanoates react also below 0 °C. These zinc-graphite reagents offer certain advantages. As a solvent, THF is a convenient and universal medium for Reformatsky reactions. Yields are high or excellent and side reactions usually observed in the Reformatsky reactions such as self-condensation of substrates, and elimination or retroaldolization of the intermediate α-bromozinc oxyester, are almost completely suppressed. In the reactions of allylic zinc bromides complete allylic rearrangement normally occurs (eq. 6.63-6.66).

$BrCH_2CO_2Si(CH_3)_3$ + cyclohexanone $\xrightarrow[\text{0 °C, 75\%}]{\text{Zn-Gr, THF}}$ 1-hydroxy-1-(CH_2CO_2H)cyclohexane (6.63)

Br–CH₂CH=CH–CO_2CH_3 + PhCHO $\xrightarrow[\text{20 °C, 85\%}]{\text{Zn-Gr, THF}}$ PhCH(OH)CH(CO_2Me)CH=CH₂ (6.64)

α-bromo-γ-butyrolactone + fluorenone $\xrightarrow[\text{0 °C, 88\%}]{\text{Zn-Gr, THF}}$ 9-hydroxy-9-(γ-butyrolacton-α-yl)fluorene (HO, O) (6.65)

α-bromo-γ-butyrolactone + $CH_2{=}\overset{+}{N}(CH_3)_2$ $\xrightarrow[\text{0 °C, 55\%}]{\text{Zn-Gr, THF}}$ α-($N(CH_3)_2$)-γ-butyrolactone (6.66)

Reformatsky reagents generated with zinc-graphite can be coupled with (*E*)-or (*Z*)-allyl acetates in palladium-catalyzed substitution reactions, to give mainly (*E*)-products (eq. 6.67).[132]

$R^1CH(Br)C(O)OR^2$ $\xrightarrow{\text{Zn-Gr, THF}}$ $\xrightarrow[\text{Pd(PPh}_3)_4\text{ (cat.)}]{R^3CH{=}CHCH(R^4)OAc}$ $R^3CH{=}CHCH(R^4)CH(R^1)CO_2R^2$ + $R^3CH(CH{=}CHR^4)CH(R^1)C(O)OR^2$ (6.67)

The zinc/silver-graphite reagent allows Reformatsky reactions at low temperatures with both α-chloro- and α-bromoalkanoates, in contrast to previously found differences for these substrates (eq. 6.68, 6.69).[130]

cyclohexanone + $BrCH_2C(O)OEt$ $\xrightarrow[\text{-78 °C, 92\%}]{\text{Zn/Ag-Gr, THF}}$ 1-hydroxy-1-($CH_2C(O)OEt$)cyclohexane (6.68)

cyclohexanone + $ClCH_2C(O)OEt$ $\xrightarrow[\text{-20 °C, 86\%}]{\text{Zn/Ag-Gr, THF}}$ 1-hydroxy-1-($CH_2C(O)OEt$)cyclohexane (6.69)

Zinc/silver-graphite has been also employed for diastereospecific Reformatsky-type reactions with chiral cyclic ketones (eq. 6.70).[133]

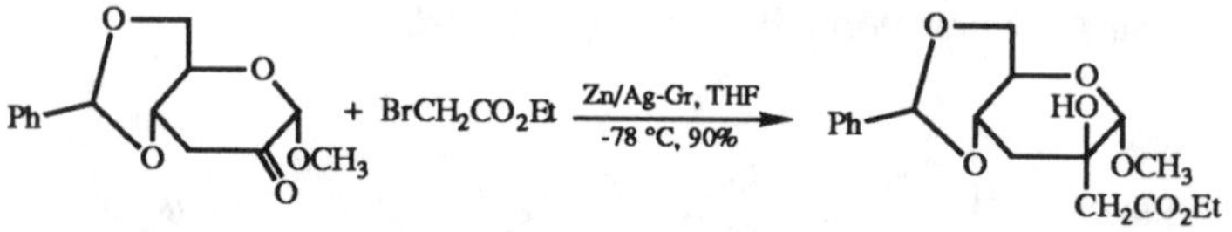

$$\text{Ph-(acetal)-ketone-OCH}_3 + BrCH_2CO_2Et \xrightarrow[-78\ ^\circ C,\ 90\%]{Zn/Ag\text{-}Gr,\ THF} \text{Ph-(acetal)-C(OH)(CH}_2CO_2Et)\text{-OCH}_3 \quad (6.70)$$

Ultrasonic radiation is beneficial to the Reformatsky reaction.[134] Yields are greater than 90% and the rate is enhanced. Thus, equation 6.71 illustrates this aspect and the β-hydroxy ester is obtained in 98% yield in only 5 min, whereas the silent process gives 98% yield in 1 h using activated zinc powder, or 95% in 5 h using the trimethyl borate-THF solvent system, and only 61% in 12 h using the classical Reformatsky reaction.

$$PhCHO + BrCH_2CO_2Et \xrightarrow[)))),\ 5\ min,\ 98\%]{Zn,\ dioxane} PhCH(OH)CH_2CO_2Et \quad (6.71)$$

By using sonication, specially activated zinc is not necessary. However, iodine and potassium iodide are effective additives, probably by suppressing enolization. The solvent of choice in this variation is dioxane, due to the better cavitation in a high boiling solvent. Ultrasonic Reformatsky reaction also gives good yields with perfluoroalkyl aldehydes.[18]

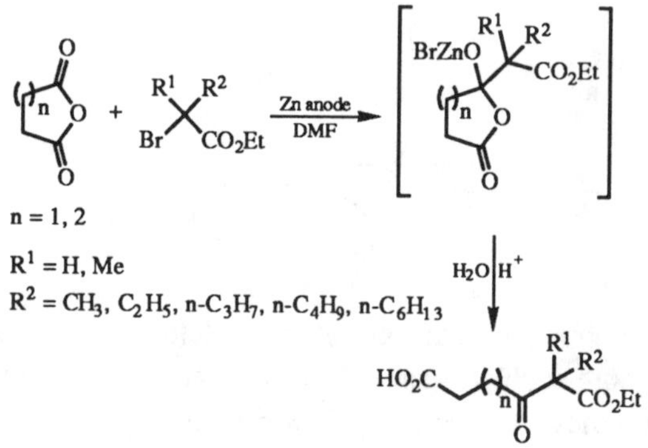

Anhydride (n) + $R^1R^2C(Br)CO_2Et$ →(Zn anode, DMF)→ [BrZnO–C(R^1R^2C–CO_2Et) lactone intermediate] →(H_2O | H^+)→ $HO_2C(CH_2)_nC(=O)CR^1R^2CO_2Et$

n = 1, 2

R^1 = H, Me

R^2 = CH_3, C_2H_5, n-C_3H_7, n-C_4H_9, n-C_6H_{13}

Scheme 6.7

Other variations have been recently introduced to extend the broad scope of the Reformatsky reactions. A series of 2-substituted 1-ethyl-3-oxoalkanedioates were obtained in good yield by an electroassisted Reformatsky reaction of ethyl 2-bromo-alkanoates with succinic anhydride.[135] The electrochemical procedure makes it possible to circumvent both the problems of zinc activation and the difficulties of the uncontrolled exothermic course of Reformatsky reactions with cyclic carboxylic anhydrides. The organometallic species were generated using a sacrificial zinc anode. The system also required a nickel cathode and tetrabutylammonium bromide as supporting electrolyte in DMF (Scheme 6.7).

One special approach, which has been scarcely exploited, involves the use of catalysts in order to perform Reformatsky reactions under milder conditions.[99] In a recent and very promising paper, Chinese chemists have reported a Reformatsky reaction catalyzed by titanocene dichloride.[136] β-Hydroxy esters were obtained in a few minutes with good yields (65-95%). The method presents important advantages over other protocols utilized in Reformatsky reactions. All reactions were conducted at room temperature, and neither anhydrous solvent (THF) nor activated zinc were required (eq. 6.72). In contrast, the only drawback of the process appears to be the low diastereoselectivity observed.

$$PhCHO + BrCH_2CO_2Et \xrightarrow[\text{THF, rt, 5 min, 90\%}]{\text{Zn-Cp}_2\text{TiCl}_2\text{(cat.)}} PhCH(OH)CH_2CO_2Et \qquad (6.72)$$

4. Metals Other than Zinc

The intermediate zinc enolates of Reformatsky reactions can be substituted by other metal enolates, which have extended the scope and selectivity of this reaction.

Magnesio-Reformatsky reactions utilizing magnesium in similar conditions to Grignard preparations, were essentially restricted to bromo *tert*-butyl esters[137,138] and bromo arylacetamides.[139] As in Grignard reactions, self-condensation is an undesirable side reaction. This inconvenience can be overcome by the extraordinary reactivity of zinc/silver-graphite.[130] Interestingly, magnesio-Reformatsky reactions can be also carried out with magnesium-graphite.[131] The high reactivity of this metal reagent allows reactions with ethyl α-halogenoalkanoates even at -78 °C to give the corresponding magnesio-ester enolate, which can react with a variety of substrates (eq. 6.73). In general, yields are moderate to good, although lower than those of classical Reformatsky reactions using activated zinc. Surprisingly, the highly reactive Rieke-magnesium was ineffective for this purpose.

$$BrCH_2CO_2Et \xrightarrow[\text{THF, -78 °C}]{\text{Mg-Gr}} [BrMgCH_2CO_2Et] \xrightarrow[\text{30 min, 71\%}]{\text{cyclohexanone, THF}} \text{1-hydroxycyclohexyl-}CH_2CO_2Et \qquad (6.73)$$

Lithium metal has been also employed in Reformatsky reactions.[140] However, lithium can cause reductive dehalogenations of haloesters and its use is strictly limited to either branched chain α-haloesters or to alkyl di- or trihaloacetates.

The preparation of Reformatsky-type reagents from cadmium powder is also possible. These reagents have been obtained from cadmium metal and *tert*-butyl α-bromoacetate in DMSO or HMPA, but in relatively low yields (20-60%).[141] Cadmium powder generated by reduction of cadmium chloride with lithium naphthalenide in DME or THF, reacts with α-haloesters to give the Reformatsky-type reagents which add to aldehydes or ketones in high yields (> 90%).[142] Reactions are slower than with other metals and an important limitation is the extreme toxicity of cadmium powder.

Tin metal in DMF or finely dispersed tin, prepared by reduction of tin(II) chloride in THF, have been utilized. The importance of tin-mediated Reformatsky-type reactions arises from the complexing ability of tin species, which are beneficial in stereocontrolled processes. Thus, α-haloketones and α-haloesters form aldols having *erythro* configuration preferentially.[143,144]

Reformatsky-type additions of haloacetonitriles to aldehydes are efficiently mediated by metallic nickel.[145] Activated nickel has been prepared in DME by the reduction of nickel(II) iodide with lithium naphthalenide. With this activated nickel, haloacetonitriles reacted with a variety of aryl and alkyl aldehydes to afford β-hydroxy nitriles in good yields (eq. 6.74).

$$RCHO + XCH_2CN \xrightarrow[\text{2) aq. HCl} \quad 59\text{-}84\%]{\text{1) Ni*, DME, 85 °C}} RCH(OH)CH_2CN \qquad (6.74)$$

This reaction is highly chemoselective and thus the addition of bromo acetonitriles to ketones gave very poor yields under similar conditions. Another advantage is the almost complete suppression of the self-condensation of aldehydes, due to the lower basicity of nickel intermediates compared to the usual zinc reagents.

Rieke-indium[146] and indium metal[147] have been also employed but require α-iodo esters for practical reaction rates.

Reaction of α-iodo esters with aldehydes or ketones in the presence of cerium metal or cerium amalgam gives good to excellent yields of β-hydroxy esters, and usually proceeds at low temperatures (eq. 6.75).[148] Halogen, nitrile, ester, and nitro groups are compatible within the ketone or aldehyde. Nevertheless, the diastereoselectivity in intermolecular Reformatsky-type reactions mediated by cerium metal is relatively low. The ratio *erythro*/*threo* is no greater than 57:43.

$$\text{cyclodecanone} + ICH_2CO_2Et \xrightarrow[\text{1 h, 75\%}]{\text{Ce, THF, 0 °C}} \text{1-hydroxy-1-}(CH_2CO_2Et)\text{cyclodecane} \qquad (6.75)$$

Lanthanum, samarium, and neodymium metals all promote this Reformatsky-type reaction, but cerium is usually employed. Curiously, the method is applied to ketones because aldehydes give preferentially pinacol products.[149]

In view of the limited and often unpredictable effects of reactive metals on Reformatsky reactions, these processes are now carried out by metal halides *via* the formation of the corresponding metal enolates.[99] Thus, titanium(II) chloride, chromium(II) chloride, samarium(II) iodide, and cerium(III) chloride or iodide among others, have been employed. Cerium(III) halides tend to favor aldol products and dehydration reactions are detected with other metal salts. From the viewpoint of the stereocontrol in Reformatsky-type reactions, chromium(II) chloride and particularly samarium(II) iodide have proved to be the most useful reagents. Chromium(II) chloride forms aldol products with high *erythro* selectivity. With samarium(II) iodide both α-halo esters and α-halo ketones can be utilized as substrates and intramolecular reactions proceed with complete stereocontrol.

5. Stereoselective Reformatsky Reactions

The Reformatsky reaction is one of the most useful methods for the formation of carbon-carbon bonds. However asymmetric reactions, which constitute a crucial target of modern organic syntheses, have received less attention. Stereoselective Reformatsky reactions represent a fascinating approach to many optically enriched molecules and have stimulated renewed interest in this old reaction. The stereochemical features of Reformatsky reactions have been previously reviewed.[150]

Although important degrees of asymmetric induction can be obtained from chiral donors or carbonyl acceptors, recent efforts to increase the stereoselectivity are focused mainly on the specific coordination of zinc or other metals with highly chelating agents.

The stereoselectivity is greatly favored by amino groups in the substrate having complexing ability.[151-154] This effect is particularly high when the nitrogen atom is at β-position to the carbonyl group. Thus in a diastereoselective Reformatsky reaction from the rigid 2-dibutylaminocyclobutanone, the *cis* product is largely favored by either zinc or lithium enolates (eq. 6.76).[154]

Bu_2N ... + $M\diagdown\!\!\diagup C(O)O^tBu$ → Bu_2N, HO, $CO_2{}^tBu$ + $^tBuCO_2$, Bu_2N, OH (6.76)

Metal	Yield (%)	cis/trans
ZnBr	70	9:1
Li	80	7:3

Similarly, other substrates containing amino or amido groups enhance the diastereoselectivity.[155-157] Again the importance of the metal enolate in these processes should be mentioned, since products obtained with a zinc enolate frequently have opposite configuration to those with a lithium enolate.[157]

Stereoselective Reformatsky reactions have been carried out with transition metal carbonyl complexes, which are usually compatible with zinc enolates under reaction conditions.[158,159] The chromium tricarbonyl complex of 2-methoxybenzaldehyde participates in a highly stereoselective Reformatsky reaction and *R,R/S,S* racemates were exclusively isolated in reactions with zinc ester and zinc nitrile enolates (eq. 6.77).[160]

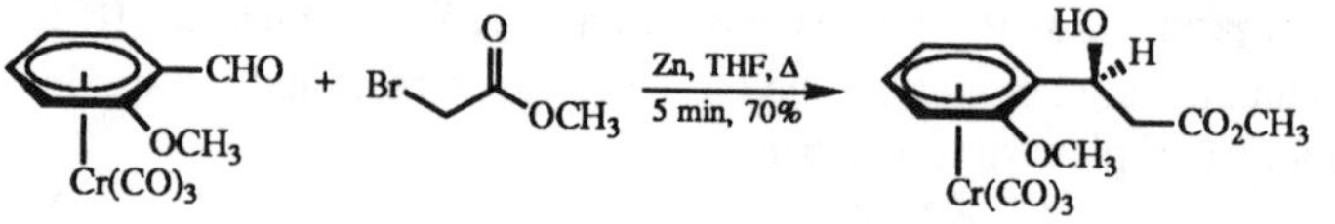

$$(\text{2-}CH_3O\text{-}C_6H_4CHO)Cr(CO)_3 + BrCH_2C(O)OCH_3 \xrightarrow[5\ \text{min},\ 70\%]{Zn,\ THF,\ \Delta} (\text{2-}CH_3O\text{-}C_6H_4)Cr(CO)_3\text{-}CH(OH)CH_2CO_2CH_3 \quad (6.77)$$

Another useful approach involves the addition of chiral ligands containing amino groups or amino alcohol units.[111,150,161] The addition of (-)-sparteine inverts the high *syn*-selectivity of zinc ketone enolates.[108] Japanese chemists have accomplished a concise study of the asymmetric synthesis of β-hydroxy esters using a Reformatsky reaction in the presence of chiral amino alcohols.[162] The Reformatsky reagent was prepared by the reaction of *tert*-butyl bromoacetate and Zn-Cu couple in THF at 90°C. This mixture was added to a THF solution of benzaldehyde and (*S*)-(+)-diphenyl(1-methylpyrrolidin-2-yl)methanol (DPMPM, 1 equiv) at 0 °C. The resulting optically active β-hydroxy ester was obtained in 91% yield with 75% ee having (*S*)-configuration. The result demonstrated that the Reformatsky reagent attacked from the *Si*-face in the presence of chiral amino alcohol. The same (*S*)-β-hydroxy ester was also obtained with a catalytic amount of (*S*)-(+)-DPMPM (0.4 equiv), but with only 44% ee (Scheme 6.8). The hydroxy group of the amino alcohol appears to be essential for the enantioselection. Very poor optical yields (1% ee) were obtained when the hydroxy group was protected as the methyl ether.

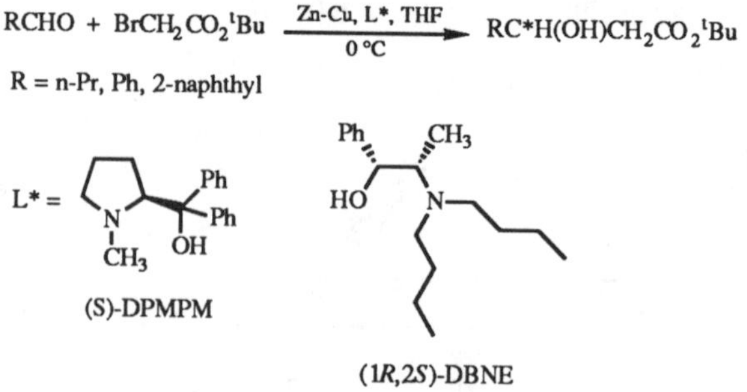

$$RCHO + BrCH_2CO_2{}^tBu \xrightarrow[0\ °C]{Zn\text{-}Cu,\ L^*,\ THF} RC^*H(OH)CH_2CO_2{}^tBu$$

R = n-Pr, Ph, 2-naphthyl

L* = (S)-DPMPM; (1*R*,2*S*)-DBNE

Scheme 6.8

Likewise, indium-induced Reformatsky reaction with stoichiometric amounts of chiral amino alcohols such as cinchonine and cinchonidine, gave optically active β-hydroxy esters in 40-70% ee.[163] The organoindium reagent prepared from ethyl iodoacetate and indium powder, was treated with benzaldehyde in the presence of a stoichiometric amount of cinchonine (eq. 6.78). The ee was estimated to be 71% and the configuration of the major enantiomer was determined to be *S*. However, the use of cinchonidine afforded the (*R*)-enantiomer predominantly (64% ee).

Ketones did not react with this indium-Reformatsky reagent in the presence of chiral amino alcohols, probably owing to the increased bulk of the chelated reagents. Furthermore, the method appears to be limited to aromatic aldehydes which gave good enantioselectivities. Curiously, zinc-Reformatsky reagent under the same conditions, gave no trace of β-hydroxy esters because zinc reagent is not compatible with the hydroxy group of amino alcohols. Other chiral ligands, quinine or quinidine, were also examined but ees. were lower.

$$ICH_2CO_2Et + PhCHO \xrightarrow[\text{cinchonine, 71\% ee}]{\text{In, THF-pentane}} PhC^{*}H(OH)CH_2CO_2Et \qquad (6.78)$$

References

1. Barbier, P. *C. R. Hebd. Acad. Sci. Paris* **1899**, *128*, 110.
2. Wagner, G.; Saytzeff, A. *Justus Liebigs Ann. Chem.* **1875**, *175*, 351.
3. Grignard, V. *C. R. Hebd. Acad. Sci. Paris* **1900**, *130*, 1322.
4. Blomberg, C.; Hartog, F. A. *Synthesis* **1977**, 18 and references cited therein.
5. Molle, G.; Bauer, P. *J. Am. Chem. Soc.* **1982**, *104*, 3481.
6. Moyano, A.; Pericàs, M. A.; Riera, A.; Luche, J.-L. *Tetrahedron Lett.* **1990**, *31*, 7619.
7. For a comprehensive treatment of Barbier-type reactions: Larock, R. C. *Comprehensive Organic Transformations*: VCH: New York, 1989; pp 553-557.
8. Pearce, P. J.; Richards, D. H.; Scilly, N. F. *J. Chem. Soc. Perkin Trans. 1* **1972**, 1655.
9. Scilly, N. F. *Synthesis* **1973**, 160.
10. Luche, J.-L.; Damanio, J. C. *J. Am. Chem. Soc.* **1980**, *102*, 7926.
11. Luche, J.-L; Petrier, C.; Gemal, A. L.; Zirka, N. *J. Org. Chem.* **1982**, *47*, 3805.
12. De Souza-Barboza, J. C.; Petrier, C.; Luche, J.-L.; *Tetrahedron Lett.* **1985**, *26*, 829.
13. Petrier, C.; Luche, J.-L.; Dupuy, C. *Tetrahedron Lett.* **1984**, *25*, 3463.

14. Greene, A. E.; Lansard, J. P.; Luche, J.-L.; Petrier, C. *J. Org. Chem.* **1984**, *49*, 931.
15. Einhorn, J.; Luche, J.-L. *Tetrahedron Lett.* **1986**, *27*, 501.
16. Kitazume, T.; Ishikawa, N. *Chem. Lett.* **1981**, 1679.
17. Kitazume, T.; Ishikawa, N. *J. Am. Chem. Soc.* **1985**, *107*, 5186.
18. Ishikawa, N.; Koh, M. G.; Kitazume, T.; Choi, S. K. *J. Fluorine Chem.* **1984**, *24*, 419.
19. Solladie-Cavallo, A.; Farkharic, D.; Fritz, S.; Lazrak, T.; Suffert, J. *Tetrahedron Lett.* **1984**, *25*, 4117.
20. Nutzel, K. *Houben Weyl-Methoden der organischen Chemie*; Georg Thieme Verlag: Stuttgart, 1973; Vol. XIII/2a, p 654.
21. Rousseau, G.; Conia, J. M. *Tetrahedron Lett.* **1981**, *22*, 649.
22. Shono, T.; Nishiguchi, I.; Sasaki, M. *J. Am. Chem. Soc.* **1978**, *100*, 4314.
23. Jackson, R. F. W.; James, K.; Wythes, M. J.; Wood, A. *J. Chem. Soc. Chem. Commun.* **1989**, 644.
24. Knochel, P.; Yeh, M. C. P.; Berk, S. C.; Talbert, J. *J. Org. Chem.* **1988**, *53*, 2390.
25. Yeh, M. C. P.; Knochel, P. *Tetrahedron Lett.* **1988**, *29*, 2395.
26. Yeh, M. C. P.; Knochel, P. *Tetrahedron Lett.* **1989**, *30*, 4799.
27. Chou, T.-S.; Knochel, P. *J. Org. Chem.* **1990**, *55*, 4791.
28. Retherford, C.; Knochel, P. *Tetrahedron Lett.* **1991**, *32*, 441.
29. Knoess, H. P.; Furlong, M. T.; Rozema, M. J.; Knochel, P. *J. Org. Chem.* **1991**, *56*, 5974.
30. Majid, T. N.; Knochel, P. *Tetrahedron Lett.* **1990**, *31*, 4413.
31. Rao, S. A.; Tucker, C. E.; Knochel, P. *Tetrahedron Lett.* **1990**, *31*, 7575.
32. Dunn, M. J.; Jackson, R. F. W. *J. Chem. Soc. Chem. Commun.* **1992**, 319.
33. Yeh, M. C. P.; Chen, H. G.; Knochel, P. *Org. Synth.* **1991**, *70*, 195.
34. Knochel, P.; Rao, S. A. *J. Am. Chem. Soc.* **1990**, *112*, 6146.
35. Knochel, P.; Rozema, M. J.; Tucker, C. E.; Retherford, C.; Furlong, M.; Achyutharao, S. *Pure Appl. Chem.* **1992**, *64*, 361 and references cited therein.
36. Zhu, L.; Wehmeyer, R. M.; Rieke, R. D. *J. Org. Chem.* **1991**, *56*, 1445.
37. Shono, T.; Ishifune, M.; Kashimura, S. *Chem. Lett.* **1990**, 449.
38. Mukaiyama, T.; Harada, T. *Chem. Lett.* **1981**, 1527.
39. Nokami, J.; Otera, J.; Sudo, T.; Okawara, R. *Organometallics* **1983**, *2*, 191.
40. Uneyama, K.; Matsuda, H.; Torii, S. *Tetrahedron Lett.* **1984**, *25*, 6017.
41. Hiyama, T.; Sawahata, M.; Obayashi, M. *Chem. Lett.* **1983**, 1237.

42. Hiyama, T.; Obayashi, M.; Nakamura, A. *Organometallics* **1982**, *1*, 1249.
43. Okude, Y.; Hirano, S.; Hiyama, T.; Nozaki, H. *J. Am. Chem. Soc.* **1977**, *99*, 3179.
44. Hiyama, T.; Kimura, K.; Nozaki, H. *Tetrahedron Lett.* **1981**, *22*, 1037.
45. Butsugan, Y.; Ito, H.; Araki, S. *Tetrahedron Lett.* **1987**, *28*, 3707.
46. Wang, W. B.; Shi, L. L.; Xu, R. H.; Huang, Y. Z. *J. Chem. Soc. Perkin Trans. 1* **1990**, 424.
47. For a review on organoantimony reagents in organic synthesis: Huang, Y. Z. *Acc. Chem. Res.* **1992**, *25*, 182.
48. Yamamoto, Y.; Yatagai, H.; Naruta, Y.; Maruyama, K. *J. Am. Chem. Soc.* **1980**, *102*, 7107.
49. Wada, M.; Ohki, H.; Akiba, K. *Bull. Chem. Soc. Jpn.* **1990**, *63*, 1738 and references cited therein.
50. Tanaka, H.; Inoue, K.; Pokorski, U.; Taniguchi, M.; Torii, S. *Tetrahedron Lett.* **1990**, *31*, 3023.
51. Araki, S.; Ito, H.; Butsugan, Y. *J. Org. Chem.* **1988**, *53*, 1831.
52. Araki, S.; Katsumura, N.; Ito, H.; Butsugan, Y. *Tetrahedron Lett.* **1989**, *30*, 1581.
53. Araki, S.; Shimizu, T.; Johar, P. S.; Jin, S.-J.; Butsugan, Y. *J. Org. Chem.* **1991**, *56*, 2538.
54. Araki, S.; Shimizu, T.; Jin, S.-J.; Butsugan, Y. *J. Chem. Soc. Chem. Commun.* **1991**, 824.
55. Araki, S.; Imai, A.; Shimizu, K.; Butsugan, Y. *Tetrahedron Lett.* **1992**, *33*, 2581.
56. Chastrette, M.; Gauthier, R. *J. Organomet. Chem.* **1974**, *66*, 219.
57. Wu, T.-C.; Xiong, H.; Rieke, R. D. *J. Org. Chem.* **1990**, *55*, 5045.
58. Yanagisawa, A.; Habaue, S.; Yamamoto, H. *J. Am. Chem. Soc.* **1991**, *113*, 8955.
59. Inaba, S.; Rieke, R. D. *Synthesis* **1984**, 844.
60. For a review of samarium(II) iodide: Soderquist, J. A. *Aldrichimica Acta* **1991**, *24*, 15.
61. For a recent and excellent review of lanthanide reagents: Molander, G. A. *Chem. Rev.* **1992**, *92*, 29.
62. Imamoto, T. *Rev. Heteroat. Chem.* **1990**, *3*, 87.
63. Imamoto, T.; Takeyama, T.; Koto, H. *Tetrahedron Lett.* **1986**, *27*, 3243.
64. Imamoto, T.; Takiyama, N. *Tetrahedron Lett.* **1987**, *28*, 1307.
65. Ruder, S. M. *Tetrahedron Lett.* **1992**, *33*, 2621.
66. Hart, H.; Chen, B.; Peng, C. *Tetrahedron Lett.* **1977**, 3121.
67. Boldrini, G. P.; Savoia, D.; Tagliavini, E.; Trombini, C.; Umani-Ronchi, A. *J. Organomet. Chem.* **1985**, *280*, 307.

68. Savoia, D.; Trombini, C.; Umani-Ronchi, A. *Pure Appl. Chem.* **1985**, *57*, 1887 and references cited therein.
69. Petrier, C.; Luche, J.-L. *J. Org. Chem.* **1985**, *50*, 910.
70. Petrier, C.; Einhorn, J.; Luche, J.-L. *Tetrahedron Lett.* **1985**, *26*, 1449.
71. Einhorn, C.; Luche, J.-L. *J. Organomet. Chem.* **1985**, *322*, 177.
72. Petrier, C.; Dupuy, C.; Luche, J.-L. *Tetrahedron Lett.* **1986**, *27*, 3149.
73. Luche, J.-L.; Allavena, C. *Tetrahedron Lett.* **1988**, *29*, 5369.
74. Luche, J.-L.; Allavena, C.; Petrier, C.; Dupuy, C. *Tetrahedron Lett.* **1988**, *29*, 5373.
75. Dupuy, C.; Petrier, C.; Sarandeses, L. A.; Luche, J.-L *Synth. Commun.* **1991**, *21*, 643.
76. Sarandeses, L. A.; Mouriño, A.; Luche, J.-L. *J. Chem. Soc. Chem. Commun.* **1991**, 818.
77. Sarandeses, L. A.; Mouriño, A.; Luche, J.-L. *J. Chem. Soc. Chem. Commun.* **1992**, 798.
78. Wilson, S. R.; Guazzaroni, M. E. *J. Org. Chem.* **1989**, *54*, 3087.
79. Chan, T. H.; Li, C. J.; Wei, Z. Y. *J. Chem. Soc. Chem. Commun.* **1990**, 505.
80. Chan, T. H.; Li, C. J. *Organometallics* **1990**, *9*, 2649.
81. Li, C. J.; Chan, T. H. *Organometallics* **1991**, *10*, 2548.
82. For a review of TMM derivatives: Trost, B. M. *Angew. Chem. Int. Ed. Engl.* **1986**, *25*, 1.
83. Ramon, D. J.; Yus, M. *Tetrahedron Lett.* **1992**, *33*, 2217.
84. Oda, Y.; Matsuo, S.; Saito, K. *Tetrahedron Lett.* **1992**, *33*, 97.
85. Gao, X.; Huang, B.-Z.; Wu, S.-H. *Huaxue Xuebao* **1991**, *49*, 827.
86. Schmid, W.; Whitesides, G. M. *J. Am. Chem. Soc.* **1991**, *113*, 6674.
87. Nokami, J.; Wakabayshi, S.; Okawara, R. *Chem. Lett.* **1984**, 869.
88. Uneyama, K.; Kamaki, N.; Moriya, A.; Torii, S. *J. Org. Chem.* **1985**, *50*, 5396.
89. Uneyama, K.; Nanbu, H.; Torii, S. *Tetrahedron Lett.* **1986**, *27*, 2395.
90. Li, C. J.; Chan, T. H. *Tetrahedron Lett.* **1991**, *32*, 7017.
91. Chan, T. H.; Li, C. J. *J. Chem. Soc. Chem. Commun.* **1992**, 747.
92. Waldmann, H. *Synlett* **1990**, 627.
93. Reformatsky, S. *Ber. dtsch. chem. Ges.* **1887**, *20*, 1210.
94. Shriner, R. L. *Org. React.* **1942**, *1*, 1.
95. Rathke, M. W. *Org. React.* **1975**, *22*, 423.
96. Gaudemar, M. *Organomet. Chem. Rev. A* **1972**, *8*, 183.
97. Diaper, D. G. M.; Kuksis, A. *Chem. Rev.* **1959**, *59*, 89.
98. Nützel, K. *Houben Weyl-Methoden der organischen Chemie*; Georg Thieme Verlag: Stuttgart, 1973; Vol. XIII/2a, p 805.
99. Fürstner, A. *Synthesis* **1989**, 571 and references cited therein.

100. For an article on zinc enolates discussing the Reformatsky and Blaise reactions: Rathke, M. W.; Weipert, P. *Comprehensive Organic Synthesis*; Heathcock, C. H., Ed.; Pergamon Press: Oxford, 1991; Vol. 2.
101. Blaise, E. E. *C. R. Hebd. Acad. Sci. Paris* **1901**, *132*, 478.
102. Bellassoued, M.; Gaudemar, M. *J. Organomet. Chem.* **1974**, *81*, 139.
103. Hannick, S. M.; Kishi, Y. *J. Org. Chem.* **1983**, *48*, 3833.
104. Dekker, J.; Boersma, J.; Van der Kerk, G. J. M. *J. Chem. Soc. Chem. Commun.* **1983**, 553.
105. Dekker, J.; Budzelaar, P. H. M.; Boersma, J.; Van der Kerk, G. J. M. *Organometallics* **1984**, *3*, 1403.
106. Orsini, F.; Pelizzoni, F.; Ricca, G. *Tetrahedron* **1984**, *40*, 2781.
107. Orsini, F.; Pelizzoni, F.; Ricca, G. *Tetrahedron Lett.* **1982**, *25*, 3945.
108. Hansen, M. M.; Bartlett, P. A.; Heathcock, C. H. *Organometallics* **1987**, *6*, 2069.
109. Dekker, J.; Schouten, A.; Budzelaar, P. H. M.; Boersma, J.; Van der Kerk, G. J. M.; Spek, A. Y.; Duisenberg, A. J. M. *J. Organomet. Chem.* **1987**, *320*, 1.
110. Dewar, M. J. S.; Merz, Jr., K. M. *J. Am. Chem. Soc.* **1987**, *109*, 6553.
111. Guetté, M.; Capillon, J.; Guetté, J. P. *Tetrahedron* **1973**, *29*, 3659.
112. Balsamo, A.; Barili, P. L.; Crotti, P.; Ferretti, M.; Macchia, B.; Macchia, F. *Tetrahedron Lett.* **1974**, 1005.
113. Erdik, E. *Tetrahedron* **1987**, *43*, 2203.
114. Darzens, G. *C. R. Hebd. Acad. Sci. Ser. C.* **1936**, *203*, 1374.
115. Darzens, G.; Levy, A. *C. R. Hebd. Acad. Sci. Ser. C.* **1937**, *204*, 272.
116. Fürstner, A. *J. Organomet. Chem.* **1987**, *336*, C33.
117. Santaniello, E.; Manzocchi, A. *Synthesis* **1977**, 698.
118. Kuroboshi, M.; Ishihara, T. *Tetrahedron Lett.* **1987**, *28*, 6481.
119. Slougui, N.; Rousseau, G. *Synth. Commun.* **1987**, *17*, 1.
120. Tamaru, Y.; Ochiai, H.; Nakamura, T.; Tsubaki, K.; Yoshida, Z. *Tetrahedron Lett.* **1985**, *26*, 5559.
121. Tamaru, Y.; Ochiai, H.; Nakamura, T.; Yoshida, Z. *Angew. Chem. Int. Ed. Engl.* **1987**, *26*, 1157.
122. Ochiai, H.; Nishihara, T.; Tamaru, Y.; Yoshida, Z. *J. Org. Chem.* **1988**, *53*, 1343.
123. Tamaru, Y.; Nakamura, T.; Sakaguchi, M.; Ochiai, H.; Yoshida, Z. *J. Chem. Soc. Chem. Commun.* **1988**, 610.
124. Tamaru, Y.; Ochiai, H.; Nakamura, T.; Yoshida, Z. *Tetrahedron Lett.* **1986**, *27*, 955.
125. Nakamura, E.; Sekiya, K.; Kuwajima, I. *Tetrahedron Lett.* **1987**, *28*, 337.
126. Rieke, R. D.; Uhm, S. T. *Synthesis* **1975**, 452.
127. Rieke, R. D. *Acc. Chem. Res.* **1977**, *10*, 301.

128. Rieke, R. D.; Li, P. T. J.; Burns, T. P.; Uhm, S. T. *J. Org. Chem.* **1981**, *46*, 4323.
129. Boldrini, G. P.; Savoia, D.; Tagliavini, E.; Trombini, C.; Umani-Ronchi, A. *J. Org. Chem.* **1983**, *48*, 4108.
130. Csuk, R.; Fürstner, A.; Weidmann, H. *J. Chem. Soc. Chem. Commun.* **1986**, 775.
131. Csuk, R.; Glänzer, B. I.; Fürstner, A. *Adv. Organomet. Chem.* **1988**, *28*, 85 and references cited therein.
132. Boldrini, G. P.; Mengoli, M.; Tagliavini, E.; Trombini, C.; Umani-Ronchi, A. *Tetrahedron Lett.* **1986**, *27*, 4223.
133. Csuk, R.; Fürstner, A.; Sterk, H.; Weidmann, H. *J. Carbohydr. Chem.* **1986**, *5*, 459.
134. Han, B. H.; Boudjouk, P. *J. Org. Chem.* **1982**, *47*, 5030.
135. Schwarz, K.-H.; Kleiner, K.; Ludwig, R.; Schick, H. *J. Org. Chem.* **1992**, *57*, 4013 and references cited therein.
136. Ding, Y.; Zhao, G. *J. Chem. Soc. Chem. Commun.* **1992**, 941.
137. Moriwake, T. *J. Org. Chem.* **1966**, *31*, 983.
138. Borno, A.; Bigley, D. B. *J. Chem. Soc. Perkin Trans. 2* **1983**, 1311.
139. Mladenova, M.; Blagoev, B.; Kurtev, B. *Bull. Soc. Chim. Fr.* **1979**, 77.
140. Villieras, J.; Perriot, P.; Bourgain, M.; Normant, J. F. *J. Organomet. Chem.* **1975**, *102*, 129.
141. Gaudemar, M. *C. R. Hebd. Acad. Sci. Ser. C.* **1969**, *268*, 1439.
142. Burkhardt, E. R.; Rieke, R. D. *J. Org. Chem.* **1985**, *50*, 416.
143. Harada, T.; Mukaiyama, T. *Chem. Lett.* **1982**, 161.
144. Harada, T.; Mukaiyama, T. *Chem. Lett.* **1982**, 467.
145. Inaba, S.; Rieke, R. D. *Tetrahedron Lett.* **1985**, *26*, 155.
146. Chao, L. O.; Rieke, R. D. *J. Org. Chem.* **1975**, *40*, 2253.
147. Araki, S.; Ito, H.; Butsugan, Y. *Synth. Commun.* **1988**, *18*, 453.
148. Imamoto, T.; Kusumoto, T.; Tawarayama, Y.; Sugiura, Y.; Mita, T.; Hatanaka, Y.; Yokoyama, M. *J. Org. Chem.* **1984**, *49*, 3904.
149. Fukuzawa, S.; Sumimoto, T.; Fujinami, T.; Sakai, S. *J. Org. Chem.* **1990**, *55*, 1628.
150. Heathcock, C. H. *Asymmetric Syntheses*; Morrison, J. D., Ed.; Academic Press: Orlando, 1984; Vol. 3, p 144.
151. Lucas, M.; Guetté, J. P. *J. Chem. Res. (S)* **1978**, 214.
152. Lucas, M.; Guetté, J. P. *Tetrahedron* **1978**, *34*, 1681.
153. Lucas, M.; Guetté, J. P. *J. Chem. Res. (S)* **1980**, 53.
154. Adlington, R. M.; Baldwin, J. E.; Jones, R. H.; Murphy, J. A.; Parisi, M. F. *J. Chem. Soc. Chem. Commun.* **1983**, 1479.
155. Ito, Y.; Terashima, S. *Tetrahedron Lett.* **1987**, *28*, 6625.
156. Ito, Y.; Terashima, S. *Tetrahedron Lett.* **1987**, *28*, 6629.

157. Van der Steen, F. H.; Jastrzebski, J.; Van Koten, G. *Tetrahedron Lett.* **1988**, *29*, 2467.
158. Hisatome, M.; Watanabe, J.; Yamakawa, K.; Kozawa, K.; Uchida, T. *J. Organomet. Chem.* **1984**, *262*, 365.
159. Hisatome, M.; Kawajiri, Y.; Yamakawa, K. *Tetrahedron Lett.* **1982**, *23*, 1713.
160. Brocard, J.; Pelinski, L.; Lebibi, J. *J. Organomet. Chem.* **1987**, *337*, C47.
161. Braun, M. *Angew. Chem. Int. Ed. Engl.* **1987**, *26*, 24.
162. Soai, K.; Kawase, Y. *Tetrahedron: Asymmetry* **1991**, *2*, 781 and references cited therein.
163. Johar, P. S.; Araki, S.; Butsugan, Y. *J. Chem. Soc. Perkin Trans. 1* **1992**, 711.

VII. CYCLIZATIONS

A. SIMMONS-SMITH-TYPE REACTIONS

1. Introduction

The Simmons-Smith cyclopropanation of olefins is another relevant application of organozinc reagents in organic synthesis.[1-5] The importance and synthetic usefulness of this reaction arise from several unique characteristics: a) the process is practically general with regard to olefin structure, b) stereospecificity, with retention of olefin geometry, c) the stereoselection in Simmons-Smith reactions, consisting of the *syn* direction of the introduction of the methylene unit, is largely caused by oxygen functionalities within the substrate. This crucial *syn*-directing effect was early recognized,[6,7] and has been widely employed in synthesis (eq. 7.1).[8-16] d) Numerous functional groups, including carbonyls, are tolerated in moderation.

$$\text{R-CH=CH-CH(OH)-CH}_3 \xrightarrow[>99\%]{Zn\text{-}Cu/CH_2I_2} \text{cyclopropane product (R, H, H, HO, H, CH}_3) \qquad (7.1)$$

Because of the importance of enantiomerically pure chiral cyclopropanes,[17] considerable effort has been devoted to the performance of asymmetric Simmons-Smith reactions. These include chiral auxiliary-directed cyclopropanations of chiral acetals,[18,19] ketals,[20,21] enol ethers,[14,22] vinyl boronates,[23] iron complexes of α,β-unsaturated carbonyl compounds,[24,25] and allyl 2-hydroxy glucopyranosides.[15,16] In any event, with a few exceptions,[14-16,22] these methods do not exploit the particular *syn*-stereoselection promoted by free hydroxy groups.

The Simmons-Smith cyclopropanation does not proceed readily unless the zinc is first activated. In general this can be accomplished by using commonly activated zinc (*e.g.* iodine-activated zinc), or by forming zinc-copper or zinc-silver couples.[3] However, poor reproducibility (yields of cyclopropanated products with a Zn-Cu couple lie in the range 12-75%) and low yields are often attributed to the difficulties in the preparation of Zn-Cu couples.

Activated zinc powder generated by the Rieke method, with reduction of zinc chloride or bromide with potassium in THF by refluxing, has been examined for

Simmons-Smith reactions.[26] Rieke-zinc reacts rapidly with diiodomethane and in the presence of cyclohexene, norcarane is obtained in 25-30% yield. This zinc also reacts with dibromomethane but in very low yield (5%). In contrast, zinc activated by reduction with lithium naphthalenide gives excellent yields. After the preparation of Zn powder, DME is replaced with ether, and cyclohexene is added followed by the addition of dibromomethane with a molar ratio of Zn/cyclohexene/dibromomethane of 1/0.5/1. The solution was refluxed to give norcarane in 94% yield (eq. 7.2).

$$\text{cyclohexene} + CH_2Br_2 \xrightarrow[\text{6 h, 94\%}]{\text{Zn*, }Et_2O,\ \Delta} \text{norcarane} \qquad (7.2)$$

Dichloromethane was unreactive under these conditions and the high reactivity of diiodomethane gave no reproducible results. It should be also mentioned that the use of Zn-Cu couple (prepared from Zn dust and Cu(I) chloride) or Zn-Ag couple in the synthesis of norcarane resulted in excellent yields (more than 90%) after, respectively, 24 h or 2 h in boiling ether.

Zinc slurries obtained from metal vapor-solvent condensations have been employed in Simmons-Smith reactions.[27] These slurries are storable as dry powders for up to seven months without loss of activity. The yields are generally low and very dependent on the solvent (eq. 7.3).

$$\text{cyclohexene} + CH_2Br_2 \xrightarrow[\text{2-4 h, solvent}]{\text{Zn*(Cu), }\Delta} \text{norcarane} \qquad (7.3)$$

Solvent	Yield (%)
Diglyme	24
Dioxane	18
THF	5
Hexane	0.2

In 1982 Repic described a sonochemical modification of the Simmons-Smith reaction using ultrasonically activated zinc, which avoids the sudden exothermic process normally associated with the reaction (eq. 7.4).[28] The metal suffers an *in situ* ultrasonic activation, and a previous chemical activation (zinc-copper, zinc-silver couples and/or the use of iodine or lithium) is not required. Furthermore, good and reproducible yields were obtained using zinc dust or even the metal in the form of mossy rods, or foil. The ultrasonic source was a simple cleaning bath operating at 50 kHz. Importantly, the method was successfully employed in a large scale preparation.[29] Cyclopropanated products were obtained in good to excellent yields (67-97%).

$$CH_3(CH_2)_7CH{=}CH(CH_2)_7CO_2CH_3 \xrightarrow[\text{)))), 91\%}]{\text{Zn, } CH_2I_2} CH_3(CH_2)_7\text{-cyclopropane-}(CH_2)_7CO_2CH_3 \quad (7.4)$$

The advantages of ultrasonically generated zinc are reproducible reaction times and yields. Likewise, the method has several advantages over the normal method of cyclopropanation as a result of changing from zinc powder to the bulk metal. Thus, there is a considerable reduction in foaming (normally associated with ethene and cyclopropane formation); the exotherm is more evenly distributed; and the reaction can be easily controlled by removing the lump of metal from the reaction mixture, which also facilitates the work-up of the reaction.

The combination of ultrasonic zinc-diiodomethane in THF has been also used for the methylenation of carbonyl compounds.[30] Reactions are conducted at room temperature in a cleaning bath (41 kHz, 53 w). Higher yields are obtained with aldehydes than with ketones. Ketones are conveniently methylenated in moderate to good yield (54-73%). Bulk zinc can be utilized, and in the absence of ultrasound, the reaction does not occur. The reagent, (iodomethyl)zinc iodide, can be also prepared from Zn-Cu couple, but methylenation yields are about 40% after 6 h in THF.

2. Simmons-Smith Reagents

The Simmons-Smith reagents can be generated by different procedures, normally utilized for the synthesis of organozinc compounds.[31,32] These methods can be grouped into three general categories.

a) Type I reagent is formed by oxidative addition of a dihalomethane to activated zinc metal or zinc-copper couple (eq. 7.5).

$$CH_2I_2 + \text{Zn-Cu} \xrightarrow[\Delta]{Et_2O} ICH_2ZnI \quad (7.5)$$

Type I reagent was first prepared by Emschwiller[33] and has been widely used due to the ease of preparation. The difficulties encountered with the initial method of preparation of the Simmons-Smith couple were soon overcome by simpler procedures, which also enabled high yields and reproducible results.[3] Recent modifications[34,35] include the use of $Zn/TiCl_4/CH_2Br_2$ and $Zn/AcCl/CuCl/CH_2Br_2$.

b)Type II reagent is formed by reaction of a zinc(II) salt with a diazoalkane, usually diazomethane or an aryldiazomethane (eq. 7.6).

$$CH_2N_2 + ZnI_2 \xrightarrow{Et_2O} ICH_2ZnI + (ICH_2)_2Zn \quad (7.6)$$

This reagent was first reported by Wittig and Schwarzenbach,[36] but has been much less utilized. It can be generated at temperatures below ambient, but the experimental disadvantages are associated with handling and the requirement of drying large quantities of diazoalkanes for practical syntheses.

c) Type III reagent is generated by alkyl-exchange reaction between an alkylzinc reagent and an 1,1-dihaloalkane (eq. 7.7).

$$CH_2I_2 + Et_2Zn \longrightarrow ICH_2ZnEt + (ICH_2)_2Zn \qquad (7.7)$$

The method is often referred to as the Furukawa procedure[37] or the modified Simmons-Smith reagent.[38] This protocol offers inherent advantages: the reagent is obtained rapidly under mild conditions at room temperature or below, tolerates a wide range of substrates, and if necessary, noncoordinating solvents can be used.

Most reagents for Simmons-Smith cyclopropanations can be satisfactorily denoted as organozinc carbenoids, ICH_2ZnI, and in general ICH_2MI with metals other than zinc. Free carbenes are not apparently formed during the cyclopropanation, and the addition probably follows a concerted path.[31] However, it is not at all clear that the reagents generated by different methods are identical. Different reaction conditions needed to generate the reagent and formation of by-products suggest the existence of different intermetallic reagents. Nevertheless, the comparable reactivity of these reagents would imply methylene transfer agents with a related structure.

Structural characterization of the cyclopropanating species is minimal, and the evidence supporting organozinc carbenoids in solution arises mainly from product formation. Simmons *et al.*[39,40] reported that hydrolysis of the reagent formed by Zn-Cu couple and diiodomethane gave methyl iodide predominantly, and by treatment with iodine, diiodomethane was regenerated. Remarkably, the authors demonstrated that the (iodomethyl)zinc species serves as the methylenating agent, discarding the existence of free carbenes as intermediates. Finally, Simmons and his associates concluded that the active reagent is $(ICH_2)_2Zn/ZnI_2$, which results from a Schlenk-type equilibrium in close analogy to other organometallics. Spectroscopic evidence for such an equilibrium was reported later[41] by NMR analysis of the reagent generated from Zn-Cu couple and dibromomethane in THF.

A recent and absolutely clarifying study has been accomplished by Denmark *et al.*[42,43] on the solution- and solid-state structures of (halomethyl)zinc reagents. These authors reported spectroscopic investigations of (chloromethyl)zinc and (iodomethyl) zinc reagents, as well as the first X-ray crystallographic analysis of an (iodomethyl) zinc compound.[43]

By reaction of diethylzinc with ICH_2Cl or CH_2I_2 in deuterated aromatic solvents (benzene or toluene) complex spectra were obtained. However, $(ClCH_2)_2Zn$-DME and $(ICH_2)_2Zn$-DME complexes were cleanly generated in benzene-d_6. The study also demonstrates that bis(halomethyl)zinc reagents are stabilized by chelation with ethers or acetone. This coordination appears to saturate the empty orbitals of zinc, which are otherwise necessary for activation of the carbon-iodine bond. The directing effect of oxygen substituents can be attributed to a complexating effect.[44]

Further information on the stabilization by ether complexation was provided by X-ray analysis of $(ICH_2)_2Zn$ complex derived from a chiral glycol ether (eq. 7.8).[43]

OCH_3 OCH_3 $\xrightarrow[0\ °C]{Et_2Zn,\ hexane}$ $\xrightarrow[-20\ °C]{CH_2I_2}$ CH_3 O Zn CH_2I CH_2I O CH_3 (7.8)

3. Metals Other than Zinc

Analogously to zinc-based cyclopropanation reagents, other metal combinations have also been employed and proposed to contain (iodomethyl)metal linkages.

Vinylmetal species derived from B, Al, Si, Ge, or Sn can be cyclopropanated with some success with the Simmons-Smith reagent.[45,46] Interestingly, aluminum-based reagents react only with isolated olefins.[47,48]

Knochel has recently demonstrated the utility of (iodomethyl)zinc iodide for one-carbon homologation in combination with organocopper reagents.[49-51]

The treatment of samarium metal with diiodomethane offers a reagent suitable for cyclopropanation reactions, and the method is a useful alternative to the normal Simmons-Smith procedures.[52,53] Probably, ICH_2SmI species are formed by oxidative metallation, and further α-elimination generates samarium(II) iodide and methylene. Alternatively, cyclopropanation can be achieved with samarium amalgam in the presence of ICH_2Cl. Thus, reaction of geraniol with Sm/Hg- ICH_2Cl gives a single diastereoisomer in excellent yield (eq. 7.9). In contrast to the Zn-mediated reaction, no by-products resulting from cyclopropanation of the isolated olefin are detected. Isolated olefins and even homoallylic alcohols cannot be cyclopropanated with this method, which demonstrates the specificity of this samarium-based system for allylic alcohols.

OH $\xrightarrow[THF,\ -78\ °C\ to\ rt,\ 98\%]{Sm/Hg-ClCH_2I}$ OH (7.9)

The enhanced chemoselectivity is also complemented by a notable diastereoselectivity.[52-54] A plausible reason is that the initiation of Sm/Hg-promoted Simmons-Smith reactions occurs at -60 °C, whereas most zinc-mediated reactions are performed in boiling ether (eqs. 7.10, 7.11). Ethylidenation reactions can also be carried out with Sm/Hg, so avoiding the use of diethylzinc for such processes.[52,53,55-57]

OH $\xrightarrow[-78\ °C\ to\ rt,\ 85\%]{Sm/Hg-CH_2I_2,\ THF}$ OH (7.10)

>95% de

$$F_3C\text{-}CH(OH)\text{-}CH{=}CH\text{-}C_6H_{13} \xrightarrow[0\ ^\circ C\ \text{to rt},\ 92\%]{Sm/Hg\text{-}CH_2I_2,\ THF} \text{cyclopropane product } (F_3C, OH, H, C_6H_{13}) \qquad (7.11)$$

The treatment of α-haloketones with diiodomethane and samarium metal at 0 °C affords cyclopropanols in good yields.[58] Esters also react under the same conditions which constitute a one-carbon homologation route to cyclopropanols (eq. 7.12).[59] The reactions seem to involve an initial nucleophilic acyl substitution reaction which gives an α-iodoketone. This undergoes reduction by samarium or samarium(II) iodide and provides an enolate, which is cyclopropanated to afford the cyclopropanol.

$$H_{17}C_8\text{-}CH{=}CH\text{-}(CH_2)_7CO_2CH_3 \xrightarrow[50\ ^\circ C,\ 70\%]{Sm,\ CH_2I_2,\ THF} H_{17}C_8\text{-}CH{=}CH\text{-}(CH_2)_7\text{-}C_3H_4\text{-}OH \qquad (7.12)$$

Electron-deficient alkenes react with active methylene dibromides and indium metal to afford cyclopropanes in 35-95% yield (eq. 7.13).[60] A similar reaction of aldehydes with dibromomalononitrile, lithium iodide, and indium also gives epoxides.

$$CH_2{=}CHCOCH_3 + Br_2C(CN)_2 \xrightarrow[94\%]{In,\ LiI,\ DMF} \text{cyclopropane } (H_3COC,\ CN,\ CN) \qquad (7.13)$$

In addition to the vast number of Simmons-Smith-type reagents,[61] cyclopropane-forming reactions can be accomplished with other metallic systems following a different pathway. Thus, one type of Wurtz reaction using sodium or sodium-potassium alloy is useful in the closing of small rings, especially three-membered rings.[62] Similarly, 1,3-dibromopropane can be converted into cyclopropane by zinc and sodium iodide.[63] Another method utilizes a *gem*-dihalide and copper.[64,65]

An interesting cyclopropanation reaction uses dichlorocarbene generated by treating carbon tetrachloride with titanium(IV) chloride and lithium aluminum hydride.[66] The adducts with olefins are obtained in 60-82% yield (eq. 7.14).

$$\text{cyclohexene} \xrightarrow[CCl_4]{TiCl_4\text{-}LiAlH_4} \text{7,7-dichloronorcarane (Cl, Cl)} \qquad (7.14)$$

As previously mentioned (Chapter IV), a low-valent titanium reagent can also convert 1,3-diols to cycloproapnes, when at least one phenyl group at the α-position is present (eq. 7.15).[67,68] In general, this reductive cyclization yields a *cis-trans* mixture along with other by-products resulting from partial or complete deoxygenation.

$$\text{Ph}_2\text{C(OH)CH}_2\text{CH(OH)R} \xrightarrow[\text{THF}]{\text{TiCl}_3\text{-LiAlH}_4} \text{1,1-diphenyl-2-R-cyclopropane} \qquad (7.15)$$

B. CYCLOADDITIONS

Numerous cycloaddition reactions are facilitated by the presence of metals. In some cases, the metal reagent generates a suitable intermediate which readily adds to any dienophiles in the reaction mixture and affords high yields of adducts. Thus, the chemistry of *ortho*-quinodimethanes (*ortho*-xylylenes) has benefited greatly from such a strategy.

ortho-Quinodimethanes are very reactive dienes in the Diels-Alder reaction and can be generated *in situ* by a number of routes, mainly by thermolytic and photolytic processes.[69,70] Intramolecular Diels-Alder reactions of these intermediates have also been widely utilized in the synthesis of natural products.[70-73]

Dehalogenation of *vic*-dihalides with activated metals constitutes a useful procedure for the preparation of *o*-quinodimethanes. Zinc-silver couple is superior to activated zinc for generation of the *o*-quinodimethane from the α,α'-dibromo-*o*-xylene. Further cycloaddition with alkenes provides an attractive route to occidol sesquiterpenes (Scheme 7.1).[74]

$$\text{3,6-dimethyl-}\alpha,\alpha'\text{-dibromo-}o\text{-xylene } (CH_3, CH_2Br, CH_2Br, CH_3) \xrightarrow[\text{DMF}]{\text{Zn-Ag}} [o\text{-quinodimethane } (CH_3, CH_3)] \xrightarrow[74\%]{CH_2{=}CHCOCH_3} \text{5,8-dimethyl-2-}COCH_3\text{-tetralin} \xrightarrow[69\%]{CH_3Li} \text{5,8-dimethyl-2-}C(CH_3)_2OH\text{-tetralin}$$

Scheme 7.1

The highly reactive metallic nickel, generated by the reduction of nickel(II) iodide with lithium in the presence of naphthalene, reacts under mild conditions with *o*-xylene derivatives in the presence of electron-deficient olefins to give substituted 1,2,3,4-tetrahydronaphthalenes in moderate to good yields (Scheme 7.2).[75] The *o*-xylylene species formed exhibits a reactivity comparable to that of a reactive diene, although the intermediacy of a biradicaloid has also been proposed.[76] The process would involve insertion of nickel into one of the carbon-halogen bonds, followed by 1,4-elimination of nickel(II) halide to give the *o*-xylylene intermediate. The authors made several attempts to trap the nickel complexes, but they obtained only *o*-

xylylene polymers.[75]

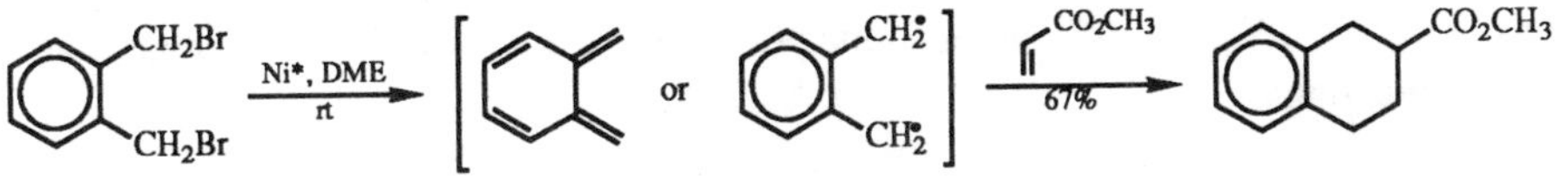

Scheme 7.2

Ultrasound induces a reaction between activated zinc (Cava's method)[77] and α,α'-dibromo-*o*-xylene to produce *o*-xylylene, which can be trapped by various dienophiles to form adducts in good yields.[78] In the absence of dienophile, dimerization of the intermediate occurs (about 10% yield), although an 80% yield of the dimer can be obtained by reaction of dibromo-*o*-xylene with lithium (1 equiv) in an ultrasonic bath. There is no reaction in the absence of ultrasound (Scheme 7.3).[79] This ultrasonic technique for the generation of *o*-xylylenes has also been employed in carbohydrate chemistry to produce compounds with ring similarities to anthracyclonones.[80]

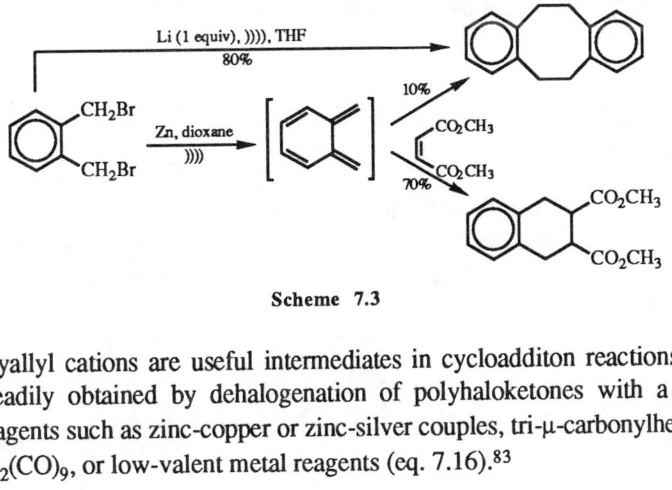

Scheme 7.3

2-Oxyallyl cations are useful intermediates in cycloadditon reactions.[81,82] They can be readily obtained by dehalogenation of polyhaloketones with a variety of reducing agents such as zinc-copper or zinc-silver couples, tri-μ-carbonylhexacarbonyl diiron, $Fe_2(CO)_9$, or low-valent metal reagents (eq. 7.16).[83]

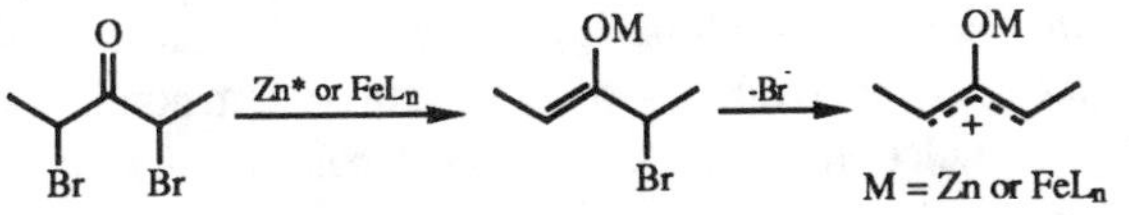

(7.16)

Thus, α,α'-dihaloketones can be dehalogenated with a zinc-copper couple on alumina.[84] The zinc oxyallyl cation can be trapped with isoprene to give a mixture of five- and seven-membered ketones (Scheme 7.4).[85] The product **1** is the natural terpene karahanaenone. The use of $Fe_2(CO)_9$ in the above reaction also results in the formation of the same three products with a somewhat greater proportion of the five-membered ketone (**3**). The naturally-occurring monoterpenoid karahanaenone (**1**) and its regioisomer (**2**) have also been obtained by another route, the reaction of α-halosilyl enol ethers with isoprene in the presence of zinc chloride,[86] which generates oxyallyl cations.

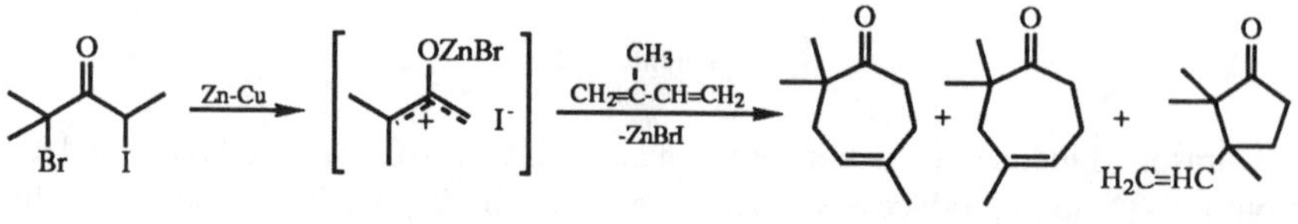

Zn-Cu; OZnBr; I⁻; CH_2=C(CH_3)-CH=CH_2; -ZnBrI; H_2C=HC

1 (54%) **2** (23%) **3** (23%)

Scheme 7.4

The [3+4]-cyclocoupling of polyhaloketones and furane represents a versatile cyclization process. The reaction is usually promoted by $Fe_2(CO)_9$,[87] but a zinc-silver couple is also effective.[88] Although yields are not so high, this variation is more economical for large scale preparations (eq. 7.17).

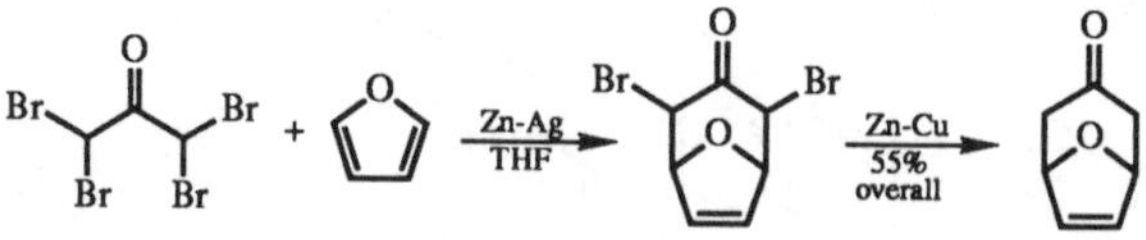

Zn-Ag, THF; Zn-Cu, 55% overall (7.17)

Another relevant alternative for the preparation of 2-oxyallyl carbocations from α,α'-dihaloketones is the use of iron-graphite (eq. 7.18).[89] The intermediates can be conveniently trapped, or can participate in cycloaddition reactions to afford a wide variety of cycloadducts.

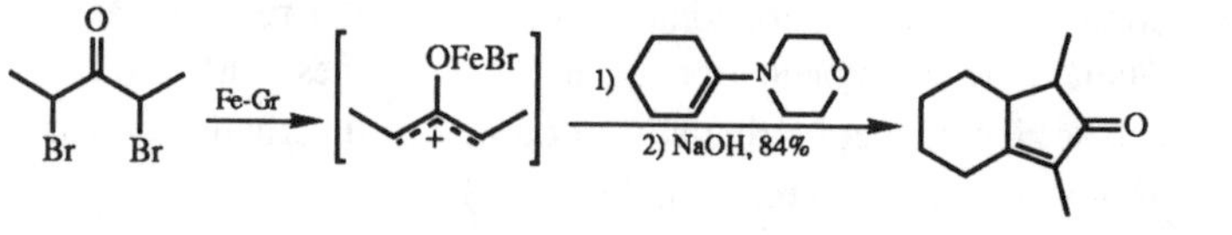

Fe-Gr; OFeBr; 1) 1-morpholinocyclohexene; 2) NaOH, 84% (7.18)

The zinc enolates are less electrophilic than the iron enolates, which is reasonably explained by the greater covalency of the iron-oxygen bond. Therefore, cycloadditions with poorly nucleophilic dienes proceed best in the presence of $Fe_2(CO)_9$ as reductant.

The zinc-promoted cycloadditon of α,α'-dibromoketones to 1,3-dienes is facilitated by ultrasound.[90] Oxyallyl cations can be generated in the presence of either zinc chloride or zinc-copper couple using dioxane as solvent. Highly hindered bicyclo[3.2.1]oct-6-en-3-ones are easily accessible by this method (eq. 7.19). The reaction in the absence of ultrasound gives only low to moderate yields and requires long reaction times.

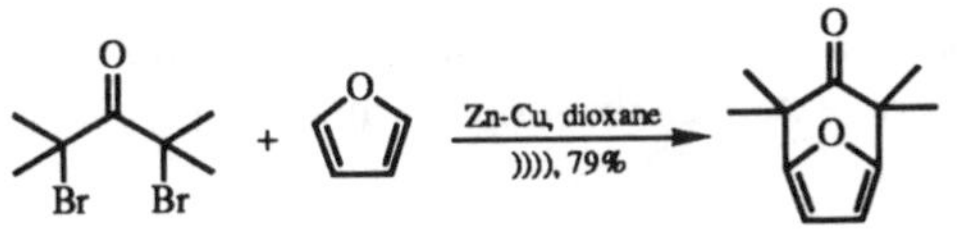

(7.19)

Similarly, sonically dispersed mercury emulsions are efficient in the reductive debromination of α,α'-dibromoketones in acetic acid to give α-acetoxyketones.[91] By using ketones as solvents the products are 1,3-dioxolan derivatives (eq. 7.20).[92]

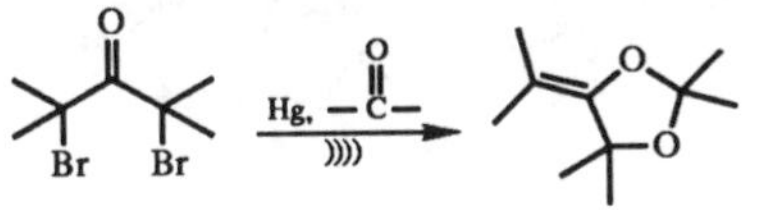

(7.20)

Metallo-ene reactions are another important and useful group of cycloaddition reactions. The intramolecular version may be highly regio- and stereoselective, thus being an efficient route to many natural products.[93,94]

As mentioned, ordinary magnesium can be activated by catalytic amounts of magnesium-anthracene complex in THF (Chapter III). This has been used in the synthesis of substituted allyl Grignard reagents in high yields.[95,96] The importance of magnesium-ene reactions starts by the discovery that allylic Grignard reagents can participate in the ene reaction under mild conditions by migration of magnesium and the concomitant formation of a new carbon-magnesium bond (eq. 7.21).[94,97]

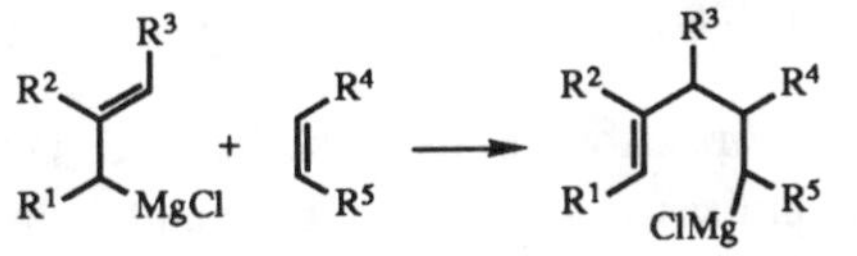

(7.21)

Thus, Scheme 7.5 highlights the synthesis of 6-protoilludene by intramolecular magnesium-ene and ketene/alkene addition reactions.[98] The key magnesium-ene/conjugate trapping step was performed by adding the chloride to magnesium-anthracene in THF at -65 °C; heating of the mixture to room temperature and further heating at 65 °C gave the non-isolated Grignard reagent. Addition of this compound to CuI-TMEDA in THF at -65 °C followed by addition of methyl-2-butynoate at -78 °C afforded the conjugate ester in 76% yield. Finally, a multistep and well-established sequence gave the desired product (Scheme 7.5).

Besides magnesium, the metallation/cyclization sequence can be performed with zinc and lithium.[94] These reactions involve the corresponding intramolecular metallo-ene additions, and a comparative study of their effectiveness has been accomplished.[99] Whereas intramolecular lithium-, magnesium-, and zinc-ene reactions are carried out in a stoichiometric fashion, Oppolzer and his associates have introduced, with very promising results, catalytic nickel-, palladium-, and platinum-ene reactions,[94] and reactions with other metals are currently under way. These processes utilize zero-valent organometallics and have extended the scope of metallo-ene reactions.

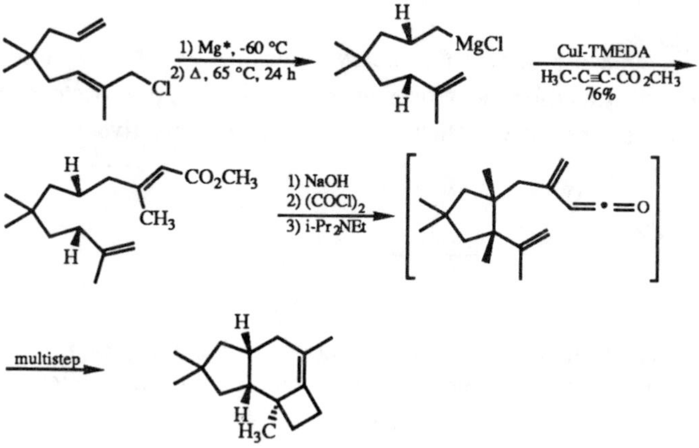

1) Mg*, -60 °C
2) Δ, 65 °C, 24 h

CuI-TMEDA
H_3C-C≡C-CO_2CH_3
76%

1) NaOH
2) $(COCl)_2$
3) i-Pr_2NEt

multistep

Scheme 7.5

C. REFORMATSKY-TYPE CYCLIZATIONS

1. Dreiding-Schmidt Reactions

A special Reformatsky-type reaction is the zinc-induced Dreiding-Schmidt reaction.[100-102] This process utilizes the alkyl 2-bromomethylpropenoates as Reformatsky donors. Apart from its interest as a useful metal-induced carbon-carbon bond-forming reaction, α-methylene-γ-lactones can be easily formed with these reagents.[103,104]

The method involves a single-step Reformatsky-type reaction in which the bromo ester is treated with zinc in anhydrous THF in the presence of the carbonyl compound (eq. 7.22).[101] Even higher and polycyclic ketones give almost quantitative yields. Formaldehyde, however, produces a mixture of products in low yield.

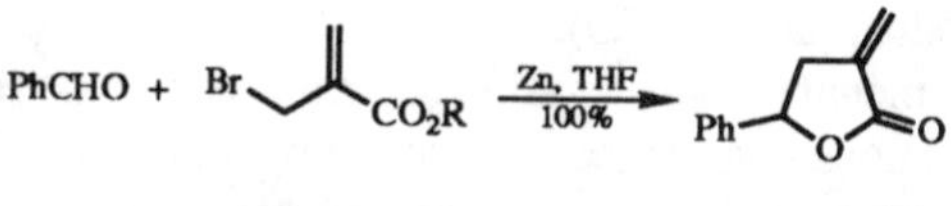

PhCHO + Br–CH₂C(=CH₂)CO_2R —Zn, THF, 100%→ (7.22)

Commercially available granulated zinc was initially employed, although further studies demonstrated that the product selectivity was markedly influenced by the degree of reactivity of the activated zinc. Thus equation 7.23, in which a carbohydrate-derived carbonyl compound is treated with ethyl 2-bromomethyl-2-propenoate, illustrates the importance of zinc activation on reaction conditions and selectivity.[105] Similarly, other carbohydrate,[106,107] steroid,[108] or alkaloid-derived[109,110] carbonyl compounds were reacted with alkyl 2-bromomethyl-2-propenoates to afford complex α-methylene-γ-lactones.

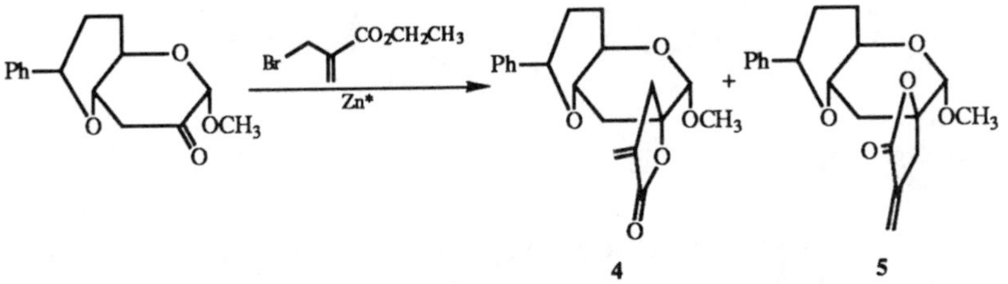

(7.23)

4 5

Activated Zn	Temp (°C)	Time (min)	4 / 5
Zn dust	66	200	54:10
Zn-Gr	0	10	84:12
Zn/Ag-Gr	-78	10	92:0

More functionalized alkyl 2-bromomethyl propenoates have been utilized[111] as well as the corresponding sulfone[112] or phosphonate derivatives.[113] While the former provide open chain products, the phosphonates give the phospholane oxides following a behavior comparable to that of carboxylic derivatives (eqs. 7.24, 7.25).

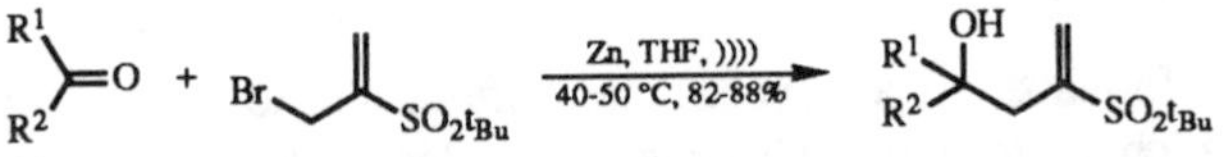

(7.24)

$R^1 = R^2 = H$, alkyl, aryl

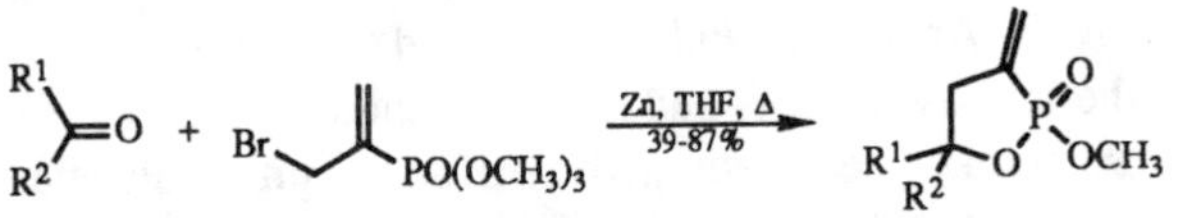

(7.25)

$R^1 = R^2 = H$, alkyl

Interestingly, lactonizations of carbonyl compounds with alkyl 2-bromomethyl alkenoates in the presence of zinc[114] or tin-aluminum couple[115] can be also carried out in aqueous media (eq. 7.26).

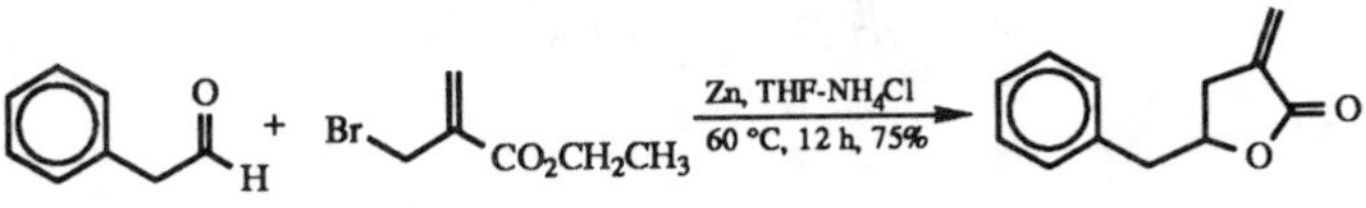

(7.26)

Chan and Li have recently introduced the combination of 2-(chloromethyl)-3-iodo-1-propene with zinc to perform a Barbier-type reaction in water (see Chapter VI).[116] The reagent reacts with carbonyl compounds to give alcohols, which can be converted into methylenetetrahydrofurans by treatment with potassium *tert*-butoxide in 2-propanol or hexane (eq. 7.27).

Zn, H_2O 45-95% ; KO^tBu 42-95% (7.27)

R^1 = alkyl, aryl
R^2 = H, alkyl

The same authors have also reported the synthesis of 2-methylene-γ-lactones by indium-promoted reaction of carbonyl compounds with methyl 2-bromomethyl-2-propenoate in water (eq. 7.28).[117] Reactions in the presence of indium metal proceed rapidly and without the need of any promoter, such as ammonium chloride or the use of sonication which are often required with zinc and tin. These reactions in aqueous media involve presumably radical species as intermediates rather than the true organometallic reagents.

PhCHO + Br–$CH_2C(=CH_2)CO_2Me$ → (In, H_2O, rt, 5 min, 96%) (7.28)

2. Intramolecular Reactions

Intramolecular Reformatsky-type cyclizations are valuable tools for the stereocontrolled formation of medium and large rings. Recent advancements have enhanced the importance of this strategy, which is especially important in the synthesis of natural products.[102] Again the kind of metal and type of activation have considerable influence on the stereoselectivity and reaction conditions.

By treatment of dibromodiketones with activated zinc, intramolecular ring closure occurs, probably *via* a biradical species. The resulting diketones can be dehydrogenated by DDQ or chloranil to heterocyclic quinones (eq. 7.29).[118]

Zn-Cu, DMSO, 50-60% ; DDQ 70-80% (7.29)

X = O, S, NCH_3

This intramolecular coupling of α-bromoketones can be extended to synthesis of cyclic 1,4-diketones. The reaction offers an attractive approach to the synthesis of substituted cyclooctanediones (eq. 7.30).[119]

$COCH_2Br$ $COCH_2Br$ Zn-Cu, DMSO NaI, $NaHCO_3$ 48% (7.30)

A synthesis of *cis*-fused α-methylene-γ-lactones can be achieved by reaction of an α-bromomethyl α,β-unsaturated ester with zinc dust at 65 °C or with zinc-copper couple at 25 °C (eq. 7.31).[120] The reaction involves an intramolecular Reformatsky-type reaction followed by spontaneous lactonization. The reaction of the starting bromoaldehyde with bis(1,5-cyclooctadiene)nickel also gives the same product in 52% yield. Thus, both metal reagents favor formation of a *cis*-ring fusion in this perhydroazulene derivative.

CH_2Br CO_2CH_3 CHO Zn-Cu, THF, 25 °C 60-62% (7.31)

Rieke-zinc[121,122] and ultrasonically activated zinc[123] promote intramolecular cyclizations, even with unusual Reformatsky acceptors (eq. 7.32).

CO_2CH_3 Br Zn, THF,)))) 25 °C, 70% CO_2CH_3 OH (7.32)

In a very interesting intramolecular Reformatsky reaction, ω-(α-bromoacyloxy) aldehydes undergo cyclization in the presence of diethylaluminum chloride (eq. 7.33).[124-126] The reaction represents an important approach to the construction of macrocycles.

Br CHO Zn, Et_2AlCl, THF 35-68% HO (7.33)

n = 9, 11, 12

Intramolecular samarium-Reformatsky reactions have also enabled the formation of medium- and large-ring molecules.[127,128] Furthermore the homogeneous reaction with samarium(II) iodide offers inherent advantages over the normal zinc-promoted Reformatsky reaction. This is particularly evident when zinc metal should be previously activated.

3. Other Zinc-mediated Cyclizations

In an efficient synthesis of γ-butyrolactones, Japanese chemists condensed dimethyl maleate with aldehydes or ketones in the presence of high purity zinc powder.[129] The resulting γ-substituted-β-carbomethoxy-γ-butyrolactones are obtained in 40-85% yield (eq. 7.34).

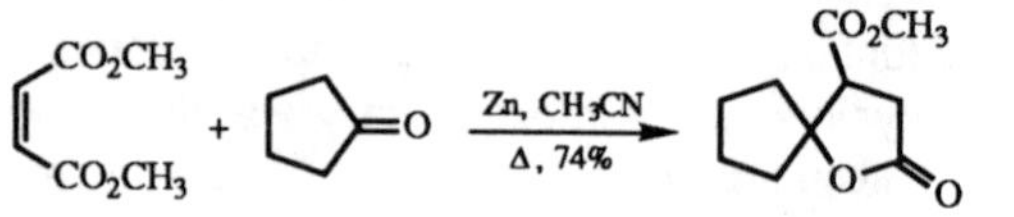

(7.34)

Likewise, activated zinc promotes a Reformatsky-type reaction (indeed a Blaise reaction) of α-bromo esters with *O*-silylated cyanohydrins to provide β-keto-γ-butyrolactones or tetronic acids (eq. 7.35).[130] THF was found to be the best solvent since in ether or ether-benzene mixtures the reaction was sometimes difficult to initiate.

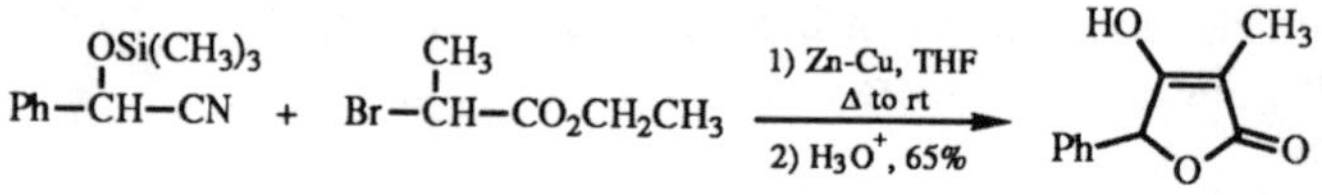

(7.35)

With a related methodology fluorinated β-keto-γ-butyrolactones or tetronic acids were prepared by an ultrasound-assisted Reformatsky-type reaction between *O*-trimethylsilylated cyanohydrins and ethyl α-fluoro(or trifluoromethyl)bromo-acetate.[131] In these conditions, reactions proceeded smoothly at room temperature using commercially available unactivated zinc powder. The reaction did not proceed in the absence of ultrasound (eq. 7.36).

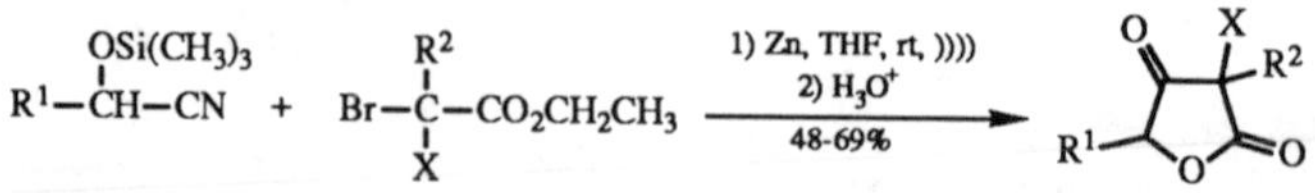

(7.36)

$R^1 = CH_3, C_6H_5, 4\text{-}CH_3OC_6H_4$
$R^2 = H, CH_3$
$X = F, CF_3$

It is known[132] that α-bromo esters and Schiff bases undergo Reformatsky-type reactions in the presence of activated zinc to form β-lactams. The so-called ester enolate-imine condensation is currently being researched and recent advances on this subject have been reported.[133,134] This reaction is markedly improved by ultrasonic irradiation and by use of dioxane as solvent.[135] β-Lactams were obtained in good yields (70-95%) in a few hours at room temperature, in the presence of activated (acid-washed) zinc granules and a crystal of iodine (eq. 7.37). Unactivated zinc granules can be also employed but give lower yields (50-70%). In the absence of ultrasound and using toluene as solvent, β-lactams were slowly obtained in the range of 25-50% yield. Nevertheless, this sonication procedure is limited to 3-unsubstituted β-lactams.[133,134]

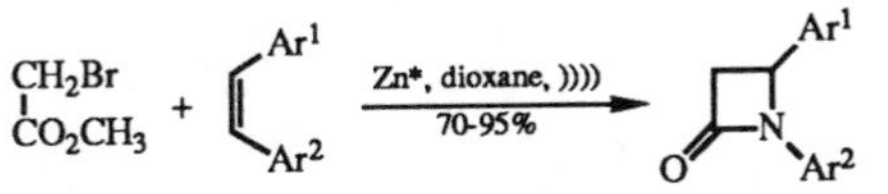

CH_2Br–CO_2CH_3 + $Ar^1CH{=}NAr^2$ —(Zn*, dioxane,)))); 70-95%)→ β-lactam (Ar^1, Ar^2) (7.37)

The [2+2]-cycloaddition of dichloroketene to alkenes is also improved by ultrasound.[136] Short reaction times, good yields, ambient temperatures, and the use of ordinary zinc dust, instead of a zinc-copper couple are significant advantages of this method (eq. 7.38). Cycloadducts from limonene and cycloalkenes, such as indene, norbornene, and 1,5-cyclooctadiene have been obtained in good yields.

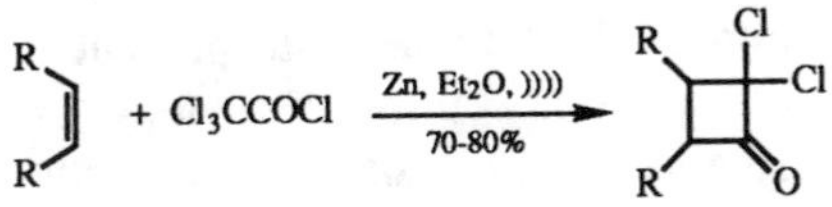

RCH=CHR + Cl_3CCOCl —(Zn, Et_2O,)))); 70-80%)→ 2,2-dichlorocyclobutanone (7.38)

A zinc-copper couple, however, induces a similar cycloaddition reaction between *N*-aryl-2-vinylazacycles and dichloroketene to afford macrocyclic lactams.[137] Subsequent exposure to certain electrophiles gives a single bicyclic product with total regio- and stereocontrol. This last step involves a simultaneous transannular cyclization-debenzylation that yields functionalized indolizidine and quinolizidine ring systems (Scheme 7.6).

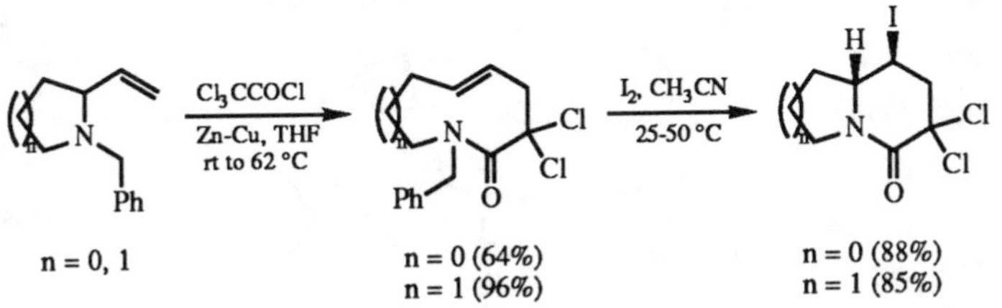

Cl_3CCOCl, Zn-Cu, THF, rt to 62 °C; I_2, CH_3CN, 25-50 °C

n = 0, 1 → n = 0 (64%), n = 1 (96%) → n = 0 (88%), n = 1 (85%)

Scheme 7.6

A combination of activated zinc and ultrasonic irradiation also promotes intramolecular cyclization of δ-iodoacetylenes to yield five-membered-ring products (eq. 7.39).[138]

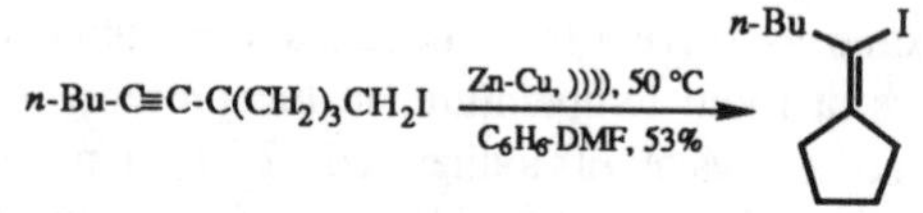
n-Bu-C≡C-C(CH2)3CH2I
Zn-Cu,)))), 50 °C
C6H6-DMF, 53%
n-Bu
I

(7.39)

Intramolecular additions of ketyl radical anions to carbon-carbon multiple bonds afford cyclized radical anions, that are further reduced and protonated.[139] These processes are not chain reactions and therefore require an excess of reducing agent (eq. 7.40).

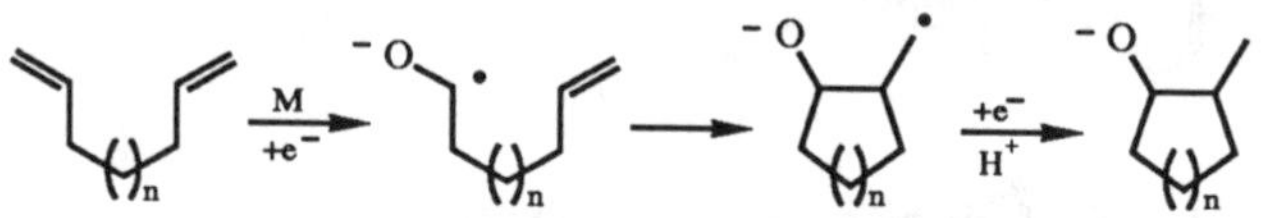
M
+e⁻
+e⁻
H⁺

(7.40)

Cyclic ketones bearing a δ,ε-unsaturated side chain at the α-position undergo five-membered ring annelation to cyclic pentanols on reaction with zinc and chlorotrimethylsilane in the presence of 2,6-lutidine.[140] The reaction probably involves reduction of the keto group to an α-trimethylsilyloxy radical, which adds to the unsaturated center. The function of 2,6-lutidine is to prevent proton or zinc-chloride-catalyzed elimination of the tertiary trimethylsilyloxy group from the intermediate. Aqueous work-up also effects hydrolysis of the silyl ether. Importantly, the bicyclic products generated with this protocol have a *cis*-ring fusion (eqs. 7.41-7.43).

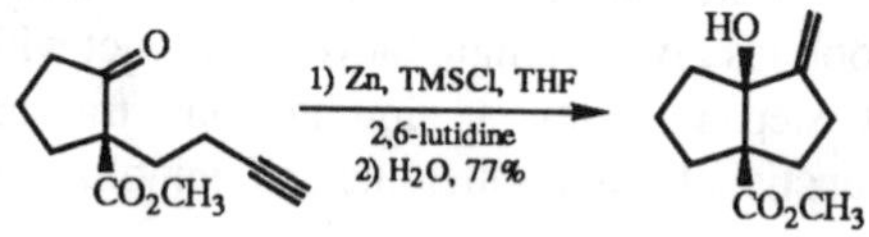
1) Zn, TMSCl, THF
2,6-lutidine
2) H2O, 77%
HO
CO2CH3

(7.41)

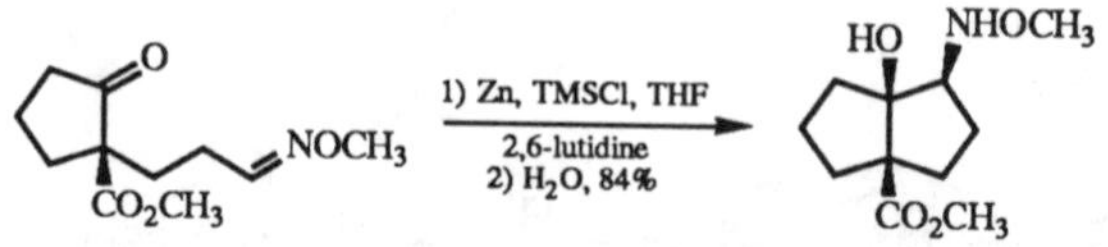
NOCH3
1) Zn, TMSCl, THF
2,6-lutidine
2) H2O, 84%
HO
NHOCH3
CO2CH3

(7.42)

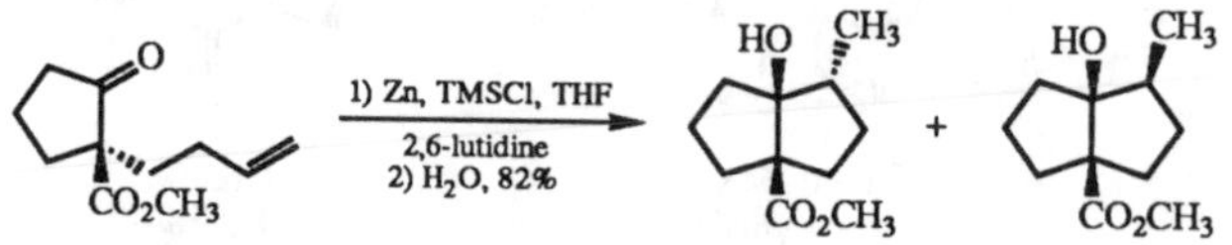
1) Zn, TMSCl, THF
2,6-lutidine
2) H2O, 82%
HO
CH3
CO2CH3
+

(7.43)

(5 : 1)

D. MISCELLANEOUS CYCLIZATIONS

Radical cyclizations under reductive conditions can be carried out with numerous metallic systems as electron donors.[139] Potassium (or sodium) in liquid ammonia is an effective reagent for reductive cyclization of 5-alkynyl ketones.[141] This reaction provides an efficient route to the D-ring of gibberellic acid (eq. 7.44).[142]

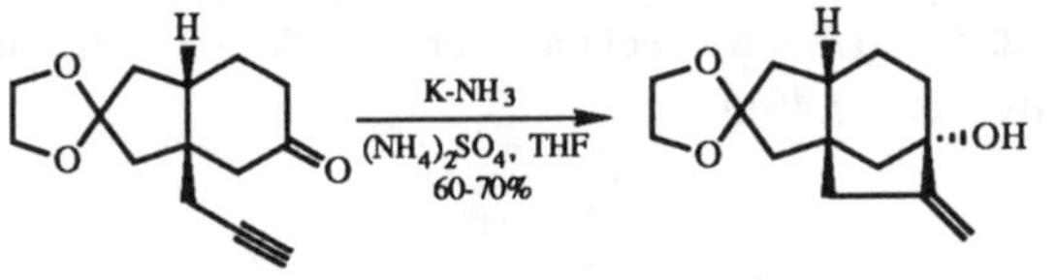

K-NH$_3$
(NH$_4$)$_2$SO$_4$, THF
60-70%

(7.44)

Sodium naphthalenide has been employed in a similar fashion for the reductive cyclization of γ-ethynyl ketones to allylic alcohols.[143] By-products resulting from overreduction are inhibited by using this reagent in THF or DME. The cyclization has been used in the synthesis of 2-methylenecyclopentanols and 2-methylene-cyclohexanols[144] as shown in Scheme 7.7.

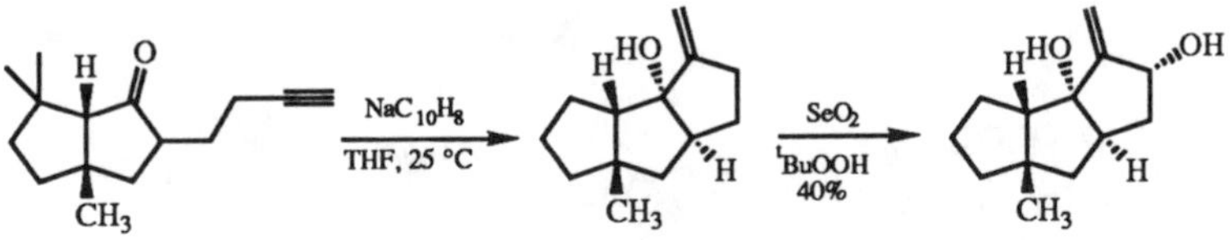

NaC$_{10}$H$_8$
THF, 25 °C

SeO$_2$
^{t}BuOOH
40%

Scheme 7.7

The Barbier reaction can be adapted to effect useful intramolecular cyclizations.[145] The reaction has been used for cyclopentenone annelation. Thus, α-allyl cycloalkanones are easily converted into iodo ketones, which on treatment in THF with magnesium activated by mercury(II) chloride are transformed into bicycloalkanols in moderate to good yield (Scheme 7.8).[146] Formation of *cis*-fused products is greatly favored. The annelation reaction gives poor yields when extended to six- and seven-membered rings.

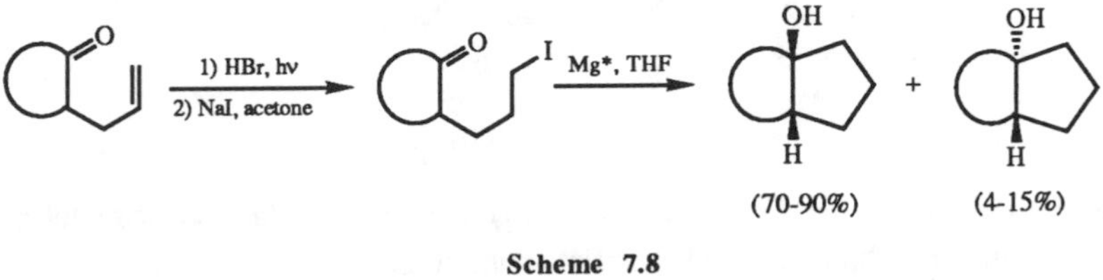

1) HBr, hν
2) NaI, acetone

Mg*, THF

(70-90%) (4-15%)

Scheme 7.8

The classical acyloin condensation, in which carboxylic esters afford α-hydroxy ketones by heating with sodium in refluxing solvents, has been used to prepare cyclic systems from diesters in boiling xylene.[147-149] Yields in the acyloin condensation can be improved by the Rühlmann procedure by running the reaction in the presence of chlorotrimethylsilane.[150] This method is normally utilized since it inhibits the competitive Dieckmann condensation. Moreover, the reaction is especially useful for the closing of four-membered rings.

The acyloin condensation of 1,4 to 1,6-diesters according to the Rühlmann version using chlorotrimethylsilane and highly dispersed sodium is markedlly simplified by sonochemical activation. Technical grade TMSCl and small cubic pieces of sodium can be used (eq. 7.45).[151]

$CO_2CH_2CH_3$ / $CO_2CH_2CH_3$ + $ClSi(CH_3)_3$ —Na, THF; 0-5 °C,)))); 82%→ $OSi(CH_3)_3$ / $OSi(CH_3)_3$ (7.45)

A fine suspension of colloidal potassium, ultrasonically generated in toluene or xylene, is a highly effective material for Dieckmann condensation of dicarboxylic esters (eq. 7.46).[152] The method is also suitable for Thorpe-Ziegler cyclizations, ketone enolization, and the preparation of Wittig-Horner reagents in good yields.

O, OCH_2CH_3, $CO_2CH_2CH_3$ —K, toluene,)))); 5 min, rt, 83%→ O, $CO_2CH_2CH_3$ (7.46)

Potassium-graphite induces formation of carbon-carbon bonds, but this reactive metal has been scarcely employed in cyclization processes. A notable exception is the synthesis of phenanthrenequinones by cyclodehydrogenation of benzil derivatives (eq. 7.47).[153] Reaction is performed with potassium-graphite in THF under very mild conditions. The same reaction with dispersed potassium gives only traces of products.

O O, R R —C_8K, THF; 20 °C→ O O, R R (7.47)

R = H (70%)
R = CH_3 (72%)

Rieke metals and highly reactive metals generated by reduction of metal halides with lithium naphthalenide,[154,155] assist numerous cyclizations and have been extensively utilized. The major portion of this important research has been accomplished by Rieke and co-workers.

Active magnesium offers a wide number of synthetic possibilities that are difficult to obtain with other metals, such as synthesis and transformations of non-Grignard organomagnesium reagents, direct metallation of 1,3-dienes and polyenes, as well as hydro- and carbomagnesiation of unsaturated compounds.[156]

Phenanthrenes are obtained in good yields (64-83%) by reaction of (Z)-2-chlorostilbenes with Rieke-magnesium in refluxing THF. (Z)-2-Bromostilbenes are reduced and isomerized to (*E*)-stilbenes by activated magnesium. The phenanthrene synthesis probably involves radical intermediates.[157] This reaction has been used for the synthesis of juncunone (Scheme 7.9), an unusual, naturally-occurring acetyl-9,10-dihydrophenanthrene.[158]

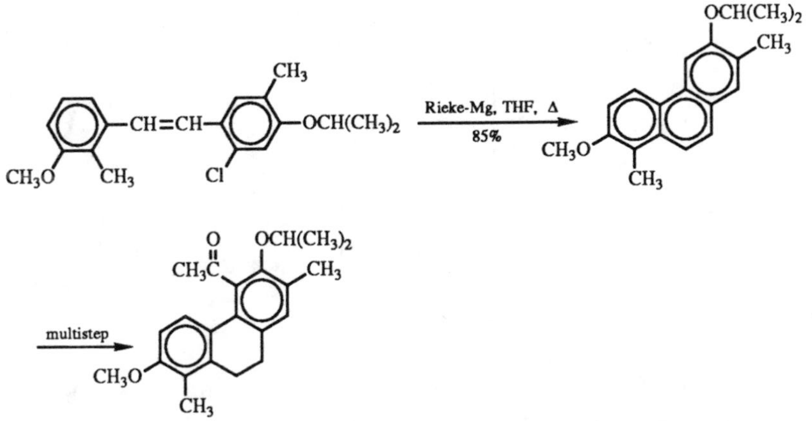

Rieke-Mg, THF, Δ
85%

multistep

Scheme 7.9

An active magnesium metal is formed from magnesium chloride by using lithium as the reducing agent in the presence of naphthalenes as electron carriers. New reactions are possible with this magnesium, which are not usually observed with normal types of magnesium. Examples are oxidative addition to carbon-oxygen bonds and the formation of multifunctional groups-containing Grignard reagents.[159] Even trimerization of nitriles is also possible with this reactive magnesium (eq. 7.48).

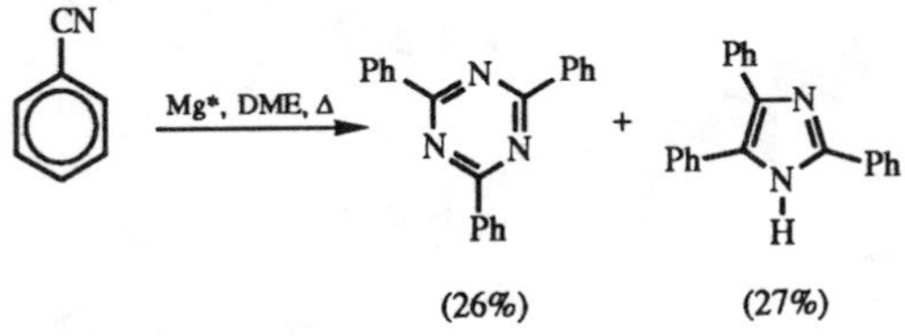

Mg*, DME, Δ

(26%) (27%)

(7.48)

The preparation of substituted 2-butene-1,4-diylmagnesium compounds can be easily accomplished by using highly reactive magnesium powder. These unsaturated magnesium reagents contain two formal Mg-carbon bonds and can act as bis-nucleophiles.[160,161] Reactions with α,ω-dibromoalkanes generate intermediates, which undergo intramolecular alkylation to afford disubstituted cycloalkanes. This strategy represents an interesting route to carbocycles (Scheme 7.10).

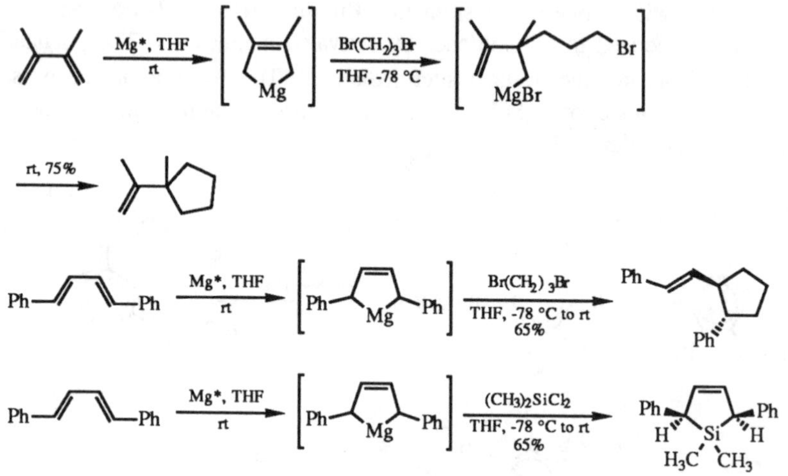

Mg*, THF rt
Br(CH$_2$)$_3$Br THF, -78 °C
Mg
MgBr
Br
rt, 75%

Ph Ph Mg*, THF rt Ph Mg Ph Br(CH$_2$)$_3$Br THF, -78 °C to rt 65% Ph Ph

Ph Ph Mg*, THF rt Ph Mg Ph $(CH_3)_2SiCl_2$ THF, -78 °C to rt 65% Ph Ph H Si H H_3C CH_3

Scheme 7.10

Additionally, the intermediate derived from reaction with alkylene dibromides or alkyl bromides can also react with other electrophiles. Thus, the addition of alkyl or aryl acid chlorides affords the corresponding ketones in good yields (eq. 7.49, Scheme 7.11).

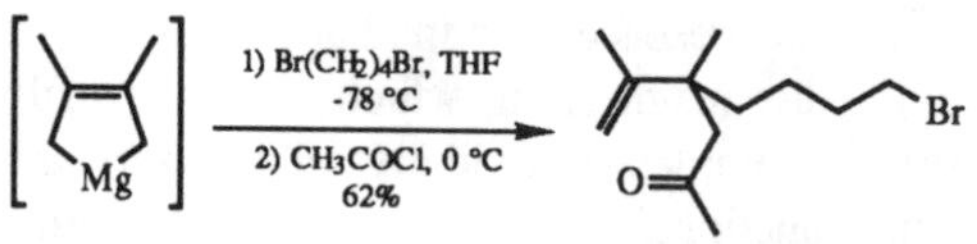

Mg 1) Br(CH$_2$)$_4$Br, THF -78 °C 2) CH_3COCl, 0 °C 62% Br O (7.49)

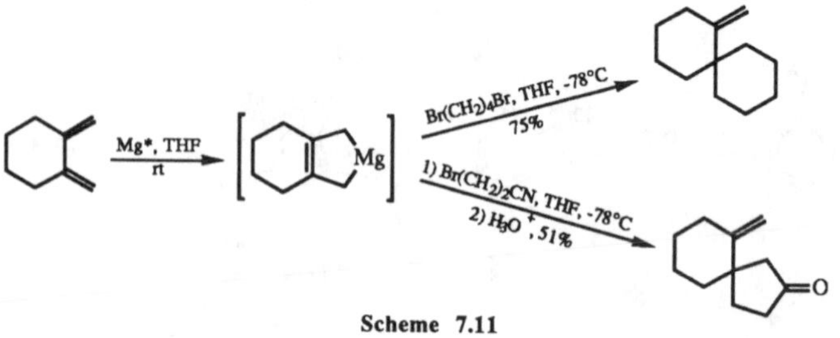

Mg*, THF rt Mg Br(CH$_2$)$_4$Br, THF, -78°C 75% 1) Br(CH$_2$)$_2$CN, THF, -78°C 2) H_3O^+, 51% O

Scheme 7.11

Similarly, magnesium complexes of 1,2-dimethylenecycloalkanes can react with carboxylic esters, providing a route to fused rings of β,γ-unsaturated ketones.[163] Esters were added to the magnesium-diene complexes at -78 °C and then the reaction mixture was gradually warmed. The formation of fused carbocycles was obtained at reflux, whereas quenching of the reaction at or below -10 °C afforded β,γ-unsaturated ketones. Experimental evidence indicates that the reaction proceeds *via* a magnesium salt of a spiro enol containing a cyclopropane ring (Scheme 7.12).

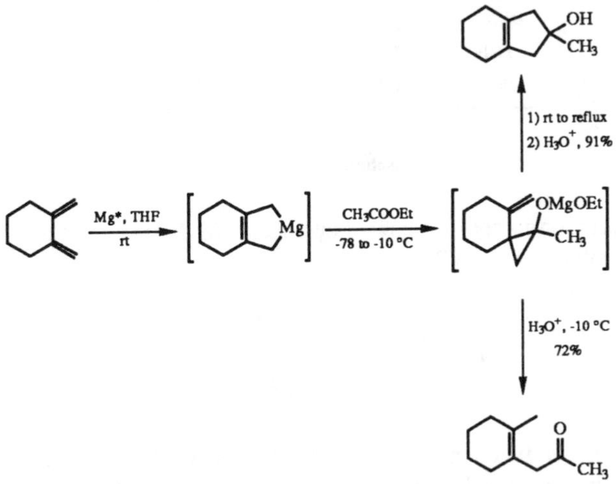

Scheme 7.12

A highly active form of copper slurry can be prepared by reduction of $CuI.PR_3$ with lithium naphthalenide. This copper metal converts aryl, alkynyl, and vinyl halides to organocopper reagents at sufficiently low temperatures to prevent the undesirable self-coupling of Ullmann reactions. Primary alkyl, allyl, and benzyl iodides, however, undergo homocoupling at 25 °C. Arylcopper reagents also give cross-coupling products with alkyl halides or acid chlorides (see Chapter IV).[164]

This active copper species reacts with epoxyalkyl halides to form an epoxyalkylcopper reagent, which undergoes an intramolecular cyclization *via* an epoxide cleavage process to afford five- and six-membered carbocycles.[165,166] The regioselectivity of this cyclization depends on chain length, substitution pattern, solvent, and the CuI-phosphine complex used to generate the active copper (Scheme 7.13). In the absence of Cu(I) salts, there is a preference in the formation of five-membered rings. The regiochemistry of the cyclization can be reversed by performing the reaction in toluene instead of THF.

6 (exo) 7 (endo)

Solvent	PR_3	Yield (%)	6 / 7
THF	PBu_3	56	1 : 6
toluene	PBu_3	45	1 : 0
THF	PPh_3	37	1 : 4
toluene	PPh_3	62	1 : 1.5

Scheme 7.13

Methyl substitution in epoxide gave a cyclohexanol derivative as a sole product regardless of the solvent (eq. 7.50).

(7.50)

Remarkably, many functional groups are tolerated in the epoxyalkyl bromides and yield highly functionalized carbocycles. This reaction can also effect cyclization to three- and four-membered rings as well as to larger carbocycles (eqs. 7.51, 7.52).

(7.51)

(7.52)

Activated calcium, generated by the reduction of calcium halides with lithium biphenylide, reacts with 1,3-dienes to give the novel calcium metallocycles, 2-butene-1,4-diylcalcium complexes.[167] These bis(organocalcium)reagents can react with several electrophiles (dihalides and dichlorosilanes) to form three-, five-, and six-membered ring derivatives in a highly regio- and stereoselective fashion (eq. 7.53).

Ca*, THF; rt — Ca — Cl(CH$_2$)$_3$Cl; THF, -78 °C; 98% (7.53)

References

1. Simmons, H. E.; Smith, R. D. *J. Am. Chem. Soc.* **1958**, *80*, 5323.
2. Simmons, H. E.; Smith, R. D. *J. Am. Chem. Soc.* **1959**, *81*, 4256.
3. Simmons, H. E.; Cairns, T. L.; Vladuchick, S. A.; Hoiness, C. M. *Org. React.* **1972**, *20*, 1.
4. Furukawa, J.; Kawabata, N. *Adv. Organomet. Chem.* **1974**, *12*, 83.
5. Zeller, K.-P.; Gugel, H. *Houben Weyl-Methoden der organischen Chemie*; Regitz, M., Ed.; Georg Thieme Verlag: Stuttgart, 1989; Vol. EXIXb, pp 195-212.
6. Winstein, S.; Sonnenberg, J.; De Vries, L. *J. Am. Chem. Soc.* **1959**, *81*, 6523.
7. Winstein, S.; Sonnenberg, J. *J. Am. Chem. Soc.* **1961**, *83*, 3235.
8. Grieco, P. A.; Oguir, T.; Wang, C.-L. J.; Williams, E. *J. Org. Chem.* **1977**, *42*, 4113.
9. Ratier, M.; Castaing, M.; Godet, J.-Y.; Pereyre, M. *J. Chem. Res.(M)* **1978**, 2309.
10. Johnson, C. R.; Barbachyn, M. R. *J. Am. Chem. Soc.* **1982**, *104*, 4290.
11. Neef, G.; Cleve, G.; Ottow, E.; Seeger, A.; Wiechert, R. *J. Org. Chem.* **1987**, *52*, 4143.
12. Ezquerra, J.; He, W.; Paquette, L. A. *Tetrahedron Lett.* **1990**, *31*, 6979.
13. Corey, E. J.; Virgil, S. C. *J. Am. Chem. Soc.* **1990**, *112*, 6429.
14. Sugimura, T.; Yoshikawa, M.; Futagawa, T.; Tai, A. *Tetrahedron* **1990**, *46*, 5955.
15. Charette, A. B.; Côté, B.; Marcoux, J.-F. *J. Am. Chem. Soc.* **1991**, *113*, 8166.
16. Charette, A. B.; Marcoux, J.-F.; Côté, B. *Tetrahedron Lett.* **1991**, *32*, 7215.
17. For a review of this subject: Suckling, C. J. *Angew. Chem. Int. Ed. Engl.* **1988**, *27*, 537.
18. Arai, I.; Mori, A.; Yamamoto, H. *J. Am. Chem. Soc.* **1985**, *107*, 8254.
19. Mori, A.; Arai, I.; Yamamoto, H. *Tetrahedron* **1986**, *42*, 6447.
20. Mash, E. A.; Nelson, K. A. *J. Am. Chem. Soc.* **1985**, *107*, 8256.
21. Mash, E. A.; Hemperly, S. B. *J. Org. Chem.* **1990**, *55*, 2055.

22. Sugimura, T.; Futagawa, T.; Yoshikawa, M.; Tai, A. *Tetrahedron Lett.* **1989**, *30*, 3807.
23. Imai, T.; Mineta, H.; Nishida, S. *J. Org. Chem.* **1990**, *55*, 4986.
24. Ambler, P. W.; Davies, S. G. *Tetrahedron Lett.* **1988**, *29*, 6979.
25. Ambler, P. W.; Davies, S. G. *Tetrahedron Lett.* **1988**, *29*, 6983.
26. Rieke, R. D.; Li, P. T. J.; Burns, T. P.; Uhm, S. T. *J. Org. Chem.* **1981**, *46*, 4323.
27. Klabunde, K. J.; Murdock, T. O. *J. Org. Chem.* **1979**, *44*, 3901.
28. Repic, O.; Vogt, S. *Tetrahedron Lett.* **1982**, *23*, 2729.
29. Repic, O.; Lee, P. G.; Giger, N. *Org. Prep. Proced. Int.* **1984**, *16*, 25.
30. Yamashita, I.; Inoue, Y.; Kondo, T.; Hashimoto, H. *Bull. Chem. Soc. Jpn.* **1984**, *57*, 2335.
31. Elschenbroich, Ch.; Salzer, A. *Organometallics*; VCH: Weinheim, 1989; p 48.
32. Wardell, J. L. *Organometallic Reagents of Zinc, Cadmium, and Mercury*; Chapman and Hall: London, 1985.
33. Emschwiller, G. *C. R. Hebd. Acad. Sci. Ser. C.* **1929**, *188*, 1555.
34. Friedrich, E. C.; Lunetta, S. E.; Lewis, E. J. *J. Org. Chem.* **1989**, *54*, 2388.
35. Friedrich, E. C.; Lewis, E. J. *J. Org. Chem.* **1989**, *55*, 2491.
36. Wittig, G.; Schwarzenbach, K. *Angew. Chem.* **1959**, *71*, 652.
37. Furukawa, J.; Kawabata, N.; Nishimura, J. *Tetrahedron Lett.* **1966**, 3353.
38. Swenton, J. S.; Burdett, K. A.; Madigan, D. M.; Johnson, T.; Rosso, P. D. *J. Am. Chem. Soc.* **1975**, *97*, 3428.
39. Blanchard, E. P.; Simmons, H. E. *J. Am. Chem. Soc.* **1964**, *86*, 1337.
40. Simmons, H. E.; Blanchard, E. P.; Smith, R. D. *J. Am. Chem. Soc.* **1964**, *86*, 1347.
41. Fabisch, B.; Mitchell, T. N. *J. Organomet. Chem.* **1984**, *269*, 219.
42. Denmark, S. E.; Edwards, J. P.; Wilson, S. R. *J. Am. Chem. Soc.* **1991**, *113*, 723.
43. Denmark, S. E.; Edwards, J. P.; Wilson, S. R. *J. Am. Chem. Soc.* **1992**, *114*, 2592 and references cited therein.
44. Beak, P.; Meyers, A. I. *Acc. Chem. Res.* **1986**, *19*, 356 and references cited therein.
45. Seyferth, D.; Cohen, H. M. *Inorg. Chem.* **1962**, *1*, 913.
46. Zweifel, G.; Clark, G. M.; Whitney, C. C. *J. Am. Chem. Soc.* **1971**, *93*, 1305.
47. Hoberg, H. *Liebigs Ann. Chem.* **1966**, 695.
48. Maruoka, K.; Fukutani, Y.; Yamamoto, H. *J. Org. Chem.* **1985**, *50*, 4412.

49. Knochel, P.; Jeong, N.; Rozema, M. J.; Yeh, M. C. P. *J. Am. Chem. Soc.* **1989**, *111*, 6474.
50. Knochel, P.; Chou, T. S.; Chen, H. G.; Yeh, M. C. P.; Rozema, M. J. *J. Org. Chem.* **1989**, *54*, 5202.
51. Knochel, P.; Rao, S. A. *J. Am. Chem. Soc.* **1990**, *112*, 6146.
52. Molander, G. A.; Etter, J. B. *J. Org. Chem.* **1987**, *52*, 3942.
53. Molander, G. A.; Harring, L. S. *J. Org. Chem.* **1989**, *54*, 3525.
54. Yamazaki, T.; Lin, J. T.; Takeda, M.; Kitazume, T. *Tetrahedron: Asymmetry* **1990**, *1*, 351.
55. Clive, D. L. J.; Daigneault, S. *J. Chem. Soc. Chem. Commun.* **1989**, 333.
56. Clive, D. L. J.; Daigneault, S. *J. Org. Chem.* **1991**, *56*, 3801.
57. Soderquist, J. A. *Aldrichimica Acta* **1991**, *24*, 15 and references cited therein.
58. Imamoto, T.; Takeyama, T.; Koto, H. *Tetrahedron Lett.* **1986**, *27*, 3243.
59. Imamoto, T.; Kamiya, Y.; Hatajima, T.; Takahashi, H. *Tetrahedron Lett.* **1989**, *30*, 5149.
60. Araki, S.; Butsugan, Y. *J. Chem. Soc. Chem. Commun.* **1989**, 1286.
61. For a comprehensive treatment of three-membered ring-forming reactions: Larock, R. C. *Comprehensive Organic Transformations*; VCH: New York, 1989; pp 71-82 and references cited therein.
62. Freidlina, R. K.; Kamyshova, A. A.; Chukovskaya, E. T. *Russ. Chem. Rev.* **1982**, *51*, 368.
63. Applequist, D. E.; Pfohl, W. F. *J. Org. Chem.* **1978**, *43*, 867.
64. Kawabata, N.; Kamemura, I.; Naka, M. *J. Am. Chem. Soc.* **1979**, *101*, 2139.
65. Kawabata, N.; Tanimoto, M.; Fujiwara, S. *Tetrahedron* **1979**, *35*, 1919.
66. Mukaiyama, T.; Shiono, M.; Watanabe, K.; Onaka, M. *Chem. Lett.* **1975**, 711.
67. Baumstark, A. L.; McCloskey, C. J.; Tolson, T. J.; Syriopoulos, G. T. *Tetrahedron Lett.* **1977**, 3003.
68. Walborsky, H. M.; Murai, M. P. *J. Am. Chem. Soc.* **1980**, *102*, 426.
69. Carruthers, W. *Cycloaddition Reactions in Organic Synthesis*; Pergamon Press: Oxford, 1990; pp 37-40.
70. Charlton, J. L.; Alauddin, M. M. *Tetrahedron* **1987**, *43*, 2873.
71. Oppolzer, W. *Synthesis* **1978**, 793.
72. Kametani, T. *Pure Appl. Chem.* **1979**, *51*, 747.
73. Funk, R. L.; Vollhardt, K. P. C. *Chem. Soc. Rev.* **1980**, *9*, 41.
74. Rubottom, G. M.; Wey, J. E. *Synth. Commun.* **1984**, *14*, 507.
75. Inaba, S.; Wehmeyer, R. M.; Forkner, M. W.; Rieke, R. D. *J. Org. Chem.* **1988**, *53*, 339.

76. Ichikawa, H.; Ebisawa, Y.; Honda, T.; Kametani, T. *Tetrahedron* **1985**, *41*, 3643.
77. Kerdesky, F. A. J.; Ardecky, R. J.; Lakshimikantham, M. V.; Cava, M. P. *J. Am. Chem. Soc.* **1981**, *103*, 1992.
78. Han, B. H.; Boudjouk, P. *J. Org. Chem.* **1982**, *47*, 751.
79. Boudjouk, P.; Sooriyakumaran, R.; Han, B. H. *J. Org. Chem.* **1986**, *51*, 2818.
80. Chew, S.; Ferrier, R. J. *J. Chem. Soc. Chem. Commun.* **1984**, 911.
81. Hoffmann, H. M. R. *Angew. Chem. Int. Ed. Engl.* **1984**, *23*, 1.
82. Mann, J. *Tetrahedron* **1986**, *42*, 4611.
83. Noyori, R.; Hayakawa, Y. *Org. React.* **1983**, *29*, 163.
84. Vinter, J. G.; Hoffmann, H. M. R. *J. Am. Chem. Soc.* **1974**, *96*, 5466.
85. Chidgey, R.; Hoffmann, H. M. R. *Tetrahedron Lett.* **1977**, 2633.
86. Sakurai, H.; Shirahata, A.; Hosomi, A. *Angew. Chem. Int. Ed. Engl.* 1979, *18*, 163.
87. Takaya, H.; Hayakawa, Y.; Makino, S.; Noyori, R. *J. Am. Chem. Soc.* **1978**, *100*, 1778.
88. Sato, T.; Noyori, R. *Bull. Chem. Soc. Jpn.* **1978**, *51*, 2745.
89. Savoia, D.; Tagliavini, E.; Trombini, C.; Umani-Ronchi, A. *J. Org. Chem.* **1982**, *47*, 876.
90. Joshi, N.; Hoffmann, H. M. R. *Tetrahedron Lett.* **1986**, *27*, 687.
91. Fry, A. J.; Herr, D. *Tetrahedron Lett.* **1978**, 1721.
92. Fry, A. J.; Ginsberg, G. S.; Parante, R. A. *J. Chem. Soc. Chem. Commun.* **1978**, 1040.
93. Ref. 69, *ibidem*; pp 261-264 and references cited therein.
94. Oppolzer, W. *Angew. Chem. Int. Ed. Engl.* **1989**, *28*, 38 and references therein.
95. Oppolzer, W.; Schneider, P. *Tetrahedron Lett.* **1984**, *25*, 3305.
96. Oppolzer, W.; Cunningham, A. F. *Tetrahedron Lett.* **1986**, *27*, 5467.
97. Lehmkuhl, H. *Bull. Soc. Chim. Fr. II* **1981**, 87.
98. Oppolzer, W.; Nakao, A. *Tetrahedron Lett.* **1986**, *27*, 5471.
99. Courtois, G.; Masson, A.; Migniac, L. *C. R. Hebd. Acad. Sci. Ser. C.* **1978**, *286*, 265.
100. Löffler, A.; Pratt, D.; Pucknat, J.; Gelbard, G.; Dreiding, A. S. *Chimia* **1969**, *23*, 413.
101. Öhler, E.; Reininger, K.; Schmidt, U. *Angew. Chem. Int. Ed. Engl.* **1970**, *9*, 457.
102. Fürstner, A. *Synthesis* **1989**, 571 and references cited therein.
103. Hoffmann, H. M. R.; Rabe, J. *Angew. Chem. Int. Ed. Engl.* **1985**, *24*, 94.
104. Petragnani, N.; Ferraz, H. M. C.; Silva, G. V. J. *Synthesis* **1986**, 157.

105. Csuk, R.; Fürstner, A.; Sterk, H.; Weidmann, H. *J. Carbohydr. Chem.* **1986**, *5*, 459.
106. Csuk, R.; Hugener, M.; Vasella, A. *Helv. Chim. Acta* **1988**, *71*, 609.
107. Rauter, A.; Figueiredo, J. A.; Ismael, I.; Pais, M. S.; González, A. G.; Díaz, J.; Barrera, J. B. *J. Carbohydr. Chem.* **1987**, *6*, 259.
108. Linding, C. *J. Prakt. Chem.* **1983**, *325*, 587.
109. Koolpe, G. A.; Nelson, W. L.; Gioannini, T. L.; Angel, L.; Simon, E. J. *J. Med. Chem.* **1984**, *27*, 1718.
110. Rebek, J.; Tai, D. F.; Shue, Y. K. *J. Am. Chem. Soc.* **1984**, *106*, 1813.
111. El Alami, N.; Belaud, C.; Villieras, J. *J. Organomet. Chem.* **1988**, *348*, 1.
112. Auvray, P.; Knochel, P.; Normant, J. F. *Tetrahedron* **1988**, *44*, 4495.
113. Collard, J. N.; Benezra, C. *Tetrahedron Lett.* **1982**, *23*, 3725.
114. Mattes, H.; Benezra, C. *Tetrahedron Lett.* **1985**, *26*, 5697.
115. Nokami, J.; Tamaoka, T.; Ogawa, H.; Wakabayashi, S. *Chem. Lett.* **1986**, 541.
116. Li, C. J.; Chan, T. H. *Organometallics* **1991**, *10*, 2548.
117. Li, C. J.; Chan, T. H. *Tetrahedron Lett.* **1991**, *32*, 7017.
118. Ghera, E.; Gaoni, Y.; Perry, D. H. *J. Chem. Soc. Chem. Commun.* **1974**, 1034.
119. Ghera, E.; Gaoni, Y.; Shoua, S. *J. Am. Chem. Soc.* **1976**, *98*, 3627.
120. Semmelhack, M. F.; Wu, E. S. C. *J. Am. Chem. Soc.* **1976**, *98*, 3384.
121. Ruggeri, R. B.; Heathcock, C. H. *J. Org. Chem.* **1987**, *52*, 5745.
122. Vedejs, E.; Ahmad, S. *Tetrahedron Lett.* **1988**, *29*, 2291.
123. Flitsch, W.; Russkamp, P. *Liebigs Ann. Chem.* **1985**, 1398.
124. Maruoka, K.; Hashimoto, S.; Kitagawa, Y.; Yamamoto, H.; Nozaki, H. *Bull. Chem. Soc. Jpn.* **1980**, *53*, 3301.
125. Tsuji, J.; Mandai, T. *Tetrahedron Lett.* **1978**, 1817.
126. Sato, A.; Ogiso, A.; Noguchi, H.; Mitsui, S.; Kaweko, I.; Shimada, Y. *Chem. Pharm. Bull.* **1980**, *28*, 1509.
127. Moriya, T.; Handa, Y.; Inanaga, J.; Yamaguchi, M. *Tetrahedron Lett.* **1988**, *29*, 6947.
128. Molander, G. A. *Chem. Rev.* **1992**, *92*, 29.
129. Shono, T.; Hamaguchi, H.; Nishiguchi, I.; Sasaki, M.; Miyamoto, T.; Miyamoyo, M.; Fujita, S. *Chem. Lett.* **1981**, 1217.
130. Krepski, L. R.; Lynch, L. E.; Heilmann, S. M.; Rasmussen, J. K. *Tetrahedron Lett.* **1985**, *26*, 981.
131. Kitazume, T. *Synthesis* **1986**, 855.
132. Gilman, H.; Specter, M. *J. Am. Chem. Soc.* **1943**, *65*, 2255.
133. Brown, M. J. *Heterocycles* **1989**, *29*, 2225 and references therein.
134. Hart, D. J.; Ha, D.-C. *Chem. Rev.* **1989**, *89*, 1447 and references therein.

135. Bose, A. K.; Gupta, K.; Manhas, M. S. *J. Chem. Soc. Chem. Commun.* **1984**, 86.
136. Mehta, G.; Rao, H. S. P. *Synth. Commun.* **1985**, *15*, 991.
137. Edstrom, E. D. *J. Am. Chem. Soc.* **1991**, *113*, 6690.
138. Crandall, J. K.; Ayers, T. A. *Organometallics* **1992**, *11*, 473.
139. For a review of intramolecular radical cyclizations: Giese, B. *Radicals in Organic Synthesis: Formation of Carbon-Carbon Bonds*; Pergamon Press: Oxford, 1986; pp 185-189.
140. Corey, E. J.; Pyne, S. G. *Tetrahedron Lett.* **1983**, *24*, 2821.
141. Stork, G.; Malhotra, S.; Thompson, H.; Uchibayashi, M. *J. Am. Chem. Soc.* **1965**, *87*, 1148.
142. Stork, G.; Boeckmann, R. K.; Taber, D. F.; Still, W. C.; Singh, J. *J. Am. Chem. Soc.* **1979**, *101*, 7107.
143. Pradhan, S. K.; Radhakrishnan, T. V.; Subramanian, R. *J. Org. Chem.* **1976**, *41*, 1943.
144. Pattenden, G.; Teague, S. J. *Tetrahedron Lett.* **1982**, *23*, 5471.
145. Leroux, Y. *Bull. Soc. Chim. Fr.* **1968**, 359.
146. Crandall, J. K.; Magaha, H. S. *J. Org. Chem.* **1982**, *47*, 5368.
147. Finley, K. T. *Chem. Rev.* **1964**, *64*, 573.
148. Bloomfield, J. J.; Owsley, D. C.; Nelke, J. M. *Org. React.* **1976**, *23*, 259.
149. Ref. 61, *ibidem*; pp 645-646 and references cited therein.
150. Rühlmann, K. *Synthesis* **1971**, 236.
151. Fadel, A.; Canet, J.-L.; Salaün, J. *Synlett* **1990**, 89.
152. Luche, J.-L.; Petrier, C.; Dupuy, C. *Tetrahedron Lett.* **1984**, *25*, 753.
153. Tamarkin, D.; Benny, D.; Rabinovitz, M. *Angew. Chem. Int. Ed. Engl.* **1984**, *23*, 642.
154. Rieke, R. D. *Science* **1989**, *246*, 1260.
155. Rieke, R. D.; Burns, T. P.; Wehmeyer, R. M.; Kahn, B. E. *High Energy Processes in Organometallic Chemistry*; Suslick, K. S., Ed.; American Chemical Society: Washington, DC, 1987; pp 223-245.
156. For an excellent survey of this subject: Dzhemilev, U. M.; Ibragimov, A. G.; Tolstikov, G. A. *J. Organomet. Chem.* **1991**, *406*, 1.
157. Brown, C.; Sikkel, B. J.; Carvalho, C. F.; Sargent, M. V. *J. Chem. Soc. Perkin Trans. 1* **1982**, 3007.
158. Carvalho, C. F.; Sargent, M. V. *J. Chem. Soc. Chem. Commun.* **1982**, 1198.
159. Burns, T. P.; Rieke, R. D. *J. Org. Chem.* **1987**, *52*, 3674.
160. Xiong, H.; Rieke, R. D. *J. Org. Chem.* **1989**, *54*, 3247.
161. Rieke, R. D.; Xiong, H. *J. Org. Chem.* **1991**, *56*, 3109.
162. Xiong, H.; Rieke, R. D. *Tetrahedron Lett.* **1991**, *32*, 5269.
163. Xiong, H.; Rieke, R. D. *J. Am. Chem. Soc.* **1992**, *114*, 4415.

164. Ebert, G. W.; Rieke, R. D. *J. Org. Chem.* **1988**, *53*, 4482.
165. Wu, T.-C.; Rieke, R. D. *Tetrahedron Lett.* **1988**, *29*, 6753.
166. Rieke, R. D.; Wehmeyer, R. M.; Wu, T.-C.; Ebert, G. W. *Tetrahedron* **1989**, *45*, 443.
167. Wu, T.-C.; Xiong, H.; Rieke, R. D. *J. Org. Chem.* **1990**, *55*, 5045.

VIII. THE BERNET-VASELLA REACTION

A. INTRODUCTION

In addition to the previous metal-mediated organic transformations, I would like to describe a zinc-induced reductive ring-opening known as the Vasella reaction[1-4] or the Bernet-Vasella reaction.[5] This useful protocol was initially applied to the field of carbohydrate chemistry and it permits a stereocontrolled ring-opening to produce highly functionalized olefinic aldehydes.

The key step in this process is the reductive fragmentation of 6-deoxy-6-halo-pyranose or 5-deoxy-5-halofuranose derivatives by using activated zinc in boiling aqueous alcohols, such as ethanol or *n*-propanol (eq. 8.1). Additionally, Vasella[4] has also reported that the fragmentation could be effected by treatment of the halogen compound with *n*-butyllithium, although the secondary alcohols derived from the addition of organolithium to the aldehyde were always found.

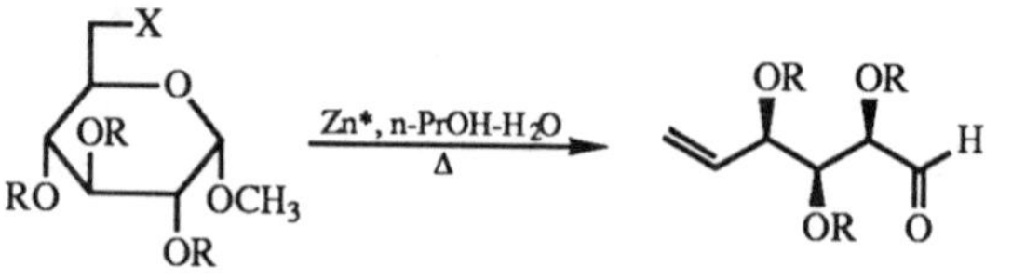

X = Br, I

Zn*, n-PrOH-H_2O, Δ

(8.1)

Both the α- and β-anomers afford the same products in approximately the same yields, and at comparable rates in most cases. These findings appear to be in agreement with Grob's fragmentation rules.[6,7] According to these rules, only β-D-anomers should fragment following a concerted antiperiplanar fragmentation whereas α-D-anomers would lead to the unsaturated aldehydes *via* the corresponding hemiacetals. These statements, however, have been discussed by Vasella, since both anomers might follow either concerted pathways or a two-step process with the intermediacy of a hemiacetal (Scheme 8.1).[4] Bernet and Vasella proposed a mechanism involving a single electron transfer (SET) as the first step, followed by the expulsion of bromide from the intermediate radical anion. A further SET process with heterolytic fragmentation yields the olefinic product. These reactions can occur on the surface of the zinc metal.

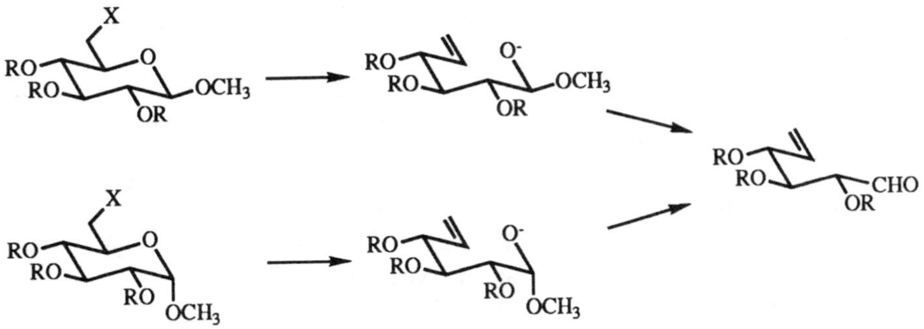

Scheme 8.1

Although the mechanism is still an obscure area in this reaction, SET processes have been discarded recently by Fürstner *et al.*[8,9] on the basis of exclusive Wurtz-type couplings of ω-deoxy halosugars with magnesium-graphite. This reagent is a very efficient single-electron donor,[10,11] and the reaction with methyl 6-deoxy-6-iodo-2,3,4-tri-*O*-methyl-α-D-glucopyranoside gave no dealkoxyhalogenation, but reductive coupling with formation of the dimeric product **1** (eq. 8.2) along with a small amount of methyl 6-deoxy-2,3,4-tri-*O*-methyl-α-D-glucopyranoside.

This failed reaction is otherwise a novel route for the preparation of *C*-disaccharides, a type of compounds in which the oxygen bridge of natural disaccharides is replaced by a carbon linkage.

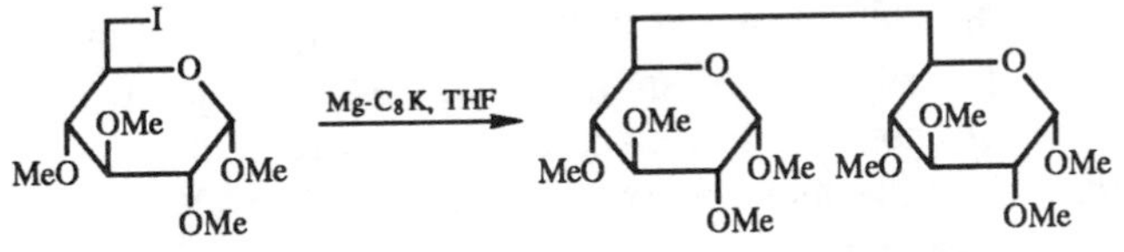

Mg-C_8K, THF (8.2)

1

B. ZINC REAGENTS

Although commercially available zinc dust or zinc powder of high purity can be utilized, the reactivity is largely enhanced by employing activated zinc. This can be simply acetic acid-washed zinc dust or better Rieke-zinc.[12-14] The less reactive zinc reagents require protic solvents and higher reaction temperatures. Unfortunately these drastic conditions are not compatible with many sensitive functional groups, and importantly the proportion of side products increases. Reactions can also proceed at room temperature for longer periods, since the temperature can affect the product distribution.[4] Rieke-zinc is a vastly superior reagent and yields products in shorter times. An important feature is that the reaction can be now conducted in anhydrous ethereal solvents such as THF or DME. A combination of Rieke-zinc in THF-methanol has been also employed.[15]

Despite these promising results and the numerous applications in organic synthesis that have appeared in the literature, the Bernet-Vasella reaction is still in its infancy. The most outstanding contributions to this topic have been accomplished by Fürstner *et al.*[8,9,16,17] who have improved and extended the scope of this zinc-mediated ring-opening by the use of potassium-graphite or zinc/silver-graphite. These metal-graphites have been successfully employed for other transformations on sugars, such as the formation of furanoid and pyranoid glycals (eqs. 8.3, 8.4), Fischer-Zach-type reductions under aprotic conditions, and reductions of *O*-alkyl- and *O*-acylglycosyl halides.[18-21]

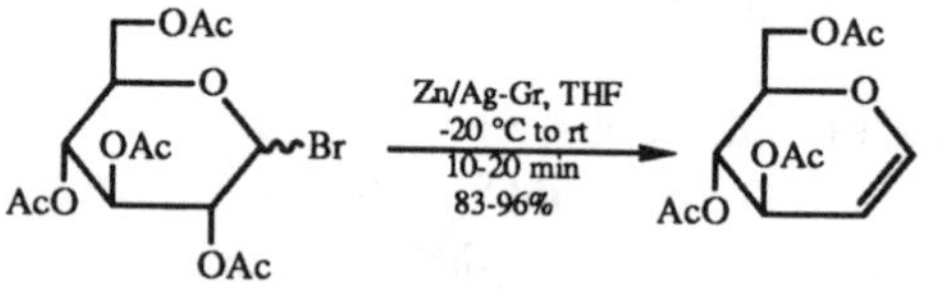

OAc O OAc Br AcO OAc — Zn/Ag-Gr, THF, -20 °C to rt, 10-20 min, 83-96% → OAc O OAc AcO (8.3)

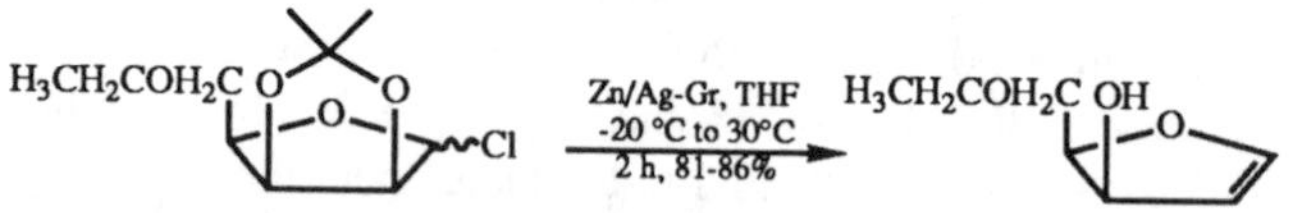

$H_3CH_2COH_2C$ O O O Cl — Zn/Ag-Gr, THF, -20 °C to 30°C, 2 h, 81-86% → $H_3CH_2COH_2C$ OH O (8.4)

The authors have extensively studied the factors influencing the reaction, functional group compatibility, and the nature of the substrate and have employed a wide number of ω-deoxy halosugars with different configurations, substitution patterns, and ring sizes. The use of either zinc/silver-graphite or potassium-graphite allows the reaction to proceed rapidly in anhydrous THF or DME at room temperature in practically all cases. Yields are good and considerably superior to those with less reactive zinc. The reagents work well even for systems where Rieke-zinc failed or gave lower yields. Furthermore, these reactive reagents inhibit almost completely the formation of side products such as enal acetals which were previously observed in zinc-induced reductions of deoxy halosugars.[1,4,22,23]

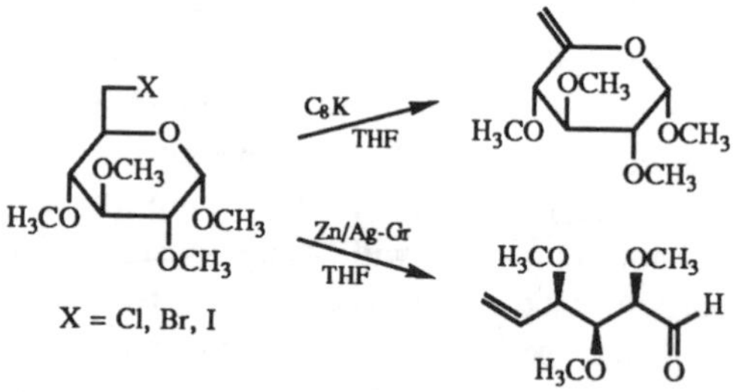

X O OCH_3 H_3CO OCH_3 OCH_3 — C_8K, THF → O OCH_3 H_3CO OCH_3 OCH_3

— Zn/Ag-Gr, THF → H_3CO OCH_3 H H_3CO O

X = Cl, Br, I

Scheme 8.2

A distinctive behavior was found between zinc/silver-graphite and the potassium-graphite laminate. While zinc/silver-graphite produces a single dealkoxyhalogenation product,with the exception of 2-*O*-alkylglycosyl halides, potassium-graphite induces dehydrohalogenation with the formation of the corresponding enol ethers (Scheme 8.2).[8]

A similar behavior was also found in the reductive elimination of derivatives **2** and **3** having the halogen at C-4 position (Scheme 8.3). Interestingly, the treatment with zinc/silver-graphite gave a similar proportion of cyclic and open-ring products. The latter was a nearly equimolar mixture of the Z- and *E*-stereoisomers (Z:*E* = 1.1:1). The complete lack of regio- and stereoselectivity in this case is not well understood. The formation of enol ethers by the action of potassium-graphite could be attributed to the Lewis base character of this reagent.[8,24] Nevertheless, the results clearly contrast with the dealkoxyhalogenations of glycosyl halides which are easily promoted by potassium-graphite.[20]

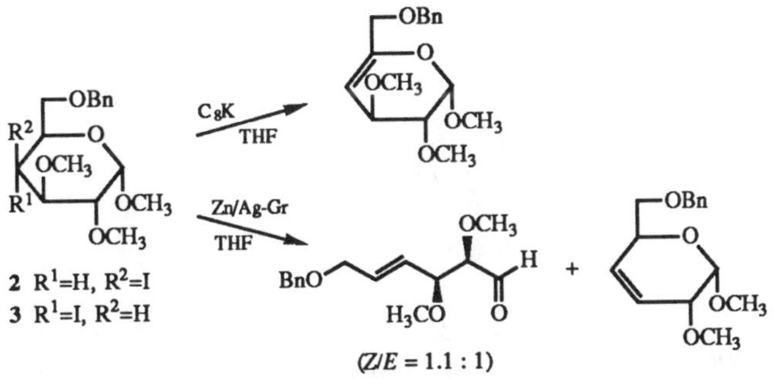

Scheme 8.3

Further considerations of the influences of the type of halogen, ring size, and functional group compatibility should be noted. The reactivity of zinc/silver-graphite toward this ring-opening dealkoxyhalogenations is in the order 6-iodo-hexopyranosides > 6-bromo-hexopyranosides > 5-iodo-pentofuranosides. 6-Bromo derivatives are also more susceptible to undergo dehalogenation. However, the reaction rate and the reduction of the carbon-halogen bond are strongly dependent on the configuration and substitution pattern of the starting sugar.[9] Very sensitive functionalities, with some exceptions, onto the sugar ring are tolerated under these aprotic and mild conditions. *O*-Sulfonyl enals are relatively unstable and decomposition is often found. The location of *O*-sulfonyl groups in the sugar ring appears to be of crucial importance. Less sterically crowded compounds can produce mixtures of products of reductive elimination and carbon-halogen bond reduction. In the presence of an azido group, the reaction results in the reduction of both the carbon-halogen bond and the azido function. This group is also incompatible under the classical Bernet-Vasella conditions using protic solvents.

The dealkoxyhalogenation process involves apparently a transition state wherein the zinc reagent is coordinated with both the endocyclic oxygen and the halogen of the sugar. In general, coordination is retarded by bulky or strongly electron-donating substituents. The protic solvents, which are essential with the less reactive zinc reagents, appear to play an important role in the Lewis acid-Lewis base interactions.

C. SYNTHETIC APPLICATIONS

The Bernet-Vasella reaction represents a simple and convenient route to cyclic and acyclic carbohydrate-derived olefinic chirons, which are versatile substrates in many synthetic strategies.[25] Thus, Fürstner and Weidmann[16] have described the preparation of the eight stereoisomeric 5,6-dideoxy-2,3,4-tri-*O*-methyl-aldohex-5-enoses in excellent yields by dealkoxyhalogenation of the corresponding 6-deoxy-6-halosugars. Four of them were prepared by this protocol and additionally, all the remaining configurations can be obtained by an aldehyde-alkene interconversion (eq. 8.5).

OCH3 CH3O OCH3 I CH3O O CH3O OCH3 CHO OCH3 CH3O OCH3 CHO OCH3 (8.5)

D-gluco D-xylo L-xylo

Moreover, zinc-induced reductive ring-openings have been applied extensively,[15,22,26-41] and Bernet-Vasella-type reactions have been key steps in various natural-product syntheses. Thus, the reductive ring-opening of ω-bromopyranose sugars was modified by Bernotas and Ganem[35] to incorporate an *in situ* reductive amination of the incipient ω-alkenyl aldehyde. The sequence of reactions involves treatment with excess activated zinc in aqueous *n*-propanol, addition of benzylamine, and reduction with sodium cyanoborohydride. The yield was superior to the two-step process in which olefinic aldehyde was isolated, and then submitted to reductive amination. The strategy was successfully employed in the synthesis of enantiomerically pure cyclic aminoalditols,[35] such as 1-deoxynojirimycin (eq. 8.6) and 1-deoxymannojirimycin, as well as in the preparation of five-carbon iminoalditols which are potent glycosidase inhibitors.[36]

Br BnO O BnO BnO OCH3 1) Zn*, n-PrOH-H2O, Δ 2) PhCH2NH2, NaBH3CN BnO NHBn BnO OBn multistep OH H N HO HO OH (8.6)

A key fragment (C10-C19) of the immunosuppressive agent FK-506 has been recently synthesized.[34] A crucial olefinic aldehyde was readily accessible through a classical Bernet-Vasella reaction using zinc in boiling aqueous ethanol (eq. 8.7). In a closely related strategy tri-*O*-acetyl-D-galactal was converted *via* a series of functional group manipulations and a Bernet-Vasella fragmentation, into a six-carbon synthon for the preparation of the C22-C27 fragment of FK-506.[37]

H_3CO OBn OCH_3 OCH_3 —I; Zn*, 95% EtOH, Δ, 95% → OBn O H H_3CO OCH_3 (8.7)

X = Br, I

Danishefsky and his associates have also reported[38] a new synthesis of the novel immunoactivator FR900483. Again, a key step was a Bernet-Vasella reaction.

Interestingly, an enantiospecific total synthesis of the aminocyclitol (-)-allosamizoline was accomplished *via* a Bernet-Vasella reaction using Rieke-zinc in THF.[39] With this highly reactive metal powder, the reaction proceeded rapidly and ring fragmentation was not observed. The furanoid enol ether **4** was isolated together with reductively dehalogenated C-6 product. The former was then converted into the desired allosamizoline (**5**) following a well-established methodology (eq. 8.8).

I, OH, HO, OCH_3, NPht; Rieke-Zn, THF, 25 °C, 1.5 h → OH, OH, NPht **4**; multistep → OH, HO, HO, O, N, $N(CH_3)_2$ **5** (8.8)

The versatility of the Bernet-Vasella reaction can be now utilized en route to products other than enals or enol ethers by appropriate choice of the sugar precursor. Ireland *et al.*[40] have reported the enantioselective synthesis of aldols and other 1,3-dioxygenated systems in a one-step procedure with zinc/silver-graphite as reagent (eq. 8.9).

R R, BnO, R^1, I; Zn/Ag-Gr, THF → BnO, OH O, R^1 (8.9)

6 R = H, $R^1 = CH_3$
7 $R = CH_3$, $R^1 = CH{=}CH_2$

8 $R^1 = CH_3$ (60%)
9 $R^1 = CH{=}CH_2$ (75%)

Very recently, six-carbon chiral synthons have been obtained from D-glucosamine in a few steps, the latter including Bernet-Vasella scission under standard conditions (activated Zn, 95% aqueous ethanol, reflux).[41] A variety of protecting groups such as alkyl ethers, silyl ethers, and esters have been utilized in different configurations on the pyran ring system (eq. 8.10).

Zn*, aq. EtOH, Δ (8.10)

10 R^1 = Ac, R^2 = Bn
11 R^1 = Bz, R^2 = TBS
12 R^1 = TBS, R^2 = TBS
13 R^1 = R^2 = Ac

References

1. Bernet, B.; Vasella, A. *Helv. Chim. Acta* **1979**, *62*, 1990.
2. Bernet, B.; Vasella, A. *Helv. Chim. Acta* **1979**, *62*, 2400.
3. Bernet, B. Ph.D. Dissertation, ETH No. 6416, Zürich, 1979.
4. Bernet, B.; Vasella, A. *Helv. Chim. Acta* **1984**, *67*, 1328.
5. This name has been kindly suggested by Prof. Vasella. Personal communication.
6. Grob, C. A.; Schiess, P. W. *Angew. Chem.* **1967**, *79*, 1.
7. Becker, K. B.; Grob, C. A. *The Chemistry of Double-bonded Functional Groups*; Patai, S., Ed.; Wiley: London, 1977; p 653.
8. Fürstner, A.; Weidmann, H. *J. Org. Chem.* **1989**, *54*, 2307.
9. Fürstner, A.; Jumbam, D.; Teslic, J.; Weidmann, H. *J. Org. Chem.* **1991**, *56*, 2213.
10. Csuk, R.; Fürstner, A.; Weidmann, H. *J. Chem. Soc. Chem. Commun.* **1986**, 1802.
11. Fürstner, A.; Csuk, R.; Rohrer, C.; Weidmann, H. *J. Chem. Soc. Perkin Trans. 1* **1988**, 1729.
12. Rieke, R. D.; Hudnall, P. M.; Uhm, S. J. *J. Chem. Soc. Chem. Commun.* **1973**, 269.
13. Rieke, R. D.; Uhm, S. J. *Synthesis* **1975**, 452.
14. Rieke, R. D.; Li, P. T. J.; Burns, B. T.; Uhm, S. J. *J. Org. Chem.* **1981**, *46*, 4323.
15. Nakane, M.; Hutchinson, R. C.; Gollman, H. *Tetrahedron Lett.* **1980**, *21*, 1213.

16. Fürstner, A.; Weidmann, H. *J. Org. Chem.* **1990**, *55*, 1363.
17. Fürstner, A. *Tetrahedron Lett.* **1990**, *31*, 3735.
18. Csuk, R.; Fürstner, A.; Glänzer, B. I.; Weidmann, H. *J. Chem. Soc. Chem. Commun.* **1986**, 1149.
19. Csuk, R.; Glänzer, B. I.; Fürstner, A.; Weidmann, H.; Formacek, V. *Carbohydr. Res.* **1986**, *157*, 235.
20. Fürstner, A.; Weidmann, H. *J. Carbohydr. Chem.* **1988**, *7*, 773.
21. Csuk, R.; Glänzer, B. I.; Fürstner, A. *Adv. Organomet. Chem.* **1988**, *28*, 85 and references cited therein.
22. Aspinall, G. O.; Chatterjee, D.; Khondo, L. *Can. J. Chem.* **1984**, *62*, 2728.
23. Aspinall, G. O.; Puvanesaràjah, V. *Can. J. Chem.* **1984**, *62*, 2736.
24. Bergbreiter, D. E.; Killough, J. M. *J. Am. Chem. Soc.* **1978**, *100*, 2126.
25. Hanessian, S. *Total Synthesis of Natural Products: The Chiron Approach*; Pergamon Press: Oxford, 1983.
26. Ferrier, R. J.; Prasit, P. *J. Chem. Soc. Chem. Commun.* **1981**, 983.
27. Ferrier, R. J.; Prasit, P. *Pure Appl. Chem.* **1983**, *55*, 565.
28. Ferrier, R. J.; Furneaux, R. H.; Prasit, P.; Tyler, P. C.; Brown, K. L.; Gainsford, G. J.; Diehl, J. W. *J. Chem. Soc. Perkin Trans. 1* **1983**, 1621.
29. Ferrier, R. J.; Prasit, P. *J. Chem. Soc. Perkin Trans. 1* **1983**, 1645.
30. Beau, J.-M.; Aburaki, S.; Pougny, J.-R.; Sinaÿ, P. *J. Am. Chem. Soc.* **1983**, *105*, 621.
31. Hanessian, S.; Ugolini, A.; Therien, M. *J. Org. Chem.* **1983**, *48*, 4427.
32. Häfele, B.; Schröter, D.; Jäger, V. *Angew. Chem.* **1986**, *98*, 89.
33. Florent, J. C.; Ughetto-Monfrin, J.; Monneret, C. *J. Org. Chem.* **1987**, *52*, 1051.
34. Villalobos, A.; Danishefsky, S. J. *J. Org. Chem.* **1989**, *54*, 12.
35. Bernotas, R. C.; Ganem, B. *Tetrahedron Lett.* **1985**, *26*, 1123.
36. Bernotas, R. C.; Papandreou, G.; Urbach, J.; Ganem, B. *Tetrahedron Lett.* **1990**, *31*, 3393.
37. Linde, R. G.; Egbertson, M.; Coleman, R. S.; Jones, A. B.; Danishefsky, S. J. *J. Org. Chem.* **1990**, *55*, 2771.
38. Chen, S.-H.; Danishefsky, S. J. *Tetrahedron Lett.* **1990**, *31*, 2229.
39. Nakata, M.; Akazawa, S.; Kitamura, S.; Tatsuta, K. *Tetrahedron Lett.* **1991**, *32*, 5363.
40. Ireland, R. E.; Wipf, P.; Miltz, W.; Vanasse, B. *J. Org. Chem.* **1990**, *55*, 1423.
41. Coleman, R. S.; Dong, Y.; Carpenter, A. J. *J. Org. Chem.* **1992**, *57*, 3732.

Index